A FIELD GUIDE TO THE
PINE BARRENS
—OF—
NEW JERSEY

Its Flora, Fauna, Ecology and Historic Sites

A FIELD GUIDE TO THE PINE BARRENS OF NEW JERSEY

Its Flora, Fauna, Ecology and Historic Sites

BY HOWARD P. BOYD

with a foreword by
V. Eugene Vivian

Illustrations by Mary Pat Finelli

Plexus Publishing, Inc.
Medford, NJ

Cover illustration: Pine warbler perched on a Jersey Pine.
By Mary Pat Finelli

Copyright© 1991
Plexus Publishing, Inc.
143 Old Marlton Pike
Medford, NJ 08055-8750
U.S.A.

All rights Reserved. No part of this book may be reproduced in any form without the written permission of the publisher.

Manufactured in the United States of America

ISBN 0-937548-18-9 (Hardbound)
ISBN 0-937548-19-7 (Softbound)

Second printing, 1991

CONTENTS

Page

Foreword. vii

Preface. ix

Pine barrens areas and pine barrens. 1
   of New Jersey
      Location. Scope. Nature of.
      Geologic history. Soils. Minerals.
      Climate. Temperature.
      Rainfall. Watersheds. Water reserves.
      Ecology. Major habitats. Fire ecology.

Development and uses of pine barrens by man 19
      Historic industries
      Agriculture

Forgotten and ghost towns. Historic sites 39
      The "Jersey Devil"

Present and future uses and preservation 61

Species descriptions. 71

"Kingdoms" of living organisms 73

Plant kingdom (Flora) 75

Flora of New Jersey pine barrens 77

Non-flowering plants:
      Algae 78
      Fungi 78
      Lichens 79
      Liverworts 88
      Mosses 88

Vascular plants 94

Higher non-flowering plants:
      Ferns and fern allies 95

    Horsetails 96
    Club-mosses 96
    Ferns 98

Flowering and seed-producing plants 104
    Trees 106
    Shrubs 118
    Sub-shrubs 134
    Vines and vine-like plants 138
    Herbaceous plants 144
        Aquatic plants 144
        Insectivorous plants and a cactus 154
        Grasses, sedges, and rushes 156
        Other plants with parallel-veined leaves 172
        Other plants with netted-veined leaves 182
        Composites 206

Flowering and fruiting table 223

Animal kingdom (Fauna) 231

Fauna of New Jersey pine barrens 231
    Mammals 232
    Birds 242
    Reptiles 272
    Amphibians 282
    Fishes 288
    Arthropods
        Arthropods other than insects 292
        Insects 296

Footnotes 391

References cited 393

Index 397

FOREWORD

By: V. Eugene Vivian

A salute to Howard Boyd who has authored a book to fill a long perceived need. "A Field Guide to the Pine Barrens" is a book that many other people will wish they had written themselves, but were dissuaded by the enormity of the task.

Mr. Boyd might have entitled his work "Everything you wanted to know about the New Jersey Pine Barrens and Much, Much More!" Anyone who obtains this book cannot be disappointed. Howard Boyd has succeeded in the formidable task of bringing together definitive and detailed answers to questions about the Pine Barrens.

The Guide is not just about plants and animals which hikers, motorists, educational groups and casual visitors might encounter on visits to the Pine Barrens. The extensive introductory sections provide interesting and informative reading about a dozen famous Pine Barrens questions and topics including:

— What and where are they?

— How and why are there efforts to preserve them?

— If there are so many forgotten towns and taverns in the Pine Barrens, why did they disappear and how can you find what's left?

— Are there really rare Pine Barren plants and animals?

— How can the Pine Barrens be a wilderness when it is located so close to the New York/Philadelphia metroplex?

— Is there a geological basis for the Pine Barrens?

— Let's get straight on the New Jersey Devil business.

By providing sought knowledge about many of these topics, Howard Boyd has provided a unique service.

But that's not all—Nowhere else can you find accurate and satisfactory information about so many groups of the wild inhabitants of the Pine Barrens. Especially noteworthy is Mr. Boyd's treatment of the insects of the pines—the insects section is most unique!

The chances are that if someone sees a lichen, moss, fern, flowering plant, mammal, bird, reptile or fish in the Pine Barrens, it will be readily located in this field guide. That's more than can be said for most similar works. All animal groups are similarly described. The special keys will be most helpful to the reader or student. The illustrations are indispensable.

This Field Guide is a must for anyone casually or seriously interested in the New Jersey Pine Barrens. I trust that the author will include me as a purchaser of one of the first copies to roll off the presses.

PREFACE

It has been longer than I care to remember since Thomas Hogan, President of Plexus Publishing, asked me when was I going to do a book on the pine barrens of New Jersey. Over the years since, I've puttered away on parts of a possible book but I've always had a major hangup: I had absolutely no ability to illustrate it. As I neared completion of a text, the people at Plexus discovered they had someone in their own organization who was well qualified to do the illustrations: Mary Pat Finelli. That made this book possible.

Apart from the popular book on *The Pine Barrens* by John McPhee, there are three good general references on the natural history and ecology of the pine barrens of New Jersey. These are Forman, 1979; McCormick, 1970; and Thomas, 1967. However, none of these meet the criteria for a descriptive field identification guide. Thomas is a good, general introductory booklet which includes listings of flora and fauna but no descriptions. McCormick is an excellent "preliminary survey" but is not an identification guide. Forman is an excellent reference text but is not practical for field identification use. Other texts pertaining to specific groups (taxa) of plants and animals are listed at the end of each major section.

This current book is an effort to bridge the gaps between existing literature and to provide one single field identification guide which anyone can carry in his or her pocket or knapsack. It can be taken and used in the field as an immediate reference to the identification of a major percentage of the flora and fauna of the pine barrens of New Jersey, without having to carry a whole library of separate field guides.

Its' primary purpose is as an initial field guide to the identification of the more common, conspicuous, and/or characteristic species of pine barrens flora and fauna. Second, it is also intended to be a compendium of brief thumbnail sketches, as they might be described by a field trip leader, on some of the more important natural and historic features of the pines. This is not intended to be a definitive, in-depth study or a major reference resource on the pine barrens. Most of the information in this field guide is already in print somewhere in the literature but this is the first effort to compile and condense this material into a single field guide to this unique, wilderness region.

The fact I have authored this field guide does not in any way even imply that I am any more knowledgeable about the pine barrens of New Jersey than many other fine naturalists, environmentalists, and conservationists who are unquestioned authorities in various aspects of the natural history of the pines. To insure the greatest possible accuracy in subject matter, a number of these individuals were invited to review sections of this book. All graciously agreed to add their expertise to this effort and I gratefully acknowledge the following reviewers.

Karl Anderson, Director, Rancocas Nature Center, New Jersey Audubon Society, reviewed the entire section on plants as well as the sections on fishes, reptiles, and mammals. I am particularly indebted to Karl for his substantial contributions to the text, especially the additions of many descriptions noted as (K.A.) in the sections on lichens, liverworts, grasses, sedges, and rushes. Joseph R. Arsenault, botanist and environmental specialist, also reviewed the entire section on plants and offered many valuable suggestions. Erwin A. Elsner, Extension Entomologist, Blueberry/Cranberry Research Center, Rutgers University, reviewed the key to the orders of insects. Joseph J. Gasior, a teacher of earth sciences at Shawnee High School, Lenape Regional High School District, reviewed much of the material on geology and water reserves contained in this book while it was included in a talk I had prepared on these subjects. Theodore (Ted) Gordon, President of the Philadelphia Botanical Club, reviewed the sections on the pine barrens, their ecology, and their development and uses by man, and the entire section on pine barrens flora, in all of which he made many valuable suggestions. Donald Kirchhoffer, a friend interested in the pine barrens, read the entire text. William W. Leap, Past President of the Camden Historical Society, reviewed the sections on development and uses of the pine barrens by man, and forgotten and ghost towns and, based on his suggestions, entire sections of the text were completely rewritten. Philip E. Marucci, Emeritus Professor of Entomology, Blueberry/Cranberry Research Center, Rutgers University, reviewed the entire section on arthropods, including insects. Brian Moscatello, Teacher-Naturalist, Rancocas Nature Center, N.J.A.S., now Director, Tenafly Nature Center, and President, Delaware Valley Ornithological Club, reviewed the section on birds. F. Gary Patterson, Associate Professor of Environmental Studies and Graduate Program Advisor for Conservation and Environmental Education, Glassboro State College, and Director, Pinelands Institute for Natural and Environmental Studies at Whitesbog, NJ, reviewed the sections dealing with the pine barrens and their ecology. V. Eugene Vivian, Emeritus Professor of Environmental Studies at Glassboro State College, and Executive Director of the former Conservation and Environmental Studies Center at Whitesbog, graciously agreed to review this work and write the foreword. Finally, Robert T. Zappalorti, Executive Director and President, Herpetological Associates, Inc., reviewed the two sections on amphibians and reptiles. To all of the above I am greatly indebted and express my sincere appreciation.

Thanks also are extended to Gay Taylor, Curator, Wheaton Museum of American Glass, for providing me access to the Wheaton Survey of Historic Glass Factories, and to the reference department, Atlantic County Library, for information on the Mays Landing brickyard.

I particularly want to express appreciation to my wife, Doris, who has very graciously pursued other interests during the many long hours, some even on our vacation trips together, while I have been researching, doing the field work, and writing this field guide. In addition, she read the entire text and made many helpful editorial suggestions.

Finally, a word is in order concerning people at Plexus Publishing. The very fine illustrations by Mary Pat Finelli are self evident throughout this book. In addition, she was the book's production manager. Loraine Page served as my very efficient editor, and Shirley Corsey served as our desktop publishing specialist. All these individuals at Plexus, as well as myself as author, were supported wholeheartedly by the publisher, Thomas H. Hogan, President and owner of Plexus Publishing, Inc. Without all of these individuals mentioned above, from Karl Anderson to Thomas Hogan, this book would not have been possible.

It has been my good fortune to have spent many very pleasant hours exploring the unique natural features of the pine barrens over the past 50 years from home bases in Philadelphia, New York City, and New Jersey, including over 30 years of residency in Camden and Burlington Counties, NJ. It is my hope that others will derive as much enjoyment and satisfaction from their association with the pine barrens as I have and that this field guide will provide an adequate reference to basic information on the flora, fauna, ecology, and other natural features, as well as some of the history of man's development and uses of the pine barrens of New Jersey.

H.P.B.

Pine Barrens Areas

New Jersey is not the only state that has a pine barrens area. There are nearly two dozen such areas in the northeastern United States. After the New Jersey pine barrens, the two largest of these are the pine barrens on Long Island in New York and on Cape Cod in Massachusetts. Other pine barrens areas are located in Maine, New Hampshire, Massachusetts, Rhode Island, Connecticut, New York, and Pennsylvania.

Most pine barrens areas stretch along the Atlantic Coastal Plain, which was laid down from 135 to five million years ago. Most are relicts left behind after the retreat of the last of the great Ice Age glaciers, some 12,000 years ago. As the climate warmed up again after the last glaciers receded, pine barrens plant and animal forms slowly moved northward into these sterile, sandy soils.

The early settlers who arrived on our shores referred to these areas as "barrens" because they were unable to raise their traditional vegetables and field crops in the sandy, acid soils of these regions.[1] Today, we know these areas are not really barren, for many forms of plant life—such as members of the pine family (pitch pine), the beech family (oaks), and the heath family (huckleberries, blueberries, cranberries)—do well in the highly acidic sandy soils. However, these areas are still called barrens, a term that is used consistently in both popular and scientific references to these areas.

Because these areas are unique, there are organizations dedicated to the study and preservation of endangered pine barrens regions. Two such organizations are the American Pine Barrens Society, based in Schnectady, New York, and the Long Island Pine Barrens Society, based near Riverhead, New York. In New Jersey, protection of the pine barrens has been mandated by governmental legislation out of which has been created the New Jersey Pinelands Commission (see page 67). This organization has adopted the term "pinelands" in referring to "that part of the pine barrens landscape (that is) protected by federal and state legislation."[2]

In this field guide, however, the term "pine barrens" is preferred for the following reasons: (1) ecologically, pine barrens refers to a considerably larger area in New Jersey (1.25 to 1.4 million acres) than is under the administration of the New Jersey Pinelands Commission (934,000 acres); (2) the nature of this book is ecological; (3) the term pine barrens is more descriptive of its background and has greater historic significance; and (4) it is a term that is more widely used outside New Jersey and some of the contents of this guide may have application outside of New Jersey. That being said, you will find that the terms "pine barrens," "pinelands," and "pines" are all used interchangeably in this book for the sake of word variation.

Pine Barrens of New Jersey

Dimensions

The pine barrens of New Jersey is one of the world's unique natural areas. Little known, largely ignored, and quickly traveled by shore-bound motorists, the mysterious barrens both attracts and repels. It is the most extensive wilderness tract along the middle-Atlantic seaboard, yet its borders lie only 25 miles from Philadelphia and 35 miles from New York City, just a few miles off center from the east coast megalopolis (see Fig. 1). Ecologically, it consists of some 2,000 to 2,250 square miles or 1.25 to 1.4 million acres of generally flat, sandy, acidic, and sterile soils which constitute a major part of the Outer Coastal Plain section of the Atlantic Coastal Plain in New Jersey.

The barrens is a broad expanse of relatively level land lying between the tidal strip along the east coast and the Delaware River valley on the west. The area lies variously from one to 10 miles inland from and roughly parallel to the seacoast, and formerly extended south nearly 90 miles from Asbury Park and Freehold in Monmouth County to Woodbine and Cape May Court House in Cape May County. It also extends, or formerly did extend, inland across southern New Jersey in places nearly 40 miles in width to New Egypt in Ocean County; Wrightstown, Pemberton, Vincentown, and Medford in Burlington County; Atco, Pine Hill, Clementon, and Blackwood in Camden County; Williamstown, Glassboro, and Newfield in Gloucester County; Elmer and Parvin State Park near Centerton in Salem County; and Bridgeton, Vineland, Millville, and Port Elizabeth in Cumberland County. There

Figure 1. Pine Barrens of New Jersey

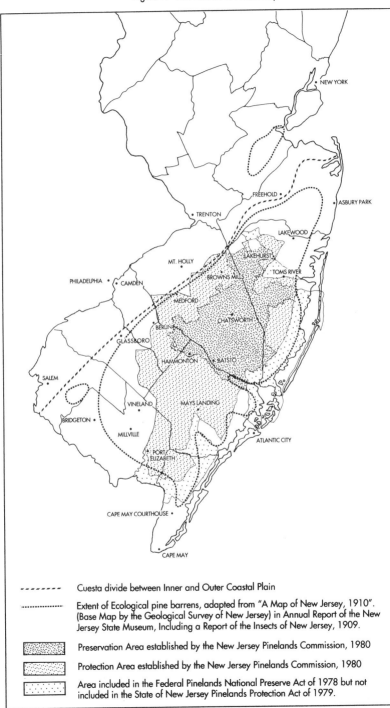

- - - - - - - Cuesta divide between Inner and Outer Coastal Plain

.... Extent of Ecological pine barrens, adapted from "A Map of New Jersey, 1910". (Base Map by the Geological Survey of New Jersey) in Annual Report of the New Jersey State Museum, Including a Report of the Insects of New Jersey, 1909.

Preservation Area established by the New Jersey Pinelands Commission, 1980

Protection Area established by the New Jersey Pinelands Commission, 1980

Area included in the Federal Pinelands National Preserve Act of 1978 but not included in the State of New Jersey Pinelands Protection Act of 1979.

also are, or were, two small pine barrens islands: one surrounding the Manalapan and South Rivers through Jamesburg, Helmetta, Spotswood, and Old Bridge almost to South River and South Amboy in Middlesex County; the other lying between Alloway and Friesburg in Salem County (see Fig. 1).

In recent years, these two "islands," as well as many localities along the northern and western edges of the pine barrens, have become obliterated as pine barrens habitat due to developmental encroachments. Today, the core area of the pine barrens is much reduced and lies mainly in the eastern 55-60% of Burlington County, together with virtually all of adjacent Ocean County (except the coast and barrier islands) and smaller adjacent segments of Camden and Atlantic Counties (see map). In large part, these areas roughly approximate the watersheds or drainage areas of: (1) the Toms River and Cedar Creek in Ocean County; (2) the Oswego and Wading Rivers and the Batsto and Mullica Rivers in Burlington County; (3) the Rancocas Creek in Burlington County; and (4) the Great Egg Harbor and Tuckahoe Rivers in Camden and Atlantic Counties.

The pine barrens is almost entirely forested, is only sparsely inhabited, lacks industrialization, and is relatively free of upland agricultural development. The relatively few people who do live here usually reside in a few small, widely scattered rural settlements, leaving between them several thousands of acres of uninhabited tracts of forests and swamps. There are places in these largely wild lands where the forests extend in every direction as far as the eye can see, in almost endless pattern, broken only by occasional small lakes, cranberry bogs, and white cedar swamps. Although much exploited by man over the past 300 years, the barrens remains a true wilderness area.

Geologic History

All of that part of New Jersey which lies southeast of a line extending from Trenton to Perth Amboy is located on the Atlantic Coastal Plain. This physiographic region includes portions of Middlesex and Mercer Counties and all of the counties further south and east. In general, the surface of this coastal plain is a low, flat, dissected plain that rises gradually from sea level along the coast to 373 feet at Beacon Hill in Monmouth County. At its northern margin, where it borders on the Piedmont Plain to the northwest, it includes a broad, shallow depression lying barely 80 feet

Geology

above sea level, extending from Raritan Bay to the Delaware River at Trenton.

At Trenton, this depression merges with the Delaware River Valley, the axis of which lies below sea level, and this valley then forms the southwestern continuation of the shallow depression. Thus, the coastal plain falls to sea level on the east, west, and south, and rises to barely 80 feet in the depression along its northern border. Over one half of this relatively flat and level coastal plain in New Jersey lies below 100 feet, and, except for the Rancocas Creek drainage, the main divide between the Delaware River slope and the Atlantic Ocean slope consists of a cuesta or ridge of low hills (Beacon Hill, Monmouth County, 373 feet; Arney's Mount, 230 feet, and Mount Holly, 183 feet, both in Burlington County) which extends southwesterly from Atlantic Highlands, Monmouth County (235 feet) in the northeast to nearly sea level in Salem County and the Delaware River lowlands in the southwest, gradually dropping in elevation from north to south. The slope to the Delaware River west of these hills is the Inner Coastal Plain. The slope to the Atlantic Ocean east of these hills is the Outer Coastal Plain.[3] This Outer Coastal Plain was formed more recently and is more sandy and less fertile than the Inner Coastal Plain. It is in this Outer Coastal Plain that the pine barrens is located. The highest elevation within the pine barrens is at Apple Pie Hill (205 feet) in Tabernacle Township, Burlington County. The next highest pine barrens elevations, between 175 and 182 feet, are found in the Forked River Mountains, which are mainly in Lacy Township, Ocean County.[4,5]

At the time of its earliest formation, the Atlantic Coastal Plain was composed of continental deposits on older, metamorphosed rocks. Then, about 140 to 100 million years ago, this coastal plain became covered by the oceans. Subsequently, all through the Cretaceous and Tertiary Periods, 135 to five million years ago, the seas covered all or most of this coastal plain, rising and receding several times. As a result of these submergences, the Atlantic Coastal Plain became overlain with layers of essentially unconsolidated, mostly permeable silts, sands, clays, greensands, and marls, each deposited by a succession of rising and falling sea levels. There are at least 15 geologic formations of Cretaceous and Tertiary age strata beneath this coastal plain, beginning with the Raritan formation at the base and ending up with the Kirkwood and, finally, the Cohansey formations on top. All these formations tilt eastward toward the ocean, the Cohansey at the rate of about ten feet per mile.

The Kirkwood formations range from the surface at their western outcrop to depths ranging from 100 to 300 feet as the formations tilt toward the east. The Cohansey formations range from the surface down varying depths averaging about 100 feet of saturated thickness but ranging to 400 feet toward the east and south. Overlying the Cohansey formations are localized deposits of Beacon Hill gravels, possibly deposited by some ancient river drainage. Due to erosion and other factors, Beacon Hill gravels today are reduced to small, scattered remnants on the tops of higher elevations such as at Apple Pie Hill.

Nearly all of these formations contain fossils of marine animals. The sedimentary record indicates this coastal plain, in Cretaceous and Tertiary times, has been deep under the ocean on several occasions, while at other times the land has been above the ocean surface. During the Miocene and Pliocene Epochs, from 25 million years ago to about five million years ago, the oceans retreated for the last time to approximately their present levels.

Finally, the Ice Age, or Pleistocene Epoch, which began about one million years ago and ended about 12,000 years ago, also left an impact upon the pine barrens. There were four major advances of ice from the arctic, at least two of which reached northern New Jersey. The last of these was the great Wisconsin Ice. The vast amounts of water trapped in the ices of these glaciers, when they covered the northern continents, caused low sea levels. During these times land south of the glaciers we now call the pine barrens would have been sub-arctic tundra and the coast of New Jersey may have been many miles east of its present shoreline.[6] Now that continental glaciers are almost non-existent, sea levels are again high. South of the terminal moraines left by the glaciers, the topography was greatly changed by sands and gravels deposited by the streams leading away from the retreating ice fronts of the glaciers. A major result in the pine barrens was the deposition and intrusion of yellow sands and gravels into parts of the otherwise sandy Outer Coastal Plain. It is generally believed that present pinelands flora and fauna developed only after the retreat of the Wisconsin Ice, or within the past 12,000 to 10,000 years.[7]

Soils

Soils of the pine barrens, developed from sandy, sedimentary deposits, are unusually porous and acid. They are so porous that rain waters drain very rapidly down through them to become part of the underground water reserves, leaving the surface very droughty and leached of nutrients.

Lying on top of the Cohansey sands formation in the pine barrens are at least 13 different major soil types. In upland areas, where the water table is generally two or more feet below ground level, these are named Lakewood, Evesboro, Woodmansie, Downer, Sassafras, and Aura. Intermediate area soils are Lakehurst, Klej, and Hammonton. In lowland areas, where the water tables generally are less than two feet below the surface, the soils are called Atsion, Berryland (including St. John's), Pocomoke, and Muck.[8]

Most of the pine barrens is located on upland sands. In general, these are sterile, highly acid (pH ranging from 3.6 to 5.5), sandy, podzol, nutrient-poor soils with only a small accumulation of humus. These soils have intensely leached upper layers lying just above lower layers of gray, illuvial soils that have been enriched by the leaching processes. They are light in texture, highly bleached, very porous, with excellent drainage, very droughty, and do not retain enough soil moisture for good vegetative development. Thus, again, the term "barrens"—an area where traditional crops of early settlers would not grow. The most common trees in these areas are pitch and short-leaf pines and several species of oaks, principally black, scarlet, blackjack, white, post, and chestnut.[9]

Although not as widespread or extensive as upland soil areas, there are more than 200,000 acres of bog and swamp lands, with overlying lowland soil types, within the pine barrens. In general, these lowland soils are quite heavy and allow for only very poor drainage, which are desirable features in the development of cranberry bogs. As a result, many of these areas have been cleared over the years and converted into cranberry production. Some of the more common trees that grow in these areas are the Atlantic white cedar, pitch pine, red maple, black gum, swamp magnolia and gray birch.[10]

Minerals

Nearly all of the sandy soils of the pine barrens contain a high percentage, often greater than 90%, of quartz or silica (SiO_2). Other than quartz, the mineral contents of the soils are very low. Chief among other minerals are the feldspars and various oxides of aluminum and iron.

Pine Barrens Areas

Climate

New Jersey's climate is basically continental in nature, characterized by cold winters and hot summers, with a range of 40° F or more in average temperature from the coldest to the warmest months. During the winter months, cold air masses move down on northwest winds from Canada to provide bone-chilling temperatures down to zero or, occasionally, even subzero temperatures Fahrenheit. However, winter temperatures average between 32-36° F. During the summer months, the circulation of air is such that warm, tropical air masses, which tend to be moist and hot, move into the state from the south and southwest. Spring and fall climates are transitional between winter and summer. No place in New Jersey escapes the tropical heat of summer and this is especially true in the pine barrens where the temperatures of the atmosphere regularly reach into the 90°-100° F range (average temperatures 72-75° F), and where the white, sandy soils often exceed 100° F.[11]

The fact that the pine barrens lies between the Delaware River Valley, the Delaware Bay, and the Atlantic Ocean probably has some moderating effect on weather and temperature conditions in the barrens. Periods of extended severe cold weather are relatively rare and periods of abnormally high temperatures seldom last more than three or four days. However, because of the presence of maritime air during the summer months, the atmosphere is definitely humid, adding to the discomfort of the high temperatures.

Precipitation and Rainfall

Precipitation in the pine barrens averages between 40 and 48 inches per year. This is fairly evenly distributed throughout the year but is somewhat concentrated during the summer months. Summer rains tend to be tropical in character with sudden, heavy downpours resulting in some runoff and erosion. Runoffs, however, are mainly local in nature. It is estimated that about one half of all rainwater drains rapidly down through the porous, sandy soil to become part of the underground water reserves, leaving the soil surfaces very droughty. The other half returns to the atmosphere as the result of evaporation and plant transpiration.[12]

The question of what impact "acid rain" may be having upon the vegetation of the pine barrens, and upon plant and animal life in its streams is currently under study by the Division of Pinelands Research, Rutgers University. Acid rain is formed when moisture in the atmosphere mixes with sulfur oxides released by burning oil, coal, and gasoline, forming sulfuric and nitric acids.

Dr. Mark D. Morgan, Assoc. Professor of Zoology, Rutgers University, Camden Campus, has been conducting studies since 1984 to measure the acidity of rain in the pinelands. In a report to the New Jersey Pinelands Commission on May 4, 1990, Dr. Morgan stated that, although the level of acid rain in the pinelands appears to be declining slightly, possibly due to the Clean Air Act of 1972, the nature and source of the acid rain is changing the type of acidity in pine barrens streams from organic to mineral. This is shown by increased levels of sulfates in the water and decreased concentrations of dissolved organic matter. Natural pine barrens acidity is created when decaying vegetation produces an organic acid that washes down through and is absorbed by the sandy soils. These organic acids leach out aluminum found in pinelands soils and bond with the aluminum making it harmless in pine barrens waters. However, the sulfuric acids in acid rain do not bond with aluminum so this mineral is washed in its pure form into streams. Morgan concluded that the pollution caused by these increased sulfates, nitrates, and aluminum may be toxic to aquatic life, but the long term effects of this pollution have not been studied and are unknown at this time.

Watersheds

There are four major watersheds in the pine barrens. Three of these tilt east to the Atlantic Ocean. The largest watershed is the basin of the Batsto-Mullica and Oswego-Wading Rivers which flow east and empty into Great Bay on the Atlantic coast, draining major pine barrens areas in central and eastern Burlington County and northern Atlantic County. Second largest is the basin of the Rancocas Creek, the only major pine barrens stream which flows west. This empties into the Delaware River, draining the western fringes of the pine barrens in Burlington and Ocean Counties. Third are the Great Egg Harbor and Tuckahoe Rivers in the southeast, both draining into the Great Egg Harbor west of Ocean City. Smallest of these four are the combined basins of the Toms River and Cedar Creek in Ocean County, both of which flow east into Barnegat Bay at and just south of Toms River.

It is an interesting and significant point that all pine barrens streams arise within the pine barrens area, originating as ground water discharge from slowly moving surface waters of the Cohansey aquifer. No streams flow into or through the pine barrens from outside the region.

Due to the generally flat terrain of the Outer Coastal Plain, all these streams have a very low gradient, or relief, and their waters have only a gradual flow toward the ocean. Due to this same generally flat terrain, there are almost no natural lakes within the pine barrens. Most existing ponds and small lakes are the result of former man-made activities such as the damming of streams to form mill ponds, canals, and raceways to power early industries.

Pine barrens streams and other bodies of water contain so-called "cedar" waters, which are quite dark, sort of reddish-brown, and somewhat tea-colored. This color is caused by their high iron content and by their absorption of vegetative dyes such as tannin. These waters also are quite acid, usually less than 5.0 pH. Yet in spite of their dark coloration and high acidity these are pure and unpolluted waters, especially in the upper reaches of these streams.

Water Reserves

The oldest, deepest, and most productive underground water reserves in New Jersey are the combined Potomac-Raritan-Magothy (PRM) aquifers (water containing sands). In some places, the lowest layer of these, the Potomac sands, may lie as deep as 6500' below the surface. It is the PRM aquifers that provide the water for many municipal water systems in southern New Jersey, particularly within the Inner Coastal Plain along the Delaware River west of the pine barrens.

Within the pine barrens, the principal upper aquifers are the Cohansey formation on top, essentially a water table reservoir, overlying the Kirkwood formation below. Together, these two formations may contain as much as 17 trillion gallons of water.[13] It is these two formations which probably supply most private, homeowner type of water systems within the pinelands, while the Kirkwood provides most of the water needs along the coast, east of the pinelands.

The waters in the shallow Cohansey aquifer frequently lie at or near the surface and it is these waters that feed the bogs, marshes, swamps, and many streams of the pine barrens. The Cohansey-Kirkwood aquifers are said to be capable of providing millions of gallons of water a day if fully developed. However, there is considerable question how much of this can be safely drawn from these aquifers without doing ecological damage to pine barrens surface waters and possible desertation of the vegetation.

All of these water reserves are due to the sandy soils, generally flat terrain, and evenly distributed precipitation which combine to provide the area with a very high rate of infiltration so that the soils absorb much of even the heaviest rainfalls, and very little is lost through runoff.

One of the major reasons why Joseph Wharton purchased so much land in the pine barrens in the late 1800s, after the bog iron, glass, and other industries became defunct, was that he planned to connect a series of shallow ponds, reservoirs, and canals and sell the water from the vast underground water reserves in the pine barrens to the City of Philadelphia as one of its main sources of municipal water. Fortunately, the New Jersey State Legislature learned of this plan and passed a bill preventing the export of any of the state's water resources outside state boundaries. It was the acquisition and protection of these same rich water reserves which prompted the State of New Jersey to purchase the 97,000-acre Wharton tract in 1954 and 1955 and establish the present Wharton State Forest.

Ecology

The pine barrens is a unique region of sandy, acid, and sterile soils made up largely of coarse sands and gravels deposited by ancient seas. These acid soils are high in iron content but low in calcium, magnesium, and potash. Water drains rapidly through layers of these porous soils to form vast underground reservoirs, leaving the surface droughty in spite of an average rainfall of nearly 45 inches per year.

The pine barrens is a region forested with pines, oaks, and cedars, with an understory of mainly heath-like shrubs. It is a region that lacks in diversity of plant forms, with a resultant impact upon animal forms. The unique features of the pine barrens ecology result in an unusual flora and fauna. A few plants and animals are known only from the New Jersey pine barrens or from similar nearby "island" habitats. Several others which occur here also occur elsewhere in a few isolated areas many miles away, while many species that are common here are rare elsewhere.

The pine barrens also is a region of acid waters in shallow, slowly moving streams of cedar waters, tinted brown by the high iron content and by organic matter leached from the bark of cedar and other trees and shrubs and from peaty bog mosses in cedar swamps.

Upland Forests

Upland forests are those where the water table generally is two or more feet below ground level. The major trees forming the canopy layer in these upland forests are pitch and short-leaf pines and several species of oaks. The dominant tree is pitch pine, *Pinus rigida*. In no other region in North America does the pitch pine cover such an extended area of country as the dominant tree.[14] Oak trees of several species—black, scarlet, white, chestnut, blackjack, and post oaks—are mixed in with the pines and have become dominant where the pines have been removed or where fires have occurred less frequently and thus have not burned off the duff from the forest floor.

Ecologists differentiate these upland forests as either pine-oak or oak-pine depending upon which type of tree is dominant. Pine-oak communities are composed of pitch, short-leaf, and Virginia pines, blackjack oak, and, in scattered areas, other oaks such as post, black, scarlet, white, and chestnut oak. These forests develop after fires or lumbering clearances and often succeed, in time, to oak-pine and finally to oak-oak.[15] In oak-pine communities, in addition to pitch and short-leaf pines, the dominant oaks are black, scarlet, white, chestnut, blackjack, and post, with black oak the most common species north of the Mullica River and southern red oak becoming prominent to the south.[16]

The natural understory of these upland forests is a variety of shrubs, the most common of which are scrub oak and members of the heath family, such as black huckleberry, lowbush blueberry, and mountain laurel.

Ground cover is provided by various lichens, mosses, bracken fern, and members of the heath family such as bearberry, and teaberry.

Dwarf Forests of the Plains

In the east-central part of the barrens, there are four tracts of specialized, upland forests known as the dwarf or "pygmy" forests. These total some 12,000 to 15,000 acres or 20 to 25 square miles. The two larger of these, located along the Burlington-Ocean County line near Warren Grove, are known as the West Plains and the East Plains. Between these two, just northwest of Warren Grove, is a smaller area known as the Little Plains. The fourth and smallest area, along and just beyond the northern border of Penn State Forest, is known as the Spring Hill, or South Plains. These tracts support a growth of unusually short, scrubby, four to 10 feet high forests of mature trees. The dominant trees are a closed-cone (serotinous) race of pitch pine whose cones open only after being subjected to very high temperatures, such as those created by fires. Over succeeding generations of evolutionary development, it appears these pines may be developing a genetic variation from more normal pitch pines.[17]

These pitch pines, together with both blackjack and scrub oaks, form a type of forest relationship indicative of repeated fires. These trees seem to be more fire resistant than others of the same species elsewhere. Considerable and rapid regeneration of these dwarf trees and shrubs takes place after fires, with multiple sprouts arising from older root crowns and even from the sides of the main trunks. Reproduction is mainly vegetative rather than seed dispersal and generation. Ground cover is provided by shrubs, most commonly black huckleberry and lowbush blueberry, and by sub-shrubs such as pyxie, broom crowberry, and hudsonia as well as several heath types like bearberry, teaberry, and trailing arbutus.

Several theories have been advanced to explain these stunted forests, but the general concensus seems to be that these are the result of a combination of factors such as infertile soils, aridity, exposure to constant and strong winds, and, perhaps most important, repeated fires which occur with greater frequency and severity in the pine plains than elsewhere. These happen perhaps as often as once every 10 to 20 years, on average.[18]

Lowland Forests

Lowland forests are characterized by water levels close enough to the surface, usually less than two feet, to exert a year-round influence upon the type of vegetation. Within the pines, three major types of lowland forests are recognized: pitch pine lowlands, cedar swamps, and hardwood swamps.

Pitch pine lowlands often occur in depressions and as narrow bands along the borders of streams and cedar swamps. These woodlands are transitional between lowland and upland vegetation types and usually have a history of human disturbances.[19] Pitch pines are the dominant trees, often in such dense stands as to exclude any oaks. Secondary trees may be a few trident red maple, black gum, and swamp magnolia. The understory growth is more varied than in upland forests with sheep laurel, staggerbush, dangleberry, black huckleberry, and sweet pepperbush being among the prominent shrubs. Sheep laurel is particularly characteristic while leatherleaf takes over near the margins of standing waters.

The herbaceous and ground cover layers are especially diverse and well developed in these forests. Turkeybeard, bracken fern, and teaberry are typical plants and the moss-lichen vegetation is especially rich[20]

Hardwood swamps often occur in areas where cedar swamps have been cut and the cedars have not regenerated, thus allowing trident red maple, black gum, swamp magnolia, and gray birch to replace the cedar. As a result, today this may often be the most common type of stream-side forest. Understory vegetation contains a variety of species including highbush blueberry, black huckleberry, dangleberry, fetterbush, sweet pepperbush, swamp azalea, leatherleaf, and sheep laurel[21]

Cedar Swamps

Many small and a few larger cedar and sphagnum swamps, or bogs, are scattered throughout the pine lowlands. Here the dominant trees are Atlantic white cedar, *Chamaecyparis thyoides*, which grow in thick but usually narrow stands along stream courses and in a few broader, lowland areas. These fine trees rise as tall, straight spires from the soggy ground and hummocks where their roots are carpeted with thick, spongy mats of sphagnum mosses.

A few shade and acid-tolerant shrubs and herbaceous plants grow out of the spongy sphagnum in these cool, dark, and humid areas, but the overriding impression is one of tall cedars reaching upward to form a dense canopy of green through which only the faintest rays of the sun can penetrate. Typical understory shrubs are highbush blueberry, dangleberry, fetterbush, and swamp azalea. Rare openings in the forest canopy may allow enough sunlight to enter to permit the intrusion of some bladderworts, sundews, pitcher plants, a few orchids, and possibly curly grass fern to become a part of the ground cover along with sphagnum mosses.

There also is a rich moss and lichen flora in these cedar swamps, most prominent of which are the thick carpet of sphagnum mosses and the bluish-gray-green "bloom" of the santee lichen that coats the bases and trunks of the cedars.

Over the years, many of these low-lying boggy areas have been cleared and transformed into cranberry bogs.

Aquatic and Semi-Aquatic Communities

Several types of aquatic communities occur throughout the pine barrens: streams, ponds and small lakes, bogs and swamps, marshes or savannas, and spongs. Many, if not most, of these areas have been disturbed over the years by man for one purpose or another.

As stated earlier (under Watersheds, p. 9), most streams are shallow, slow moving, contain "cedar" water, and all rise within the pines. However, over the years, most of these have been dammed up by man at one or several points along their courses to form ponds and small lakes for industrial purposes (see page 10). Some of today's ponds and small lakes are former cranberry bogs or reservoirs for bogs.

Shallow ponds that form in small depressions may remain so only through wet seasons. During these times, they may provide ideal habitats for sphaghum mosses and bladderworts and be surrounded by thick growths of leatherleaf. Deeper ponds and small lakes may be filled with yellow pond lilies, fragrant white water lilies, and other deep-rooted and floating leaved plants, while many shallow water plants thrive along their margins.

Savannas are sedge and grass marshes or meadows. Some of these may be abandoned cranberry bogs or they may occur as small openings along streams. Spongs are lowland areas that may formerly have been small ponds or savannas that have become overgrown with a dense intrusion of shrubs, particularly leatherleaf, sheep laurel, and highbush blueberry.

Fire Ecology

The pine barrens is a region of fires, perhaps the single most important factor in shaping its vegetation. From an ecological point of view, and from as far back in time as we have knowledge, fires have played a major role in maintaining the New Jersey pine barrens in the condition our early colonists and pioneer ancestors first found them. Before man, lightning strikes caused widespread fires. In more modern times, fires may have been deliberately set by the early American Indian inhabitants as a hunting technique to drive deer, to improve visibility, and to facilitate foot travel.[22,23]

The highly flammable vegetation burns readily, providing tinder for frequent and often widely spreading fires. These fires, fueled by the resins of the pitch pines, repeatedly burn off the tree and other plant tops as well as the duff on the dry forest floor, but few plants are actually killed. Most plants, including the dominant pitch pine with its thick bark, are quite fire resistant and soon either resprout from the roots or produce new growth from the larger limbs or even from the main trunks.

The dry, acid, and infertile soils, in combination with repeated fires, seem to eliminate the richer flora of the surrounding areas, leaving behind only those characteristic pine barrens species that can withstand these harsh conditions.

Without fires, over a period of years the pine barrens could become an oak forest. The winged seeds of the pines are extremely small and light and are hidden deep down between the bracts of pine cones. When the cones open, the seeds are released and, wind blown, fall upon the ground. If the ground is covered with a heavy layer of dead leaves and needles, these seeds cannot reach down through this duff to the ground to take root and send up new pine trees.

On the other hand, the seeds of the several species of oaks, the other major trees in the pine barrens, are the relatively heavy acorns. When they drop to the ground in the fall of the year, because they are so heavy they can work their way down through the duff to the earth on the forest floor and take root. Next spring, new oak trees shoot up through the duff. In time, there could be a gradual transition from pine woods to oak-dominated woodlands, and we might not any longer have the New Jersey pine barrens.

Fire is a leading factor in this story. Fires contribute to the ecology of the pine barrens by burning the duff off the ground leaving the earth exposed, so when the light pine seeds reach the ground they can take root and sprout new pine seedlings, thus enabling the pines to maintain their traditional dominance.

It is estimated that on average fires need to burn over an area not less than once in every 20 to 40 years in order to maintain a pine sub-climax and, consequently, a pine barrens. In short, fires are essential for the continuation of New Jersey's pine barrens.

On the other hand, from a human, non-ecological point of view, wild fires in the pinelands can be very destructive. One of the worst and most memorable fire storms in the pine barrens occurred during the period of April 20 to 24, 1963, when a whole series of wild fires burned over 190,000 acres, caused $8.5 million in property damages, and resulted in the deaths of seven persons.

Because of continually increasing human population and developments, wild fires now must be contained as quickly as possible so as to avoid destruction of property and loss of lives. An alternative is the practice of controlled burning, a program in which low-intensity fires are deliberately set and controlled during periods of least flammability. The purpose of this is to burn the tinderous duff off the forest floor in order to try to prevent future wild fires. Controlled burning has been practiced in the pine barrens by the New Jersey Division of Parks and Forestry since about 1947.

Development and Uses of the New Jersey Pine Barrens By Man

Prior to the arrival of immigrants from the Old World in the 17th century, American Indians of the Lenni Lenape, or Delaware, tribe inhabited much of southern New Jersey. These natives located most of their settlements along river banks and on the inner coastal plain rather than in the pine barrens, but they used the pine barrens for hunting and some fishing and for traveling back and forth to the coast for fishing and shell collecting. They dried the meat and left behind, along the coast, piles of shells called "middens." Recent research has revealed that the Indians left other marks on this land as they hunted, trapped, and fished. Archaeologists have discovered over 1,000 Indian sites in and surrounding the pine barrens.

Later, as the Indians succumbed to the diseases of the newly arrived settlers, their numbers became so reduced, and their health and living conditions so poor, that in 1758 all remaining Indians south of the Raritan River were encouraged to live in a 3,258-acre reservation called Brotherton, at a place called Edgepillock near the present village of Indian Mills, which is in Shamong Township, Burlington County. This is generally recognized as the first Indian reservation in the United States.[1] However, it was not successful. In 1801 most of the Indians left to join the Oneida Indians in New York State.[2]

When Old World immigrants arrived in New Jersey in the mid-1600s and early 1700s, they settled first along the coastal bays and inlets of the Hudson, Hackensack, Passaic, and Raritan river valleys in northern New Jersey and along the Delaware River valley and inner coastal plain south of Trenton. Even 100 years later, in 1765, the area of the present pine barrens on the outer coastal plain was classified as "unsettled."[3] Instead, the area of the pine barrens was used largely for lumbering and hunting, with some limited use as a range for cattle.

As the early settlers moved in and began to take charge in the pinelands, they saw the forests, the waters, and the sand as raw materials to be used for making a living and started thriving in-

Development & Uses

dustries related to these natural resources. Thus began modern man's exploitation of the pine barrens, which might be characterized as having been heavily industrialized from 1765 to about 1865.

The story of man's uses and development of the pine barrens is based largely on the rise and fall of a succession of local industries. Chief among these, successively, were: timber cutting and sawmills; charcoal burning; bog iron furnaces and forges; glass factories; and paper mills. These provided employment and drew people together into small villages and towns that existed as long as the industries prospered. Whenever the industries shut down for any reason—fire, exhaustion of natural resources, or economics—the people had to seek other employment in order to live. If a village was entirely dependent on one industry, people had to move to other places, leaving behind many deserted and now forgotten communities (see page 39).

Today, most of those old industries and many of their surrounding communities are gone, and the pine barrens has more or less reclaimed its own wild, little developed, sparsely populated condition. Yet man's destructive influence on the pine barrens is still being felt. Over the past 225 years, there probably is not a single acre of pine barrens terrain that has not been burned, cut over, or otherwise disturbed repeatedly.

Man's past and present effects on the pine barrens is particularly noteworthy in regard to rare, threatened, and endangered plant life. Fairbrothers in his excellent chapter "Endangered, Threatened, and Rare Vascular Plants"[4] asks why some species become rare. He then suggests some answers. A complete reading of this reference is recommended, but the following provides a brief summary.

Some species become rare as the result of natural events such as sea-level rise, glaciation, and warm-dry periods. However, the growing number of threatened and endangered plant species in the pine barrens is largely the result of human disturbances. The following sections of this field guide will briefly relate the story of human activities in the pine barrens since the late 1600s in industries such as lumbering, pitch and tar making, charcoal making, bog iron production, glass factories, brick making, and others. All of these had an impact, mainly adverse, on the natural vegetation of the pine barrens.

This process continues even today. The development of cranberry bogs and the setting out of cultivated blueberry fields, though relatively compatible with pinelands preservation, have certainly altered the natural vegetation. The destruction of all ground cover as the result of "mining" operations is especially disastrous to rare, threatened, and endangered plants.

Occasionally the flora of the pine barrens benefits from man's activities. Some of his disturbances provide new habitats, which then allow for the expansion or establishment of rare plants. Examples of these may be certain orchid species, curly-grass fern, and certain rushes and sedges.

Fairbrothers concludes that "...knowledge of individual requirements of species and habitat maintenance clearly is essential to assure the continued existence of the endangered and threatened plant species in the Pine Barrens. However, the habitat of endangered species must be protected even before we can comprehend fully the ecosystem of which they are a part." This translates into the urgent need to preserve all habitats which sustain threatened and endangered species.

Lumbering

One of the very first industries was the cutting of the virgin woodlands and the construction of sawmills to turn the timber into lumber for the construction of homes, shipbuilding, and multiple other purposes. This very substantial lumbering industry operated uncontrolled and indiscriminately in the virgin pine and oak forests and cedar swamps of the barrens.

Beginning in the very earliest decade of the 1700s and continuing through most of the 18th century, sawmills driven by water power and operated by people living in small surrounding communities dotted the pine barrens. Mounier, in his "Study of Water Powered Sawmills in the Pine Barrens of New Jersey," records 158 of these sawmill sites.[5] A few of these sawmills continued to operate in the pines until the early 1900s. As a result, the virgin pitch pines, timber oaks, and stands of Atlantic white cedar were virtually wiped out as they were harvested for use as sills, beams, framing, plaster-lath, and flooring in building construction. Oaks were particularly prized in shipbuilding. The industry operated at such a pace that as early as 1749 Benjamin Franklin decried the reckless and wanton slaughter of the woods and urgently advocated conservation and intelligent forestry.[6] In the same year, Peter Kalm, a student of Linnaeus, expressed fears that the Atlantic white cedar was being extirpated.[7,8]

In the late 1700s to mid-1800s, after the stands of saw-log size became non-existent, medium and smaller second and third growth trees were cut and recut over 20- to 30-year cycles principally to make charcoal, which was used to fire furnaces in the manufacture of bog iron and glass. This wood was also used for domestic fuel, for the manufacture of small boxes and crates, and as fuel for steam navigation and locomotives. Today, some wood is still being cut for domestic fuel and for pulpwood.

Especially prized were the great virgin Atlantic white cedar trees which grew in extensive cedar bogs, mostly in narrow belts along streams. White cedar is relatively decay resistant, non-resinous, and straight grained. For these reasons it was used for roof shingles and clapboards in house construction, and for fence posts, rails, and cooperage. In time, these cedar bogs became over-harvested after having been cut two and three times. And because cedar tree stands regrow more slowly, in approximately 60- to 100-year cycles, this phase of the lumber industry died out before the end of the 19th century for lack of marketable stands.[9]

Due to pre-settlement windfalls, many of these decay resistant and valuable cedar trees, some up to two and one half feet in diameter, had fallen into and been long buried 10 to 20 feet in the thick muck of Great Cedar Swamp and other swamps along the bay shores of Cumberland and Cape May Counties. Beginning as early as 1800 and continuing to some extent up until 1935, these buried trees were raised or "mined" and split into shingles and rails.[10] Because of their superior quality, shingles from these raised trees were used on the roof of Independence Hall in Philadelphia. Today, medium sized cedars are much desired for fencing, rustic furniture, and other round products.

Pitch, Tar, and Turpentine

Simultaneously with the lumbering industry, starting in the late 1600s and continuing into the early 1800s, the pitch pine forests were used as the resource for the distillation of turpentine and of tar and pitch, which were needed in shipbuilding. Slits were made in the sides of pitch pine trunks to encourage the flow of oleoresin, which exudes from the trees. This was gathered and distilled into turpentine and rosin. Unionville, near Mount Holly, was known as Turpentine until 1855.

Tar, often called "pine-wood tar," was distilled by using stumps, roots, and other waste materials left by the pitch pine lumbermen. These were stacked, covered with earth, and fired in much the same manner as charcoal burning (see below). The tar would run out at the bottom where it was collected.

Pitch was a residue of tar distillation. It is highly adhesive and water repellent and was used for caulking the seams of boats and ships. It also became a component, along with tallow, of shoemaker's wax.

Grist Mills

Perhaps equal in importance to the sawmills for the earliest settlers were the early grist mills. Dating from the early 1700s, these were erected along streams, especially near the better agricultural areas and near areas of population. Here, early farmers brought their grain to be ground into flour, and purchased feed for their cattle, horses, and livestock. Here, also, inhabitants obtained flour and similar supplies for their table. The number of known grist mills at one time or another within the pinelands is approximately 39.[11] Two still standing grist mills are at Batsto, still in operating condition, and at Kirby's Mill, in Medford Township, Burlington County, which is in the process of being restored by the Medford Historical Society.

Charcoal

As the lumbering industry turned to second and third growth trees, another industry—the manufacture of charcoal—developed because it could use both cordwood and smaller trees and even branches. Charcoal is partially burned wood from which water, volatile gasses, tars, resins, and impurities have been driven off, leaving a residue of practically pure carbon. This carbon, or charcoal, burns at a much higher temperature than wood, and was used to fire blast furnaces in the manufacture of bog iron and glass. All iron furnaces of the period had to have a substantial acreage of forest land "appurtenant to the works"[12] to provide timber for the manufacture of charcoal. Although the first manufacture of charcoal in New Jersey seems to have taken place at the Tinton Falls Iron Works, or Shrewsbury Furnace, in Monmouth County sometime between 1674 and 1746, the charcoal industry really did not take off until the rise of the bog iron industry in the mid-1700s.

Charcoal was manufactured by slowly burning wood in oxygen-deprived charcoal "pits," which are not really pits at all but stacks of wood from 10 to 20 feet high piled on the ground in a mound like an inverted cone with its top cut off, or like a rounded beehive. These stacks then were covered with sods and earth called "floats"

Development & Uses

and were fired by lighting the pile through an opening near the base or by dropping burning kindling down through a center hole on top, which then was sealed shut. A slow fire was then maintained for a week to 10 days. Except for vent holes, the earth covering kept out air so the wood did not burn to ashes. But the heat from the combustion changed the wood to a high carbon charcoal by driving off most of the water content as well as the volatile gasses, tars, and resins which escaped through small holes near the top. After cooling, the stacks were uncovered. Using long-handled rakes with long teeth, the charcoal was raked out from the "pits," after which the cooled charcoal product was hauled to the "coal house" or to market.

The clear cutting methods used to obtain the wood necessary to operate these charcoal pits wiped out vast acreages of pine barrens woodlands—another example of man's exploitation of the barrens. It was generally recognized that a functioning bog iron furnace needed a minimum of 20,000 acres of woodlands in order to have enough timber, converted into charcoal, to "fire" the furnace. A common practice was to divide the 20,000 acres into 1,000-acre sections, one of which was cut each year to produce the fuel for that season's "blast," with the hope that at the end of the 20-year cycle the first section would have regrown a sufficient crop of new wood that it could be recut and the cycle continued. This hope, however, was not realized and may have been a factor in the demise of the charcoal and bog iron industries. The charcoal industry continued shipping schooner and wagon loads of charcoal to Philadelphia and New York City until approximately 1848 to 1865, with only a few small charcoal operations continuing into the 20th century.

Ted Gordon documents two of these enduring charcoal makers in "The Last of the Oldtime Charcoal Makers and the Coaling Process in the Pine Barrens of New Jersey." He writes of George Crummel of Jenkins who "in his eighties, continued to make charcoal right up to the time of his death in 1960 or 1961." Crummel was "the last of his breed in Burlington County." There was also Herbert Payne of Whiting, Ocean County. Gordon states that Payne "prevailed until August, 1974 when, storehouse full of 'coal,' he laid down his saw for another season." Gordon witnessed and photographically documented that burning which proved to be Payne's final one, due to subsequent illness and a stroke. Thus Payne, whose career as a collier had "spanned some four decades" proved to be "the last of the old-time charcoal makers in the pines."[13]

Bog Iron

The best known and most important of all early pine barrens industries was the "mining" of bog iron "ore" and the manufacture of bog iron in furnaces constructed for the purpose throughout the pine barrens. Although it is known that the American Indians were the first to use bog iron ore by mixing the reddish ore with bear grease to make paint for their faces, it remained for the early settlers to recognize the economic potential of bog iron and to capitalize on it.

In simple terms, bog iron ore is formed by the chemical action of decayed vegetative matter in streams upon iron salts in, beneath, and surrounding stream beds. Strata of marls, or greensands, in the soil contain a soluble form of iron. In many very slowly moving streams, marshes, and other sluggish waters, vegetative "cedar" waters percolate down through these marl beds. These acid waters produce a chemical reaction which picks up the iron, in solution, and carries it to the surface. Here, in contact with the air, and aided by a bacterium such as *Leptothrix ochracea,* it oxidizes. The resulting oxides, precipitated as a reddish floc or sludge, settle and are deposited in the beds of slowly moving waters. As deposits pile up, usually mixed with sand, mud, and decayed vegetation, they harden into thick, rocky "ironstone" ore beds, the layers of ironstone sometimes running from 18 inches to 2 feet in thickness. Chemically, bog ore is a low grade variety of limonite, rarely exceeding 40% in iron content.[14] Given ideal conditions, a bog cleared of iron ore may replenish itself in about 20 to 30 years.[15] Thus it is a renewable resource.

The pine barrens possessed an abundance of the three necessary elements for the manufacture of iron: native raw "ore" in the form of "ironstone" or bog iron; water to provide power; and wood from the extensive forests to provide fuel. Even the clam and oyster shells used as a flux, or reducing agent, were readily available from the nearby coastal areas. The shallow, sluggish sections of the Wading and Mullica River systems were especially rich in "ironstone" resources.[16] In general, iron works were constructed where streams could provide enough power to operate air blast bellows for the furnace and hammers for the forges. The adjacent forests provided an adequate supply of wood for conversion into charcoal, the principal fuel used to fire the furnace.

A typical furnace was a large, brick, double-walled chimney or hearth, enclosed by a large shed called the casting or molding house. The chimney itself stood about 20 feet high and was about 20 to 24

Development & Uses

feet square at its base. Its purpose was to reduce the raw bog iron to crude pig iron and castings. Furnaces usually were built near a hillside or elevation and were filled or "charged" from the top. The smelting process consisted of loading, or "charging," the furnace by wheeling measured loads of ore, charcoal, and flux (clam or oyster shells) across a bridge from the hillside, dumping them in alternate layers into the furnace, and then firing the furnace underneath with charcoal or other available wood. The furnace was fanned to extremely high temperatures by a constant air blast from bellows powered by water. Furnaces were started up each spring after the ice was gone and water from the dammed streams could flow through spillways to turn waterwheels. These powered giant bellows which kept the furnaces "in blast" until the next winter freeze when they went "out of blast."

The melted iron was heavy and was tapped off from the bottom and run into "pigs" in sand molds. The waste products, or "slag," were light, floated on top, and were drawn off. The end product of these furnaces was the pig iron which later was cast into pots, pans, kettles, skillets, firebacks, stoves, sash weights, iron pipe, and during the war years into cannons and cannon balls.

In forges, which were more refined versions of furnaces, the pig iron was remelted and refined into wrought iron to be pounded into shape as tools, horseshoes, and rims for wagon wheels. Some furnaces were so constructed as to combine the functions of both furnace and forge. These were called bloomeries.[17] Rolling and slitting mills further processed the wrought iron into sheet iron, nails, and iron rods.

The first iron works in New Jersey was founded at Tinton Falls, near Shrewsbury, Monmouth County, about 1675. One of the oldest iron works in southern New Jersey was at Mount Holly. This iron works was in existence in 1715.[18] However, this was short lived. It was to be many years before the iron industry would benefit from the demands for iron products created by the French and Indian Wars of 1756-1763 and the Revolutionary War for Independence, 1775-1783. Nearly half of the more than 30 furnaces and forges in the pine barrens and in southern New Jersey were built between 1760 and 1800,[19] many during and shortly after the Revolutionary War. In just the two years 1765 and 1766, Charles Read of Philadelphia established four bog iron furnaces in the pines: Atsion in 1765; and Taunton (Medford Township), Aetna (Medford Lakes), and Batsto in 1766.

The bog iron industry reached its maximum development in the early 1800s, especially between 1812 and 1840, and then began to decline markedly after about 1840. It was almost completely gone by 1865. The primary causes of this decline were: (1) the bog ore did not prove to be as readily renewable as anticipated. This resulted in depletion of bog iron ore beds to such an extent that furnaces and forges were forced to import raw materials;[20,21] (2) the decimation of the forests around existing furnaces so that adequate supplies of wood were no longer available to produce charcoal; (3) the discovery and mining in Pennsylvania of anthracite coal, a more efficient fuel than charcoal; (4) the discovery further west of valuable deposits of iron ore of much better quality than the New Jersey bog ores; and (5) the western expansion of railroad transportation. So, the iron industry moved west to use better grade ores in anthracite fired furnaces.[22]

The industry left behind many remnants as reminders of its past activities in our pine barrens. Chief among these are: the many small ponds and lakes created by the dams constructed to provide water power for furnaces, forges, and mills; vast, clear cut acreages of once heavily forested lands; and many abandoned villages and small towns that were located adjacent to the furnaces, where people who had worked in these operations had lived.

Glass

Sand basically is silica, and silica is the basic ingredient in glass, so with such an abundance of sand as a natural resource on the outer coastal plain of southern New Jersey, and given the abundance of wood and charcoal to fire the furnaces, it was only natural that the manufacture of glass would become a major industry of the pine barrens.

With few exceptions, such as the Lebanon Glass Works, most early glass works in 19th century New Jersey were established in southern portions of the pine barrens, generally south of the Mullica River.[23] The first successful glass factory in New Jersey was established in 1739 at Allowaystown, originally Wistarburgh for its founder, Caspar Wistar, in Salem County. Soon other centers became established, some of the earliest in pine barrens territory being: Port Elizabeth, 1779 (on the extreme southern edge of Pinelands National Reserve); Waterford, 1822; Estellville, 1825; Medford, 1825; and Winslow, 1829.

Pinelands and South Jersey glass was known as "green glass" and

Development & Uses

it is generally stated that the very first mason jars for canning were made in the pines near Greenbank, possibly at Crowleytown in Burlington County.[24] However, this is not certain, for one John Fowler who, at 14 years of age, worked in the Tansboro Glass factory in Camden County, is quoted as stating that the first mason jars were invented and made in the Tansboro plant by a Leon Mason, one of the co-owners of the business.[25]

Most of the early glass factories enjoyed only a fleeting prosperity and were plagued by devastating fires that periodically burned the glass "houses" to the ground. Most pine barrens glass factories operated successfully only a relatively short while, from approximately the mid-1820s into the 1880s with a few operating until around 1920.[26]

Brick and Tile Manufacturing

Brick making as an industry in New Jersey dates back to 1665 at Elizabethtown in Monmouth County, just north of the pine barrens.[27] One of the better known brickyards in southern New Jersey operated near Mays Landing in Atlantic County from 1892 to 1931 and used clay from a site east of Babcock's Creek. In 1916, bricks for the Traymore Hotel in Atlantic City were fired in these kilns, some of the ruins of which are still visible today.[28]

Articles such as tile pipes, drains, and pottery were produced from clay deposits found at pine barrens sites such as Old Half Way and Union Clay Works in Ocean County. One of the better known manufacturing operations in the pines took place around 1900 in Ocean County at the Pasadena Terra Cotta Co. plant, or the E.N. and J.L. Townsend & Co. plant of Wheatlands, now in ruins.

Sand and Gravel "Mining"

Another industry based on the use of natural resources in the pine barrens is the "mining" of sands and gravels out of open surface pits. Although these resources are valuable in road construction and several industries, most notably the glass industry, the extraction of these products is viewed by many as wanton exploitation that leaves a blight upon the landscape. At the very least, the stripping of surface vegetation, the digging of deep pits, the resulting erosions, and the upsetting of the water table levels create unnatural disturbances to the ecology of the pine barrens. There are 75 active "mining" sites involving more than 15,000 acres of lands within the Pinelands National Reserve.[29]

Tanning and Currying

An abundance of wildlife provided early settlers, trappers, and hunters with plenty of furs and skins. As a result, tanning and currying was another industry that came into being because of necessity and demand.

The treatment of leather was accomplished by soaking hides in a decoction of tannin that came from the bark of oak trees, although sumac also was used. Sometimes trees were cut down just for the bark alone. An example of an early community where a tannery was located, and from which it took its name, is Tansboro, near Berlin, in Camden County, where a tannery operated as early as 1800.

Following the tanning, leather had to undergo another process, called currying, before it could be used. This meant the leather was prepared by scraping, cleaning, stuffing, beating, smoothing, and coloring. This was done either by the tanner or by others whose only skill was currying leather.[30]

Paper

Another early, but rather localized industry in the pine barrens was the manufacture of paper. Apparently, the actual manufacture of paper was restricted to only three locations: Harrisville (1835-1891); Pleasant Mills or Sweetwater (1861 or 1862-1915); and Weymouth (1866-1887). At Atsion, a mill was constructed for the manufacture of paper between 1851 and 1854 but was never used as such. It was converted to a cotton mill in 1871.

Little is known about the Weymouth or Pleasant Mills/Sweetwater operations but apparently the operators of the latter mill used almost anything they could obtain as raw materials. With linen and cotton fiber in short supply, the mill owners turned to straw, bark, hemp, cat-tail stalks, ground wood, and salt marsh grass.[31]

Much more is known about the Harrisville operations due to the fine research by Fowler and Herbert and by Dellomo. Although the paper mill at Harrisville was located in the pine barrens, its main raw material for the manufacture of its principal product, a brown "butcher-type" paper, was salt marsh hay carted by horse and wagon rom the coastal marshes. This salt hay was supplemented by a combination of other materials such as old rags, rope, scrap paper, and bagging. The pine barrens site was chosen for several reasons: to make use of the water power from the Oswego River; because of the nearby and abundant supply of wood to fire the boilers needed in the manufacturing process; and because there was a ready work force still residing in the village from the previous slitting mill operations.

Cotton

As with paper manufacturing, the manufacture of cotton yarn in cotton mills was a very limited and localized industry in the pine barrens. Apparently, this was restricted to four locations: Pleasant Mills or Sweetwater (1821-1856); Atsion (1871-1882-3); Retreat (Factory) (about 1815 to 1830); and Shreveville, now Smithville, in Burlington County, (1831-1858).

The Shreveville operation did not manufacture cotton yarn from raw cotton. Rather, the enterprise consisted of a cotton spinning and weaving plant, a spool cotton manufactory, and a complex for printing cotton yard goods.

The Pleasant Mills-Sweetwater and Atsion operations were more typical cotton mills and at both sites the two industries of cotton and paper manufacture seemed able to utilize, at different times, the same physical plants for the production of both of these products. At Pleasant Mills the original building was erected as a cotton mill in 1821-1822 and, after a fire, was rebuilt in 1861 as a paper mill. At Atsion the basic factory building was constructed as a paper mill about 1852 but was never used as such. It was converted into a cotton mill in 1871. This was a successful operation for a little over a decade. Cotton was brought up to Atsion by rail from the south, processed into cotton yarn, and then shipped by rail to markets in Philadelphia and New York City. At one time, the mill employed about 170 hands and turned out 500 pounds of cotton yarn a week.[32,33] The former paper/cotton mill building burned to the ground on March 27, 1977 (see page 50).

Taverns

Taverns, or wayside inns, or "hotels," were an important part of the early life of pine barrens residents and travelers. Sometimes called "jug taverns" from the practice of travelers on horseback stopping to get a saddle bottle or jug filled before proceeding on their way, these were welcome houses to both weary travelers and local workers, and often served as social centers for nearby communities. Many of these taverns were prominent landmarks and played important roles in the development and in the folk history of the pine barrens. A few of the more notable taverns in the pines were: Eagle Tavern, near Speedwell (1798-pre-1849) (see page 47); Half

Moon and Seven Stars Tavern at Flayatem (Flyatt) on the old Tuckerton Road (pre-1808-late-1860s) (see page 48); Quaker Bridge Tavern at Quaker Bridge on the old Tuckerton Road (1808-1809 to post-1849) (see page 50); Washington Tavern, later Sooy's Tavern at Washington, on the old Tuckerton Road (1775-post-1854) (see page 51); and Bodine's Tavern at Wading River on the old Tuckerton Road (early 1780s-late 1830s) (see page 53).

By the mid-1800s, when the early industries, principally iron, which attracted tavern keepers to locate in the pines, failed, and furnaces and forges fell into disuse, and people moved away to seek other employment, and when the railroads began to replace the stage coaches, these taverns began to close their doors and pass into history.[34]

Realty Developments

Much of the land in the pine barrens has changed hands many times, often as the result of land speculations. A great many land development schemes involving thousands of acres have been launched and promoted as offering good homesites or as investment properties, only to fail after too few sales forced abandonment. As a result, ownership of many parcels of land is in doubt since some deeds may never have been recorded and some of the same parcels were then sold to unsuspecting new owners.

Some of the more notable early land development or planned community speculative schemes that have failed are: Hanover Farms in the mid-1800s at Upton Station; the Forest Colony between 1830 and 1850 at Retreat; Cedar Crest after 1865 at Bamber; Fruitland in 1862 at Atsion; Raleigh in 1885 at Atsion; and Paisley from 1895 to the early 1900s at old White Horse in Tabernacle Township. Of these and several others, perhaps the most notorious was Paisley.

Paisley was a "get-rich-quick" scheme to develop 1,400 acres of piney woods about halfway between Tabernacle and Chatsworth. Lots were advertised with glowing inducements and promises. The developers even published a promotional newspaper called the *Paisley Gazette* from "the corner of Fifth and Main Streets, Paisley, Burlington County, New Jersey." Only a few lots were ever sold and only a few houses were ever built in Paisley, none of which exist today.[35,36] Instead, along the double curves in Route 532, where Paisley was to have become "The Magic City," there is only the building of a hunting club, and a typical pine barrens white sand road takes off south toward Apple Pie Hill.

Bucking the many failures, however, are a couple of notable successes among early realty developments. Lakewood was developed in the 1880s as a health resort from the village of Bricksburg that surrounded the old Washington Furnace or Bergen Iron Works, and Medford Lakes was developed in the late 1920s and during the 1930s as a summer colony on the site of the old Aetna Furnace, Oliphant's Mill, and cranberry bogs.

Sphagnum, Peat, Plants, and "Greens"

Bog or sphagnum mosses are plants that cover extensive areas in low-lying bogs and cedar swamps throughout the pine barrens. Sphagnum is a plant that can absorb water up to many times its own weight, thus it is a very valuable plant as a ground cover helping to keep the soil and roots of plants moist. For the same reason, it is valuable as a packing material around the roots of nursery plants to keep them moist while being shipped, and as "peat moss," used in the cultivation of plants that require a constant supply of water around their roots. Sphagnum mosses are so absorbent that it is said that the Indians dried and used them as diaper material for their babies.

Sphagnum mosses are reputed to have antiseptic and curative powers and researchers have found that they do indeed contain a bacterium that produces an effective antibiotic. Apparently these mosses were used by the Indians to heal sores and it is well known that during World War I, sphagnum mosses were gathered for use in army hospitals.

Generations of pine barrens residents have gathered sphagnum mosses, which have then been baled and shipped to florists, gardeners, nurserymen, and others. A few old-timers still carry on a little of this activity even today as a supplemental source of income.

Sphagnum is a plant that grows without roots. The upper stems of the plant continue to grow year after year, while the lower stems die off and become compressed, along with leaves and stems from other bog plants, into peat. Although limited quantities of this have been dug from time to time, peat from sphagnum bogs in New Jersey is of an inferior grade and the digging of peat has never become a major pine barrens industry.

Other exploitation of pine barrens vegetation has resulted from extensive collecting. Many native plants such as spurge, wild indigo, colic root, false sarsaparilla, sassafras, wild cherry, bearberry, wintergreen, horse mint, and others have been gathered for drug use.

Drug houses have depended on these wild plants for many years.

"Greens" and pine cones have been gathered for sale as decorations, especially at Christmas time. Branches and wreaths of ground pine, mistletoe, white cedar, pine and holly trees end up for sale each Christmas in metropolitan areas.

Unfortunately, other native plants such as swamp azalea, laurel, and other shrubs, as well as many rare orchids and other wild flowers are dug up each year and removed from bogs and pine habitats for transplanting elsewhere. This is not only illegal, but, because of their very critical and narrow requirements for damp, sandy, highly acidic soils, these plants stand no chance of survival outside of their natural environment. Yet the collection and transporting of these plants has been so extensive that some are now endangered. Their protection is vital for their continued survival.

Agriculture in the Pine Barrens

Most pine barrens soils are sterile, acid, sandy, porous, and light in texture, and do not retain enough moisture for good vegetative development. Only in scattered and generally more southern portions of the barrens where soils are heavier than usual have commercial vegetable and fruit farms developed and been maintained in communities such as Tabernacle, Shamong, Winslow, Hammonton, Folsom, Nesco, Elwood, Egg Harbor City, Vineland, and Millville. On typical pine barrens lowland soils (sandy, acidic, nutrient poor, with high, almost surface level water tables) only native cranberries and blueberries have been successful agricultural products on a commercial scale.[37]

Today, the cultivation and harvesting of cranberries and blueberries are the two most important, and perhaps the only, industries of any substance in the core area of the pine barrens. Together, the cultivation of these two crops cover approximately 11,000 acres (3,000 cranberry, 8,000 blueberry) of pine barrens land. In addition, since 10 acres of upland is desired as watershed for every acre of cranberry in cultivation,[38] cranberry and blueberry farmers utilize and, in a broad sense, protect over 100,000 acres, a sizable portion of New Jersey's pinelands.

Cranberry Agriculture

Cranberry, *Vaccinium macrocarpon,* is a vine, a member of the acid favoring heath family, and a native North American plant that grows wild in low fields, meadows, bogs, and along streams. It was here when the first immigrants from the Old World arrived on these shores, and the native American Indians used cranberries as a food, for medicinal purposes, and as a dye. It was the Indians who introduced cranberries to the early settlers. At first, the colonists also treated cranberries as just a wild and edible fruit which was picked for home use, but around 1816-1820, cranberry cultivation was begun on Cape Cod in Massachusetts.

The first cranberry cultivation in New Jersey was started between 1825 and 1840, with plants taken from the wild, possibly at a bog of one John Webb near Cassville, Ocean County, or possibly at the bog of one Benjamin Thomas, located on the edge of Burr's Mill pond, about eight miles from Pemberton in Burlington County.[39] Nearly all of New Jersey's cranberry bogs have been, and still are, located within the pine barrens.[40]

It is said that the name "cranberry" was given to this vine by the early settlers because they imagined that in its full flowering stage its stem, curled back pinkish white petal-like lobes, and downward protruding yellow stamens resembled the neck, head, and bill of a crane. Hence the early name, "crane-berry," now cranberry.

Areas for growing cranberries are especially located and prepared for their cultivation. Usually, these are situated in lowland areas where the soils are relatively heavy with poor drainage, near stream headwaters and small streams where the waters can be channeled into reservoirs for the constant supply of water needed during the growing season. The area to be "bogged" is cleared of all vegetation, leveled, and ditched with irrigation or drain channels. The main ditch is channeled down through the center (usually) of a bog, with several side ditches, parallel to each other, leading away from the main ditch. Another ditch is dug all around the outside edges of the bog connecting with all the side ditches and the main ditch. The purpose of all these ditches is to maintain the correct water level to keep roots damp by adding water during dry seasons and to carry rain water away from the plants during excessively wet periods. Thus, in reality, cranberry bogs are actually low fields raised slightly above water level rather than true bogs. Finally, a dam or dike is built (raised) all around the outside edges of the bog into which gates are installed to control the flow of water into and out of the bog, as well as to control the level of water in the bog. The tops of these dams or dikes also serve as service roadways for the bogs.

Before planting, a cover of sand is spread over the bed of the bog and this gradually mixes and blends in with the muck and organic material. The bog is then planted with cuttings from "seed" cranberry vines which are spread over the bog and pressed into the soil. In time, these plants produce long, woody runners from which grow short, upright shoots, four to six inches long, which bear the flowers and the fruit. It takes from three to five years for the vines to grow enough to spread out and cover a new bog with sufficient growth to produce a profitable crop of berries. With care, a cranberry bog will bear fruit indefinitely.

An adequate supply of pure, unpolluted water is essential for cranberry culture which is one reason why a cranberry farmer wants so much upland watershed surrounding his bogs. Not only do the plants themselves need a soil with an adequate and controlled moisture content, afforded by irrigation, in order to grow well, but water is needed for other reasons. Because cranberry vines have a very shallow root system, cranberry bogs are flooded throughout the winter months to protect the roots of the plants from freezing. In the spring, after the last hard frost, the water is drained off in time for an early bloom in June. The plants then grow, blossom, and produce fruit during the summer and early fall with just the right amount of moisture around their roots, the levels of water controlled by the height of boards in the cranberry gates. On rare occasions, flooding may be used during the summer as a last resort control of insect infestations.

By late September and throughout October bogs are again flooded, this time each bog individually, in sequence, for the wet harvesting process, a method of harvesting pioneered in New Jersey in the mid-1960s. This is the most colorful time in the cranberry cycle. Cranberries are very buoyant, and when the bogs are flooded the buoyancy of the ripe berries lifts the vines off the bed of the bog. This allows the mechanical cranberry machines, called "beaters," controlled by workers, to strip the cranberries from the vines, and the bright red berries pop up to the surface of the water. Here they are contained by booms, guided onto loading conveyers, dropped into large crates on truck beds, and driven to the processing plants for cleaning and shipment to market.

After harvesting is completed, waters are again drained off and the plants allowed to grow in the controlled moisture soil until the advent of freezing weather when they are once again flooded for winter. Many but not all cranberry farmers in the pine barrens are members of the Ocean Spray cooperative which takes and processes the fruit. A few cranberry farmers dry-pick their berries and this is the fruit that housewives purchase in stores and supermarkets. In the early years, New Jersey produced the largest volume of cranberries in the United States, but today New Jersey is third to Massachusetts and Wisconsin in cranberry production.[41] In 1989, New Jersey cranberry growers had more than 3,000 acres under cultivation and production averaged just under one hundred 100 lb. barrels per acre.[42]

Blueberry Agriculture

Wild blueberries, *Vaccinium corymbosum,* also were an important food crop for the native Indians and the early settlers who followed them, but the development and commercialization of cultivated blueberries is a relatively recent development.

In large measure, this is due to the foresight and energy of one of the more famous native pine barrens residents, Miss Elizabeth C. White. In 1910, Miss White, then in her late thirties, came upon Bulletin No. 93 entitled "Experiments in Blueberry Culture" by Dr. Frederick V. Coville, a botanist with the Bureau of Plant Industry, United States Department of Agriculture. Miss White and her father offered, and Dr. Coville accepted, the use of Whitesbog's facilities for him to continue and expand his experimental studies. Miss White then decided to augment his work by introducing native, wild, "swamp huckleberry" plants selected for having the largest sized berries it was possible to find out in the swamps and woods of the pine barrens. She advertised all over the area for these and made up, for the use of searchers, packages containing three 2 oz. jars, an aluminum gauge, wood plant labels to tag bushes, and typewritten instructions. Although Miss White advertised for "huckleberries," this was simply the common term used by "piney" natives for all of the blue to black berries that grew wild out in the swamps and woods. For the real differences between huckleberries and blueberries, refer to page 130.

Upon receiving reports on the location of spotted bushes, Miss White personally traveled, either on horseback or by horse and buggy, to the site and supervised the digging and transplanting of the bushes which then were set out at Whitesbog. Approximately 100 bushes were found by this method and many were named for their discoverers, all proud "pineys." By cross pollinating, hybridizing, and breeding combinations of Dr. Coville's experimental varieties with Miss White's native stock, the propagation and cultivation of blueberries became a reality as a successful commercial crop as early as 1916.

It should be noted that another of Miss White's innovations was the first use, in 1916, of cellophane as a covering on blueberry boxes to enhance their sale, a technique that revolutionized the fruit packing process. She also was the founder of the Tru-Blue Cooperative to which most blueberry farmers in the pine barrens belong.

Blueberries are members of the same heath family as cranberries, and all cultivated blueberries in New Jersey are varieties of the highbush blueberry. Bushes are started from cuttings and are planted in rows in low, moist, acid soils composed of muck or peat mixed with sand, on lands usually lying somewhat above the lower cranberry bogs and where the water table is within 18 inches of the surface. Blueberries blossom in May and the berries ripen from mid-June into early August, depending upon the variety. Berries used to be picked by hand and many still are, but a more modern method is to use a mechanical blueberry picker that straddles the rows and shakes the fruit from the bushes. New Jersey farmers have nearly 8,000 acres under cultivation. In annual production New Jersey is second in the nation in pounds produced.

Forgotten and Ghost Towns, Historic Sites, and Folklore

It may be difficult for today's casual visitor to the pine barrens to realize that this entire tract once was dotted with sizable communities, some as large as a thousand people. These communities were built around sawmills, gristmills, iron furnaces and forges, glass factories, paper and cotton mills, and agricultural operations. It may also be hard to believe that ocean-going vessels moved upstream on the Mullica River to anchor within a mile of Batsto. But all this is true. Much of it came to a halt in the late 1800s, and the clock began to turn back in the pine barrens. Old towns began to die, old roads became overgrown, forests regrew and thickened, and cedar streams once again flowed freely toward the ocean.

Company Towns

Many old pine barrens industries such as the iron furnaces and forges at Allaire, Atsion, and Batsto, the paper mill at Harrisville, and later, the cranberry bogs at Double Trouble and Whitesbog, to name only a few examples, were really industrial and community centers in which all buildings, including the surrounding village homes for the workers, were built and owned by the owners of the operations.

Many of these operations really were "proprietorships" or "plantations" or "company towns." In a sense, these resembled old feudal establishments in which the workers lived near their jobs, labored long hours, were paid in company script, and they and their families obtained their daily needs at the company store—to which they often were indebted long before pay days. Owners sometimes provided schools up to the fifth grade, with school masters, for the workers' children and built small churches or chapels for Sunday worship conducted by itinerant preachers. In short, these were isolated, self-sufficient communities entirely dependent upon the success or failure of the base industry, the productivity of the workers, and the benevolence as well as the business ability of the owners, who often purchased and sold these operations in rapid succession,

Figure 2. Location of Forgotten and Ghost Towns and Historic Sites described in text.

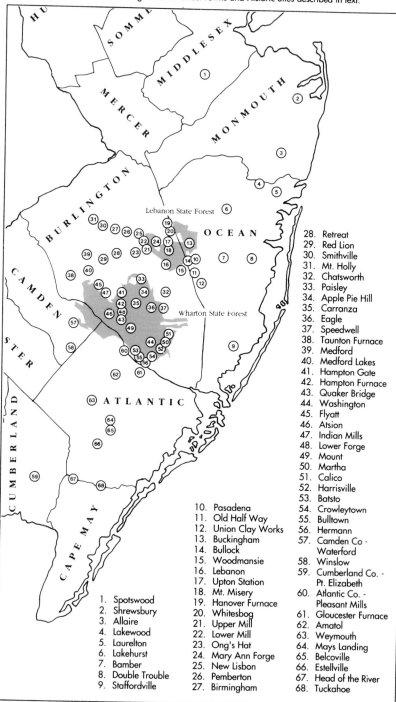

1. Spotswood
2. Shrewsbury
3. Allaire
4. Lakewood
5. Laurelton
6. Lakehurst
7. Bamber
8. Double Trouble
9. Staffordville
10. Pasadena
11. Old Half Way
12. Union Clay Works
13. Buckingham
14. Bullock
15. Woodmansie
16. Lebanon
17. Upton Station
18. Mt. Misery
19. Hanover Furnace
20. Whitesbog
21. Upper Mill
22. Lower Mill
23. Ong's Hat
24. Mary Ann Forge
25. New Lisbon
26. Pemberton
27. Birmingham
28. Retreat
29. Red Lion
30. Smithville
31. Mt. Holly
32. Chatsworth
33. Paisley
34. Apple Pie Hill
35. Carranza
36. Eagle
37. Speedwell
38. Taunton Furnace
39. Medford
40. Medford Lakes
41. Hampton Gate
42. Hampton Furnace
43. Quaker Bridge
44. Washington
45. Flyatt
46. Atsion
47. Indian Mills
48. Lower Forge
49. Mount
50. Martha
51. Calico
52. Harrisville
53. Batsto
54. Crowleytown
55. Bulltown
56. Hermann
57. Camden Co - Waterford
58. Winslow
59. Cumberland Co - Pt. Elizabeth
60. Atlantic Co - Pleasant Mills
61. Gloucester Furnace
62. Amatol
63. Weymouth
64. Mays Landing
65. Belcoville
66. Estellville
67. Head of the River
68. Tuckahoe

thus resulting in numerous turnovers in both ownership and management.

When these historic industries shut down for good, the dependent villages also began to crumble as people began to leave to seek employment elsewhere. As years passed, more and more families moved away and the old communities dwindled into oblivion, leaving behind only deserted ruins and memories. Most of these have reverted gradually into their earlier primitive condition. Today, there may be 100 or more old ghost or forgotten towns in the pine barrens of southern New Jersey, most hardly discernible in the regrowth of mature forests, save for a lingering catalpa or sycamore tree or Norway spruce, or an occasional small stand of ebony spleenwort (fern), almost sure giveaways of past human habitation.

The stories of old, historic industries and villages in the pine barrens are fascinating. Together, these provide a comprehensive picture of natural and human existence in the pine barrens over many past generations. In a small and nature-oriented book as this, it is possible to mention only some of these, with only a few, brief, synoptic statements on each. Such a listing of historic sites follows. Because of its size, and the concentration of bog iron furnaces and forges, and glass works or factories in the Batsto-Mullica and Oswego-Wading River watersheds, the Wharton tract, now Wharton State Forest, has the greatest number of these old sites. The second greatest number is found in and surrounding Lebanon State Forest. Both forests are, for the most part, located in Burlington County.

Historic Sites

A. Sites in Middlesex, Monmouth, and Ocean County

Middlesex County

SPOTSWOOD SAW AND GRISTMILLS (from 1750) and SPOTSWOOD IRON FORGES (pre 1762-1772 and 1763-1780)—On Manalapan River, now South River. Spotswood is located within the northern pine barrens "island" (see page 4). "Several sawmills and gristmills" operated here, as well as two forges which were combined into one operation in 1772.

Monmouth County

SHREWSBURY or MONMOUTH FURNACE, or TINTON or TINTERN FALLS IRON WORKS (pre 1673-pre 1746)—Probably on Hockhockson Brook, a tributary of North Shrewsbury, or Navesink River. Just north and outside of pine barrens but listed because it was first iron works in New Jersey.

Forgotten & Ghost Towns

PALMER'S SAWMILL (from 1750) and WILLIAMSBURG FORGE or MONMOUTH FURNACE, or HOWELL FURNACE or HOWELL WORKS or HOWELL IRON WORKS, or ALLAIRE (pre 1803-1846).—On Manasquam River in Wall Township. A major iron works on northern fringe of pine barrens. In 1822, James P. Allaire purchased the former Williamsburg/ Monmouth/Howell iron works and successfully operated it until 1846. During this period, Allaire developed a complete, self-sustaining community or "iron plantation" with iron works, blacksmith shop, carpenter and wheelwright shops, enameling furnace, dwelling houses for 200 workers and their families, store, church, school, bakery, and other buildings needed for the business and for community living, most of which are still standing and in use today. Approximately 500 persons lived in the village which now is a fine state park.[1,2]

Ocean County

THREE PARTNERS' (SAW)MILL (from 1786) and WASHINGTON FURNACE, or BERGEN IRON WORKS (1814-1818 or 1830, and 1832-1833 to 1854 or 1865)—On south branch of Metedeconk River. Principal product of iron works was water pipe sold and shipped to New York. Site now part of present village of Lakewood.

BUTCHER'S FORGE OR WORKS (1808 - 1840s)—On south branch of Metedeconk River about five miles east of Bergen Iron Works. Principal product was iron water pipe for New York. Site now present village of Laurelton.

DAVID WRIGHT'S SAWMILL (pre 1793) and FORGE (1789-1797), and FEDERAL FORGE or FURNACE, or FEDERAL WORKS, or DOVER or MANCHESTER FURNACE (1795-pre 1855)—Two forges were built on the Horicon branch of Toms River, about eight or nine miles inland from the Toms River landing. These were merged in 1797. Site now part of present village of Lakehurst.

LOWER or MARTHA FORGE, or PHOENIX FORGE (pre 1815-pre 1855)— On a branch of Toms River, below Horicon Lake, a short distance east of present village of Lakehurst. After a fire in 1816-17, was rebuilt and renamed Phoenix. Operated in conjunction with Federal or Dover Furnace.

DOVER FORGE (1809-1868)—On Dover Forge pond on middle branch of Cedar Creek, four miles downstream from Bamber. After original furnace was destroyed by fire, owner Joseph Austin, 2nd, moved to and operated Hampton Furnace until 1828 when he dismantled the latter, carted it to Dover, rebuilt it and put it back into operation. Site along northern edge of present Double Trouble State Park.

FERRAGO FORGE or FURNACE, or BAMBER FORGE (1810-1865)—On

Bamber Lake on middle branch of Cedar Creek about four miles upstream from Dover Forge. Named for Latin word for iron (*ferrum*). Built and operated by John Lacey and family until 1843. This accounts for name of present Lacey Township and for Lacey Road over which iron products were hauled to Forked River for shipment to New York. Site now present village of Bamber.

POTTER'S SAWMILL (from 1765) and DOUBLE TROUBLE CRANBERRY BOGS (1860s -1964)—On Cedar Creek, in Lacey and Berkeley Townships, between Toms River and Forked River. Thomas Potter of Good Luck fame may have been first operator of a sawmill to harvest Atlantic white cedar along Cedar Creek. Water-powered sawmills operated at Double Trouble as far back as 1765 and a sawmill dating back to 1906 is still standing on the property. Double Trouble is best known as a village and company town based upon cranberry agriculture. Most original buildings are still standing. Site now a state park.

WESTECUNK or WEST CREEK, or STAFFORD FORGE (1797-1838-9)—On north branch of West Creek, just west of present village of Staffordville. Extensive buildings were erected and a substantial and successful business was conducted here for nearly 40 years.

B. Sites Within and Adjacent to Lebanon State Forest
Ocean County

PASADENA or WHEATLANDS (pre 1878 - early 1900s)—Beside old N.J. Southern R.R. tracks, Manchester Twp., near Burlington Co. line. Only remains at old village site are broken foundations and some non-native trees. Interesting ruins of old tile factory hidden in pines half mile south of village site. Beck (1936) called this the Pasadena Terra Cotta Co. but Ted Gorden has pieces of pipe lettered "E.N. & J.L. Townsend & Co." of Wheatlands. Area later became site of major fruit farm.

OLD HALF WAY and UNION CLAY WORKS (mid 1800s - early 1900s)—Two small industrial village sites in Manchester (Old Half Way) and Lacey (Union Clay Works) Twps., two and four miles, respectfully, southeast of Woodmansie. Natural deposits of Cohansey clay were "mined" to depths of 10 to 40 feet to extract the clay. Clay from Old Half Way was hauled by "donkey" engines over a narrow gauge railroad to Woodmansie on the N.J. Southern R.R. Most was then shipped to Philadelphia for manufacture into drain and sewer pipe. Clay extracted at Union Clay Works was manufactured on site into sewer pipe. Today, both old sites and their surrounding villages, as well as the trails to them, are so overgrown with vegetation they are difficult to reach. All that remains are large, deep, abandoned pits partially filled with water, bluish-green from the color of the clay.[3]

Forgotten & Ghost Towns

BUCKINGHAM SAWMILL (1880-approx. 1895)—Located along old Philadelphia and Long Branch R.R., later the Pennsylvania R.R., from Camden to Toms River, about half-way between Whitesbog and Pasadena, or halfway between Mt. Misery and Whitings, in Manchester Twp. Sawmill based on "cedar swamping." Only remains at site of old village are old foundations and Norway spruce trees facing abandoned R.R. right-of-way.

Burlington County

LEBANON GLASS WORKS (1851-1866)—Located southeast of Pakim Pond, near camping areas, within Lebanon State Forest. Plant built in cluster of cedar trees for which town was named. Main plant product was window panes but workers also turned out fancy bottles and glass walking canes. Town contained about 20 houses and nearly 200 men were employed in operation. Only visible remains are a mound covering foundations of the old glassworks and a slight rise where there used to be a small cemetery.

UPTON or UPTON STATION (mid-1800s - mid-1900s)—A station and siding on former Philadelphia and Long Branch R.R., later the Pennsylvania R. R. from Camden to Toms River, in Pemberton Twp., on northwest side of Rte. 70, about half way between Mt. Misery and Whitesbog. Receiving and shipping point for both villages. Home of "Rattlesnake Ace," Asa Pittman. Few scattered houses remain along abandoned R.R. right-of-way.

MT. MISERY or PETER BARD'S MILL (1715? - early 1900s)—Old village sited about three quarters of a mile southeast of Rte. 70, on border between Pemberton and Woodland Twps., on northern edge of Lebanon State Forest. Settled by French Huguenots who may have named village for its hill site (100') and for the difficulties of living there. Became major center for lumber industry. Village inn ("hotel"?) a stage coach stop. Remains of old mill dam and site of old sawmill still discernable. In the 1930's, a Civilian Conservation Corps (C.C.C.) camp was located here and this is now the site of a religious conference center.

HANOVER FURNACE (1791-2 - 1863-4)—Major iron works and later charcoal operation on north branch of Rancocas Creek, a few miles upstream and east of Browns Mills, and less than two miles north of Whitesbog, Pemberton Twp. Principal products were pig iron, sad(flat)irons, fireplace backs, water pipe, and cannon-balls (War of 1812). Only stray bricks and scattered slag remain of furnace and "iron plantation" village of nearly 200 homes and 1,000 persons[4] on 24,000 acres of land.

WHITESBOG or WHITE'S CRANBERRY BOGS (1857-1967-8)—Whitesbog village is located one half mile north of Rte. 530, Pemberton Twp., its access road one mile west of the intersection of Rtes. 530 and 70. Whitesbog has a double ancestry. In 1857, James A. Fenwick developed his first commercial cranberry bogs on 108 acres along Cranberry Run. (He had experimented with a small bog at "Skunk's Misery" near Pemberton in 1854.) In 1866, Joseph J. White developed 30 acres of bogs near Rake Pond, New Lisbon. Ultimately, from the marriage of J. J. White to Mary A. Fenwick, a daughter of James Fenwick, Whitesbog developed into the largest cranberry and blueberry operation in the pine barrens. The first cultivated blueberries in the world were developed here between 1911 and 1916 (see Blueberry Agriculture, page 36). Whitesbog is the only old community in the Lebanon area where many original buildings are still standing, with a few still in use. A visit will provide an overview of a 19th and early 20th century cranberry village as a company town based on agriculture. This was sold to the State of New Jersey in 1967-8 and is now a part of Lebanon State Forest.[5,6]

UPPER MILL (from 1720)—Named for mill that once operated along McDonald's branch of Rancocas Creek, about one half mile north of Rte. 70, on northwestern edge of present community of Presidential Lakes. Nothing visible today except low mound on which part of small village stood.

ONG'S HAT (early 1800s)—Small wayside settlement in Southampton Twp. on west side of road about one and 1/2 miles north toward Pemberton from Four Mile Circle. Famous for its odd name based on an old pine barrens yarn. Nothing remains of Ong's Hat except present day roadside food stand.

MARY ANN FORGE (1827 - pre 1877)—Small forge on Mt. Misery Run about three miles southeast of New Lisbon and four miles southwest of Hanover Furnace which supplied this forge with pig iron to convert into bar iron and wagon axles and tires. There are practically no visible remains.

LISBON or NEW LISBON MILLS (pre 1810 - post 1831) and FORGE (1810 - pre 1831)—Located on south side of Rancocas Creek, west of old road from New Lisbon to Cedar Bridge. Forge was small and short lived. Site of sawmill, gristmill, and forge was center for present village of New Lisbon, possibly named by Portuguese workers who came here to cut timber.

THE MILLS (pre 1787 - post 1815) and FORGE (pre 1787 - pre 1811) at NEW MILLS—A sawmill, gristmill, forge, blacksmith shop, dwelling houses, and sundry other buildings were located at or near the main north branch of Rancocas Creek. Site now part of present village of Pemberton.

BIRMINGHAM FORGE (1800-1814)—Located on north branch of Rancocas Creek, a few miles west of New Mills, now Pemberton. In an effort to obtain a more adequate wood supply, the owners moved half their forge fires to Retreat (see next page).

Forgotten & Ghost Towns

RETREAT FORGE (1808-1814) and RETREAT FACTORY (1830s - 1840s)—The owners of Birmingham Forge moved two of their four forge fires to Friendship Creek, a branch of the Rancocas, in Retreat, two miles southeast of Vincentown. In the 1830s - 40s, Retreat boasted a thread and hosiery mill, sawmill, blacksmith shop, store, a number of residences, and a 22-room mansion, the Cushman House.[7]

RED LION (1700s - 1900s)—Small settlement just off intersection of Rts. 70 and 206 (Red Lion Circle) best known for its old Red Lion Inn and a good "piney" yarn.

PARKER'S MILLS (from 1776), SHREVEVILLE (1831-1858), and SMITHVILLE (1865-1975-6)—On north branch of Rancocas Creek, two miles east of Mt. Holly, Access from Rte. 38. After early sawmill and gristmill, site became major industrial complex and company town for over 135 years, first as cotton spinning, weaving, spool cotton and cotton yard goods printing operation (Shreveville) and then as a plant to produce woodworking and other machinery (Smithville). Latter noted for its high-wheeled "Star" bicycle and its famous "Bicycle Railway." Site now a fine county cultural and heritage park.

MT. HOLLY MILLS (early 1700s) and MT. HOLLY IRON WORKS (1730-1778)—One or more sawmills, gristmills, and a fulling mill, along with a major iron works, were located along the south branch of Rancocas Creek, near the south end of Pine Street, Mt. Holly. The iron works, which supplied needed war materials such as camp kettles, cannon, and cannon shot for the Continental Army, was destroyed by the British while in retreat from Philadelphia to New York. (Neither Mt. Holly nor Smithville, above, are strictly within the pine barrens but both are adjacent to and closely associated with considerable pine barrens history.)

C. Sites Within and Adjacent to Wharton State Forest

Burlington County

UNION WORKS or FORGE (1799-80 - early 1840s), CHATSWORTH and WOODLAND COUNTRY CLUBS (1890s - 1920s), and BUZBY'S GENERAL STORE (1865 - present)—Union Forge operated on the west branch of the Wading River, about one mile west of Chatsworth on present Rte. 532, by the dam at Union Forge Pond, now Chatsworth Lake. This small forge made bar iron hauled from the furnace at Speedwell. In the late 1880s, an Italian prince by the name of Ruspoli and his American wife built a palladian villa for themselves on the shores of the same pond, then called Lake Shamong. This led to the development of the Chatsworth, later the Woodland, Country Club that catered to European royalty and nobility and American wealth. From this, the village name was changed from Shamong to Chatsworth in

1900-01. Only remaining evidence of this period is former Shamong Hotel, now the White Horse Inn, beside abandoned R.R. tracks in village. Diagonally across from White Horse Inn is Buzby's General Store, a pine barrens landmark since 1865.

WHITE HORSE (1785 - ?) or PAISLEY (1882 - early 1900s)—On the site of an earlier White Horse Tavern, Paisley was a realty scheme that failed. It was to have become "The Magic City" in the pines along present Rte. 532, halfway between Tabernacle and Chatsworth. A few structures were built but nothing remains today. Only a hunting lodge is now located here and a white sand trail leads south to Apple Pie Hill. (See page 31)

PINE CREST SANITARIUM (early 1900s)—An enterprising doctor envisioned building a tuberculosis sanitarium on top of Apple Pie Hill in Tabernacle Twp. and providing his patients with "health water." This was never realized.

CARRANZA MEMORIAL (1928 - present)—On the Carranza Road, Tabernacle Twp., within Wharton State Forest, is a memorial to Cap't. Emilio Carranza, a young Mexican aviator who crashed at this site on July 13, 1928 while returning from New York to Mexico trying to complete a good will flight to this country. Every year, on the Saturday nearest July 13, a ceremony is conducted at the site by members of the Mt. Holly Post, American Legion, with Mexican and American officials and representatives, to honor Capt. Carranza.

EAGLE and EAGLE TAVERN (1798 - pre 1849)—Eagle was a small crossroads settlement along the southern edge of Tabernacle Twp. between Speedwell and Friendship bogs. A tavern had been licensed here since 1798 and from 1826 to 1849 it was operated by James McCambridge. It was a popular place, frequented by travelers and by workers and residents of nearby Speedwell furnace. Only remaining evidence of old Eagle is a small fenced cemetery in the pines with a single stone bearing the inscription "C.W. - 1839" for Charles Wills who had resided in Eagle.

RANDOLPH'S or SPEEDWELL MILL (pre 1775) and SPEEDWELL FURNACE (1783-1839 or 1842)—Located on the west branch of the Wading River, near the intersection of Rte. 563 and the road west to Eagle and Friendship bogs, about four miles south of Chatsworth. Before the Revolution, Daniel Randolph built a sawmill here then sold the site to his brother, Benjamin, a maker of fine furniture, who built a furnace here in 1783. Benjamin operated the furnace until 1829 then sold it to Samuel Richards. Its principal product was pig iron, some of which was hauled to Union Forge at Shamong (Chatsworth). Speedwell is also noted for a small log cabin schoolhouse which served children of both white settlers and a few remaining local Indians. Today, Speedwell is a well known launching site for canoe trips down the west branch of the Wading River.

Forgotten & Ghost Towns

TAUNTON FURNACE and FORGE, or TANTON or TINTERN FURNACE (1766-7 - post 1830)—Located on Haines or Haynes Creek, a tributary of the south branch of Rancocas Creek, about three miles south of Medford. Built by Charles Read, Sr., it consisted of a small furnace, forge with two or three fires, coal house, dwelling houses, sawmill, and 8,000 acres of timber. Its principal product was pig iron as well as cannon and shot for the Continental Army. After the iron works were abandoned, some of the property was converted, about 1855, into cranberry bogs.

AETNA MILLS (post 1675 - post 1862) and ETNA or AETNA FURNACE (1766-7 - 1773-4)—Two sawmills and a gristmill operated for many years on tributaries of the south branch of Rancocas Creek, three miles southeast of Medford. A furnace, built by Charles Read, Sr., was located between the present Stokes Road and Lower Aetna Lake. Its principal products were pig iron, bar iron, and iron ware such as kettles, flat irons, and firebacks. When Charles Read Jr. took over this furnace, he shut it down. Soon after, Shinn Oliphant built a gristmill and two sawmills on the property and operated these until 1790 when he transferred them to Thomas Ballinger. During the mid 1800's, some of the lakes were converted into cranberry bogs. All are now part of the Borough of Medford Lakes.

THE TUCKERTON ROAD (early 1700s - mid 1800s)—The Tuckerton Road was a main artery through the pines from Philadelphia and Camden to Tuckerton and Little Egg Harbor. From an early horseback trail and service road to haul timber, charcoal, bog ore, iron and other products, it became, in the 1800s, a stage coach route with stops to change horses and for passengers to get refreshed at numerous taverns, inns, and "hotels" along the route. East of the Medford-Jackson Road, the route divided in two: the upper or northern route via Hampton Gate; the lower or southern route, also called the Atsion Road, via Atsion and Quaker Bridge. The two joined at Washington for the final leg to Tuckerton.

FLAYATEM or FLYAT or FLYATT (late 1700s) and the HALF MOON AND SEVEN STARS TAVERN (pre 1808 - late 1860s)—The small settlement of Flayatem, or Flyat, or Flyatt, was located at the intersection of old Tuckerton Road with Jackson Road north from Berlin and Atco, now Oak Shade Road, Shamong Twp. It is said to have been founded by three deserters from Washington's army. The Half Moon and Seven Stars Tavern was opened by John King prior to 1808 at the northwest corner of that intersection. It catered to travelers on the stage road and several notations in the *Martha Furnace Diary* attest to Martha workers visiting the Half Moon. Today, nothing remains of either Flyatt or the Half Moon except a slight elevation where the hostelry once stood.

UNKNOWN MILL (pre 1795), and HAMPTON GATE and HAMPTON FURNACE and FORGE (pre 1795 - pre 1828 or 1850)—Hampton Gate was simply the gateway to Hampton Furnace. It was located beside the old Tuckerton Road and a lane ran to the furnace, built about 1795, and furnace community near the site of an existing sawmill, the "Unknown Mill," on the Batsto River in Shamong Twp. within Wharton State Forest. Around 1810-12, a forge with six fires was also built here The principal products of Hampton Furnace and Forge were flat and round bar iron, wagon axles and tires, sledge hammers, hollow ware, and castings for mould boards. After Samuel Richards purchased the property in 1825, records are fuzzy regarding its continued operation. It may have been dismantled by 1828 and moved to Dover Furnace by Joseph Austin.[8] It may have been destroyed by fire between 1825 and 1828 and not rebuilt. It may have been rebuilt in 1829 and operated until 1850[9]. In 1892, it was purchased by Joseph Wharton along with his purchase of Atsion. Nothing remains of the old gate and almost nothing remains of the old furnace and forge.

LOWER FORGE—An iron forge was located on the Batsto River, in Shamong Twp., below Hampton Furnace and above Quaker Bridge, within Wharton State Forest.

BROTHERTON (1758-1801) and INDIAN ANN (1805-1894)—In 1758, New Jersey purchased 3,285 acres of the Edgepillock Tract, renamed it Brotherton, and encouraged all Indians south of the Raritan River to move to this reservation. A Presbyterian minister, John Brainerd, became superintendent in 1762 and for 15 years he taught the Indians religion, trades, and agriculture. A church and a school were built and both a sawmill and a gristmill were established. It was from these two mills that the present village of Indian Mills in Shamong Twp. was named.

Although most Indians left the area in 1801, some remained and a few returned. One of these latter fathered a daughter, possibly born in 1805, who became known, when she grew up, as Indian Ann (Roberts) throughout Shamong and Tabernacle Twps. She lived on Dingletown Road, farmed her land, wove baskets, peddled berries, vegetables, and baskets from door to door, and smoked a long-stemmed clay pipe. She died in December 1894 and is buried in the Methodist cemetery in Tabernacle. A stone marker was placed over the grave of this last of the local Lenni Lenape Indians by the Burlington County Historical Society.

ATSION FORGE or IRON WORKS (1765 - 1846-8) and PAPER and COTTON MILL (1851-4 - 1882)—Atsion is on the Atsion or upper Mullica River, along Rte. 206, Shamong Twp., Burlington County, within Wharton State Forest. The name derives from the Lenni Lenape name of Atsayunk for the area. Atsion was cut out of the forests in 1765 by Charles Read, Sr. At first the forge was used to convert pig iron, hauled from Batsto, into bar iron. Read sold his interests in 1773 to a partnership which operated the works for nearly 30 years. These owners constructed a gristmill, three new sawmills, and their own furnace thus making Atsion independent of Batsto but by 1805 these partners faced financial difficulties. By 1823, foreclosure left Atsion a neglected property, the iron works deserted, and the whole in "formidable ruins." In 1823-4, Samuel Richards acquired the furnace.

Atsion reached its peak under Richards. At its height, it consisted of 74 dwellings, a church, store, school, and factory buildings consisting of furnace, forge, gristmill, and three sawmills with a work force of over 120 men. The ironmaster's home, built in 1826, is considered to be a fine example of Greek Revival architecture. Richards donated, by deed, the land on which the church was built and still occupies, together with the adjacent burial ground, and stipulated that it was to be used by "all religious denominations professing Christian religion."[10] The property contained 128,000 acres including Atsion, West Mill, Hampton Furnace, and the Skit Mill tract on the Skit branch of the Batsto River. Following Richards' death January 1, 1842, operations at Atsion continued for a short while but were completely abandoned by 1848. In 1849, Richards' daughter, Maria, married a William Fleming who built a massive stone paper mill on the site between 1851-54, but there is no evidence this ever operated as such.

In 1862, the property came into the hands of William Patterson who tried to turn it into an agricultural operation and realty development called the Fruitland Improvement Company, but this failed. In 1871, the estate passed to Maurice Raleigh who enlarged the empty paper mill and converted it to a cotton mill which operated for a little over a decade (see page 30). After Raleigh's death in 1882, the cotton factory closed and his heirs set up the Raleigh Land and Improvement Company to establish a planned community in the pines but this also failed. In 1892, Atsion was sold to Joseph Wharton. Its rail facilities became a central shipping point for the products of Wharton's agricultural operations and the old cotton mill became a sorting and packing house for cranberries. On March 27, 1977, the long vacant, old cotton factory burned to the ground.[11] The former company or village store which Samuel Richards built at Atsion in 1827 is now the State Ranger's office.

QUAKER BRIDGE (1772-post 1849)—This was a small settlement beside the Batsto River, on the old Tuckerton Road, Washington Twp., Burlington County, four miles east of Atsion, in Wharton State Forest. It was named for a bridge the Quakers built in 1772 in order to provide a safe crossing over the river to their yearly meeting in Tuckerton. The site became a favorite stopping place on journeys to and from the shore so, in 1808-9, a tavern opened here which became an important stage coach stop on the Tuckerton Stage Road. This tavern continued to operate until destroyed by fire sometime after 1849. Nothing remains of this village today except a slight rise on which the tavern and surrounding buildings were located overlooking the Batsto River.

There is an interesting old legend connected with stage coach days at Quaker Bridge: the tale of the White Stag which was said to haunt Quaker Bridge and which would appear, usually on a stormy, windy, rain-swept night, to warn of danger ahead. Quaker Bridge has a claim to fame in that a rare and extremely tiny fern, curly grass fern, *Schizaea pusilla,* was first discovered near here in 1805 or 1808. Quaker Bridge also is a well known launching site for canoe trips down the Batsto River to the lake at Batsto.

MOUNT (pre 1846-1870s)—East of Quaker Bridge, after passing through Penn Swamp, the old Tuckerton Stage Road forks left toward Washington, next passing through the completely ghost-like village of Mount, Washington Twp., in Wharton State Forest. About all that is known of this small village is that a tavern was in operation here in 1846 and likely earlier. Today, Mount is simply an open, grassy crossroads situated on a slight rise, completely surrounded by woods.

WASHINGTON and WASHINGTON TAVERN or SOOY'S TAVERN (1775-post 1867)—This old ghost town once was a small village built around a famous pine barrens tavern. Located in the center of Washington Twp., Burlington County, it stood at the intersection of the old Tuckerton Road and the road from Speedwell to Green Bank, within Wharton State Forest. Another road ran from Washington to Batsto, thus it was an important crossroads. The old Washington Tavern may have dated back to the start of the American Revolution. It certainly was operating during the Revolution for it was one of the important recruiting stations in lower Burlington County. Later, it became known as Sooy's Tavern after its popular proprietor, Nicholas Sooy. In addition to becoming an important coach stop for the Tuckerton and Egg Harbor stages, this tavern was a very popular spot for woodsmen, teamsters, and iron workers from Atsion, Batsto, and Martha. The *Martha Furnace Diary* records several instances when elections, town meetings, and other business matters took place at Sooy's. The tavern operated until at least 1854 while records for the village disappear after 1867. Today, the only evidences of past human activity are the old foundation walls of Joseph Wharton's cattle and horse barn and pit silo, destroyed by fire around 1913, and a few old cellar holes across the road.

OSWEGO SAWMILL (1741-post 1860) and MARTHA FURNACE (1793-1844-5)—Martha Furnace was founded on the same site where the Oswego Sawmill had been in operation since 1741, 52 years before Isaac Potts built his bog-iron furnace which he named for his wife, Martha, and which went into blast September 29, 1793. The site is located on the lower Oswego, earlier called the Swago River,[12] also known as the east branch of the Wading River, barely two miles north from present Rte. 679, on the east side of Harrisville Lake, nearly opposite the old Harrisville paper mill ruins, Bass River Twp, Burlington County, in Wharton State Forest. Potts retained ownership for only five years. Following a series of owners, it came into the hands of Joseph Ball and Samuel Richards in 1808. By this time, Martha consisted of a furnace with requisite buildings, stamping mill, blacksmith shop, sawmill, gristmill, store, school, hospital, an owner's or ironmaster's home, a work force of about 60 hands, about 50 dwellings for 400 persons, and around 30,000 acres of land. Much of the pig iron produced at Martha was shipped downstream to the Wading River Forge and Slitting Mill as there was no forge at Martha. Other than pig iron, major products were cast stoves, fire backs, sash weights, kettles, and hollow ware.

One of the most valuable sources of information existing today about Martha Furnace operations is contained in the *Martha Furnace Diary*, kept by company clerks from March 30, 1808 to May 16, 1815. This contains extremely interesting insights into the operations of an early 19th century blast furnace and intimate glances into the day to day lives of the people who lived and worked in the bog iron villages of the New Jersey pine barrens. This furnace prospered until the fires were extinguished in 1844 or 45, after which the site was used for the production of charcoal until 1848. A small sawmill continued to operate until at least 1860. The property was then sold to Francis Chetwood, a real estate speculator who, based on a proposed new railroad, promoted a realty scheme to sell lots just north of Martha in a town to be called Chetwood but neither the railroad nor this town materialized. The site of Martha Furnace, along with nearby Calico, was purchased by Joseph Wharton in 1896.[13,14]

CALICO (1808 - post 1834)—Calico was a small residential village barely two miles northeast of Martha in Wharton State Forest. Likely most of Calico's residents were employed by or were dependent upon the operations of nearby Martha Furnace. It has been suggested[15] that the name Calico refers to calico bush, another name for mountain laurel, which grows in profusion at the site. Today, nothing remains of Calico except a few cellar hole depressions shaded by a grove of catalpa trees which, not being native to the pine barrens, could only have been brought and planted there by former residents.

THE SKIT MILL or BELANGEE'S SAWMILL (1750 or 60 - 1820s), THE WADING RIVER FORGE AND SLITTING MILL (1795-1823), McCARTYVILLE (1832-5 - 1851), and HARRISIA or HARRISVILLE (1851-6 - 1898 or 1914)— This ghost town was located on Rte. 679 on the lower Oswego, or east branch of the Wading River, just upstream from its junction with the west branch, Bass River Twp., Burlington County, in Wharton State Forest. The first development here was either a sawmill and/or a "skit mill," a combination saw- and gristmill[16], built around 1750 or 1760, on either the east (Oswego) or the west branch of the Wading River about a half mile above their junction. This mill remained in operation for about 75 years. The next development was the Wading River Forge and Slitting Mill built by Isaac Potts in 1795. Potts had built Martha Furnace two years earlier and he probably built his forge and slitting mill to process Martha's pig iron into bar iron and sheet iron which was then cut or "slit" into nails. The forge, slitting mill, and a rolling mill were destroyed by fire in 1823 following which the property changed hands several times until, in 1832, it came into the hands of William McCarty.

McCarty rebuilt the operations as a paper mill to produce brown, "butcher" type paper (see Paper on page 29). At the same time he developed a substantial village including the paper mill, gristmill, sawmill, machine shop, blacksmith and carpenter shops, dwellings for the owner, the manager, and workers and their families, and a 550-acre farm, all of which became known as McCartyville. This operation prospered for about 15 years until financial difficulties and a devastating 1846 fire destroyed the mill. Although rebuilt and paper manufacturing resumed, the operation was sold in 1851 to four Harris brothers for whom the town was renamed Harrisia or Harrisville.

Richard Harris was the key figure and he expanded the paper mill and other operations into a thriving community. Harrisville became known as a model village of neat houses behind white picket fences lined up along streets lighted by ornamental iron lamps, the gas supplied by its own gas plant. The paper mill operated for over 40 years but declined in the late 1880s and by 1890 had run into financial trouble, had defaulted on mortgages, the factory burned down, most village residents had moved elsewhere, and the property was sold at a sheriff's sale in 1891. In 1896, the property was purchased by Joseph Wharton. Fire destroyed the old mill and surrounding village in November 1914, and after junkmen and vandals had cleaned out anything of value, the few remaining ruins still can be seen today almost across from and just below Harrisville Lake.[17,18]

BODINE'S TAVERN (early 1780s-late 1830s)—This tavern was located near the east end of the Wading River bridge, about one half mile south of Harrisville and one half mile west of present Rte. 679. This was where the old Tuckerton Road crossed the Wading River and was the last stop for stages going from Philadelphia to Tuckerton and the first stop on their return trips. Numerous references in the *Martha Furnace Diary* show this tavern also was much frequented by iron workers of the early 1800s. As at Sooy's, military exercises, elections, and even court sessions were held at Bodine's.

BATSTO or BATSTOW FURNACE (1766-1848 or 1858) and GLASS WORKS (1846-1867)—Batsto is the most complete and best preserved of old historic sites in Wharton State Forest. It is located on Rte. 542, on the Batsto River about a mile north of The Forks where the Batsto joins the Mullica River in Washington Twp., Burlington County. The name Batsto comes from the Swedish "Batstu," pronounced "baat-stoo," meaning bath house or steam bath and was used by the Indians for "bathing place." Batsto's iron furnace was built in 1766 by Charles Read, Sr. but in quick succession there were several other owners until, in 1784, Batsto was acquired by William Richards. Over the years, the Batsto furnace was rebuilt at least twice: in 1786 and again in 1829. In 1778 a slitting mill was added and in 1781 a forge was built on Nescochague Creek.

Batsto became a principal arsenal for the Continental army from 1770 to 1784 and again during the War of 1812, casting both cannon and cannon balls for the army and, during the Revolution, salt evaporation pans. During the Revolutionary War, Batsto was considered so important that men working there were exempt from military service.

Batsto reached its height between 1812 and 1830 under William Richards and his son, Jesse. In addition to military supplies, major products were stoves, fire backs, sash weights, kettles and hollow ware, fencing, and iron water pipe. A glass works was built in 1846 but it made only flat glass for window panes and trapezoidal panes of glass for gas street lamps. With the ironmaster's mansion on the hill, workers' houses across the river, and industries in between: iron furnace, forge and slitting mill, gristmill, sawmills, glass works, and brickyard, Batsto once was a company town of nearly 1,000 persons.

Batsto's furnace fires were extinguished in 1848 and the furnace dismantled in 1858. The glass works failed in 1867. Finally, more than half the town was destroyed by fire on February 23, 1874. Batsto then was purchased by Joseph Wharton in 1876. Thwarted from realizing his original purpose in acquiring so much pine barrens property (see Water Reserves, page 11), Wharton made major repairs, enlarged the mansion, erected an 80 foot tower over it, and became "lord of the manor" and its agricultural operations. He experimented with raising sugar beets and engaged in livestock breeding as well as lumbering. After Wharton died in 1909, his entire tract of around 97,000 acres was acquired from his heirs by the State of New Jersey in 1954-55.[19,20,21,22,23,24]

CROWLEYTOWN GLASS WORKS (1850-51 - 1866 or 1874)—Two miles southeast of Batsto, on Rte. 542, along the north side of the Mullica River, there is a small state park and picnic area known as Crowley's Landing, a part of Wharton State Forest. This was the site of Crowleytown in Washington Twp., Burlington County. Samuel Crowley operated one (or two?) glass works here and it is said the first mason jars were made here or in nearby Green Bank but this is disputed (see Glass, page 28). Crowley had 400 acres of land here and, in addition to his glass plants which produced mainly bottles and glass for druggists, the town boasted stores, a hotel, and houses for employees. Prior to 1858, Crowley sold his glass works but after a fire in 1866, the works were abandoned and the factory "blown down" in 1874.

BULLTOWN GLASS WORKS (1858-1870)—Bulltown was located about two miles inland from the Mullica River, in an almost straight line north from Hermann City, on small Bull's Creek, Washington Twp., Burlington County, within Wharton State Forest. After Samuel Crowley sold his glass works at nearby Crowleytown, he built a small glass works at Bulltown in 1858. This had only one furnace and five "pots" and produced bottles. By 1870, it had failed and was closed.

HERMANN CITY GLASS WORKS (1870-1873)—One mile southeast of Crowleytown, on Rte. 542, Washington Twp., Burlington County, along the north banks of the Mullica River, Hermann City once was planned to become a major city built around the glass industry. First construction took place in 1869-70 and soon there were some 40 to 70 houses, stores, hotel and tavern, river landing, and glass works which produced bottles and fancy items as decorative pieces and enclosures or "shades" to cover display items as stuffed birds. This came to an abrupt halt in just three years due to the financial panic of 1873. The old hotel, the only recently remaining building, burned to the ground in 1987.

Atlantic County

PLEASANT MILLS or SWEETWATER (1707-1900s)—Just across the Mullica River from Batsto, on the Nescochague Creek and Lake, on Rte. 542, Mullica Twp., is the village of Pleasant Mills. First known to early settlers as a Lenni Lenape summer village called Nescochague, it was settled by the Scotch in 1707, was the site of a sawmill as early as 1739, and was established as a plantation in 1762 by Elijah Clark. The Clark home later became the setting for the Kate Aylesford mansion in Charles Peterson's best selling novel of that name published in 1855. By some accounts, it was from this romantic novel that the name Sweetwater was drawn. Pleasant Mills became associated with Batsto when it was sold in 1787 to former Batsto ironmaster, Joseph Ball.

The Pleasant Mills - Sweetwater area and the nearby Mullica River was the locale where a noted Tory outlaw by the name of Joe Mulliner led a band of refugees who became renegades and brigands and terrorized the Mullica River area of the pines. Joe was captured dancing in a tavern one night in 1781, tried in Burlington, hung, and buried along the banks of the Mullica.

The small church in Pleasant Mills, built in 1808, is the outgrowth of a log cabin church built in 1758 by Elijah Clark. In its early days, it was an important pulpit for itinerant preachers. It also was, and still is, the church for the Batsto settlement. The headstones in the graveyard beside the church tell an interesting story of declining resources in the later years at Batsto.

At the outflow from Lake Nescochague, a cotton mill named the "Pleasant Mills of Sweetwater," possibly, by other accounts, the first use of the Sweetwater name, was constructed in 1821 or 22, but it was destroyed by fire in 1855-6. It was rebuilt as a paper mill in 1861 but again burned to the ground in 1878. The present structure was built as a paper mill in 1880 and was operated until 1915, with feeble revival attempts until 1925, when it was dismantled. It was remodeled and used briefly as a summer playhouse around 1951-53. Part of the structure is now a private home amid interesting old ruins.

GLOUCESTER FURNACE (1813-1848)—This iron works was located on Landing Creek, a branch of the Mullica River, within the present limits of Egg Harbor City, Atlantic County, but about six miles northeast of city center, nearly opposite Lower Bank in Burlington County. The furnace was established in 1813[25] by John Richards. In 1833 (1834?), a notation in the *History and Gazetteer of New Jersey* described Gloucester (now part of Egg Harbor City) as a post town containing a furnace, grist and sawmill, store, tavern, and a number (25) of dwellings for the 60 (125?) workmen and their families totaling some 300 persons. The furnace annually produced about 800 tons of iron, chiefly castings, and had 25,000 acres of land annexed to it.[26] The furnace closed in 1848.[27]

D. Sites in Camden, Atlantic, and Cumberland Counties

Camden County

WATERFORD WORKS GLASS FACTORY (1822-24 - 1892)—This glass works was located in a village that became known as Waterford Works after the name of the plant. It was in Winslow Twp., Camden County. A glass factory was built here between 1822 and 1824 and operated successfully for about 60 years until it closed in 1880 and burned in 1892[28]. It was noted for the high quality of its products, especially its Waterford flasks.

WINSLOW GLASS WORKS (1829-post 1884)—This glass works was located in Winslow Village, Winslow Twp., Camden County. Established here in 1829, it manufactured hollow ware and became famous for its beautiful flasks made in 1836-37. In 1851, a new owner added a gristmill and sawmill to the operations which continued until 1892.[29]

WINSLOW JUNCTION—About one mile from Winslow village is the now quite deserted Winslow Jct. crossing of three early railroads: the New Jersey Southern R.R., later the Central R. R. of New Jersey, recently abandoned by Con-rail, the Camden and Atlantic R. R., and the Philadelphia and Atlantic City R. R., the latter two to later become the Pennsylvania-Reading Seashore Lines.

Atlantic County

AMATOL (1918-1919)—Amatol was a World War I munitions (shell production and loading) depot and boom town that developed "overnight" toward the end of the war and was abandoned just as quickly after the November 1919 armistice. It was located east of Hammonton, north of the White Horse Pike, on a 6,000-acre tract between East Hammonton and

Elwood, Mullica Twp., Atlantic County. Planned for a maximum of 25,000 persons, it had reached a population of 5,500 and the plant was 90% completed when it closed. Between 1926 and 1933, the site was used as an auto speedway and testing ground. Today, the site has returned to low oak woods and the only standing reminder of Amatol is the former administration building on the White Horse Pike which was remodeled and is now a state police barracks.[30]

WEYMOUTH FURNACE and FORGE (1802-1865) and PAPER MILL (1866-1887)—Weymouth, located on the Great Egg Harbor River, Hamilton Twp., Atlantic County, on Rte. 559 from Hammonton to Mays Landing, became an iron works community around 1801-02. The furnace produced a variety of items from cannons and cannon balls for the War of 1812 to stoves and, especially, water pipe much of which was used in Philadelphia. Records indicate the furnace produced about 900 tons of castings annually and the forge, which had four fires and two hammers, made about 200 tons of bar iron. There also was a grist- and sawmill, company store, dwellings for the families of 100 or so workers, and 85,000 acres of land. A meeting house, built in 1807, is still being used and many of the gravemarkers in the adjoining cemetery were cast at Weymouth. In 1808, the operation was purchased by Samuel Richards. In 1846, a wood tramway was built from the works to a wharf on the Great Egg Harbor River at Mays Landing to make it easier and quicker to haul products to port for shipment to Philadelphia and New York. Weymouth forge burned to the ground in 1862 and three years later the furnace also was destroyed by fire. The next year, Stephen Colwell built a paper mill on the site and this operated until 1887 when it was shut down and abandoned. Following a series of fires and vandalism, only stone arches, a ruined chimney stack, and some foundations from the old mill still stand by the cedar stream in a county park.

MONROE FORGE or WALKER'S FORGE (1816 or 1820-21 - 1853)—Named for James Monroe, President of the United States, but built by Lewis M. Walker in 1816[31] or 1820 or 21[32] this forge was located on the South River, a tributary of the Great Egg Harbor River, about three miles west of Mays Landing, in Weymouth Twp., Atlantic County. In addition to his forge, Walker also built a sawmill and was successful for many years, employing up to 100[33] men. The forge continued in operation until Walker died in 1853.

BELCOVILLE (1917-18 - 1919)—Belcoville was a World War I munitions (shell-loading) plant and workers' community. It was located about one mile southwest of Mays Landing, on the southeast side of Rte. 50. At the time it was closed down the community and plant consisted of 205 dwellings and 84 factory buildings, with 1,800 employees.[34] Remaining ruins may be seen within same Estell Manor park as next site.

ESTELL GLASS WORKS (1825 or 1834 - 1877)—Within the relatively new Estell Manor (Atlantic) County Park at Estellville, on Rte. 50, about three miles southeast of Mays Landing, are the interesting old stone walls and ruins of this glass works built by John Scott in 1825[35] or by Daniel and John Estelle in 1834[36] This was a large and successful operation for nearly 50 years, employing in 1844, as many as 80 men. The works was abandoned in 1877.

ETNA or AETNA or TUCKAHOE FURNACE and FORGE (1816-1832)—This furnace was located on the north bank of the Tuckahoe River, southeast of Head of the River and south of the Head of the River Road, Estell Manor, about four miles west of Tuckahoe. It was built between 1815 and 1817 and was operated until 1832. A forge connected with this furnace was located by a pond about two miles from the furnace. Around the furnace were a sawmill, gristmill, manor house for the manager, and between 40 and 50 workers' homes. During its most active period, there were about 200 men employed around the furnace, forge, and mills.[37] The stack of the furnace was 25 feet high with three foot thick walls but today nothing can be found of this tremendous structure except a few old bricks and footings. The nearby Head of the River church, built in 1792 and surrounded by its typical burial ground, is still standing, but services now are conducted only once a year on the second Sunday in October

Cumberland County

BUDD'S IRON WORKS or CUMBERLAND FORGE (1785-1800) and FURNACE (1810-1840)—A furnace and forge operated for many years on Manumuskin Creek, Maurice River Twp, Cumberland County, about five miles north of Port Elizabeth. These were built by Eli Budd and his son, Wesley. After several changes in ownership, they were sold to Edward Smith who improved the ironworks and operated it until 1840 when it was abandoned due to lack of wood to make charcoal.[38]

COHANSIE IRON WORKS (1753 - pre 1789)—Twenty-two years before the American Revolution, there was a small iron forge in operation on Cedar Creek, Fairfield Twp., Cumberland County. A 1753 advertisement describes the forge and property and offers it for sale. Nothing else is known about this forge or how long it operated.[39]

The Jersey Devil

The people whose families have remained in the pine barrens for many generations proudly call themselves "pineys." By and large, these are rugged individuals who not only work hard but love to spin a good yarn. One of the best known of many old pine barrens legends revolves around the "Jersey Devil." A simplified version of only one of the many variations of this story follows.

It seems that back around 1735 a certain Mrs. Leeds of Leeds Point or Estellville, both in Atlantic County, was about to give birth for the 13th time. Not wanting the child, she is alleged to have complained in anger, "I don't want no baby. I hope it's a devil." So, when the child arrived on a violent, wind-swept, stormy night, it was just that—a boy with horns, the face of a horse, a snakelike body, cloven hoofs, bat wings, and a long, forked tail. As midwives fled in terror, the demon quickly grew in size and strength, stretched its wings and, with a screech, rose and flew out the window into an adjacent swamp, its harsh cries still heard over the violence of the storm.

It is reported that the devil was seen in the area for several years, winging its way from farm to farm, raiding chicken coops for food, causing the milk of cows to go sour, and generally disrupting farm life throughout the area. Over the next 200-plus years, reports of Jersey Devil sightings and depravations spread throughout southern New Jersey from Freehold and Trenton south to Cape May, as people told and retold its story. Right up to the present day, variations and additions to this yarn have continued to grow, far beyond the ability to relate in this brief recounting. For further information, refer to Vivian, 1969, McCloy and Miller, 1976, and McMahon, 1980.

Present and Future Uses and Preservation

In addition to agriculture, some of the principal uses of the New Jersey pine barrens are recreational: camping, canoeing, hiking, hunting, and the observation and photography of nature subjects. Another important use is valuable scientific research.

State Forests and Parks

There are a number of state forests and parks within the area of the pine barrens. The two largest of these are:

WHARTON STATE FOREST—A 108,000-acre tract of core area pine barrens located mainly in Burlington County, with smaller segments in Atlantic and Camden Counties. Starting around 1873, and continuing for nearly 30 years, Joseph Wharton, a Philadelphia philanthropist and founder of the Wharton School at the University of Pennsylvania, began to purchase large tracts of forest lands in the pine barrens of southern New Jersey, including old, abandoned bog iron furnaces, glass factories, paper mills and old ghost towns (see previous section on Wharton area historic sites.) When Wharton's original purpose to sell water to the City of Philadelphia was prevented (see Water Reserves, page 11), he then operated his 97,000 acres as an agricultural estate. After Wharton died in 1909, several attempts were made to have the state purchase the Wharton Tract but this did not materialize until 1954-55. The forest is now a well developed recreational facility.

LEBANON STATE FOREST—Contains about 32,000 acres of upland pine and oak forests and cedar swamps in Burlington and Ocean Counties. It is named for the old, short-lived Lebanon Glass Works (1862-1867) formerly located within its borders (see Lebanon Glass Works, page 44). Pakim Pond, the center of the recreational area in the forest, was formerly an old cranberry bog reservoir and the name "Pakim" comes from a Lenni Lenape word meaning cranberry. The deserted village of Whitesbog and its surrounding cranberry bogs and blueberry fields are a relatively recent (1967-8) addition to Lebanon State Forest. The forest is a well developed recreational facility.

Present & Future Uses

Other state forests and parks are:

ALLAIRE STATE PARK—In Monmouth County, contains 3,035 acres and includes the deserted village of Allaire and the site of the former Howell Iron Works. The village of Allaire is significant as an historical museum because most of the buildings necessary to an operating bog iron furnace and forge community during its heyday in the mid-1800s are still standing and open for visitation. The park is well developed and the former village is the centerpiece of one of New Jersey's nicest small state parks, to which a visit to observe early American industrial and community life as it existed over a century ago can be a very rewarding experience (see page 42 for further details).

PENN STATE FOREST—Contains 3,300 acres on the Oswego River in Burlington County. The 90-acre Lake Oswego is a former cranberry bog reservoir. Fine upland forests of pine and oak, and white cedar swamps surround the lake. Where the waters of the lake spill over the dam is the starting point for one of the most popular of all pine barrens canoe trips, from the lake, down the Oswego River, to the lake at Harrisville.

Throughout much of the 1960s there was considerable speculation and discussion of purported plans by federal, regional, state, and municipal authorities to build a jetport in the pine barrens. One of the sites seriously considered was Penn State Forest, where it was proposed to drain and fill in Lake Oswego to become the main runway for the jetport. Fortunately, there was such widespread opposition to this proposal that it never got off the ground. Although the park is classified as developed, it lacks in on-site supervision and maintenance.

DOUBLE TROUBLE STATE PARK—Is the site of an early sawmill (1765) and, later, a major cranberry operation (1860s to 1964). It contains nearly 2,000 acres on Cedar Creek in Ocean County. It consists mainly of old cranberry bogs, a former typical cranberry village, cedar swamps, and adjacent uplands, all purchased by the State of New Jersey in 1964 and 1975 for water conservation and recreation. This park has not yet been developed.

GREEN BANK STATE FOREST—Consists of 1,800 acres of mixed pine and oak woodlands and cedar swamps in southeastern Burlington County. It is mostly undeveloped.

BASS RIVER STATE FOREST—Contains 9,000 acres in southeastern Burlington County. The 67-acre Lake Absegami is the center of recreational activities. Absegami derives from the Lenni Lenape Indians and means "body of water with the shores visible." The name originally was applied to the bay between the mainland and the islands of Atlantic City, now called Absecon. Park is well developed.

BELLEPLAIN STATE FOREST—Contains 6,500 acres in northern Cape May County. The 26-acre Lake Nummy is an abandoned cranberry bog surrounded by pine, oak, and cedar forests in better stands than in the core area of the pine barrens due to better soil conditions and less damage by fire. Lake Nummy was named for the last Lenni Lenape Indian sachem (chief) to rule in this section of New Jersey. The remains of Sachem Nummy are said to have been buried on Nummy Island near Hereford Inlet, between Stone Harbor and Wildwood. Park is well developed.

PARVIN STATE PARK—Has 1,100 acres in Salem County and is often described as one of the finest state parks in New Jersey. Situated on the edge of a former pine barrens "island," it contains considerable pine barrens terrain, along with its general woodland habitat. The park is especially rich in both flora and fauna and is well developed.

In addition to the above, the state owns or otherwise controls 20 wildlife management areas and other sites as recreation areas and game farms within the confines of the pine barrens.[1]

Hiking

The single hiking trail in the pine barrens is the Batona, a wilderness path that threads through pine and oak woodlands, alongside cedar streams and swamps, and past old forgotten habitations. From north to south it stretches 50 miles from Ong's Hat and Carpenter's Spring in Lebanon State Forest to Batsto in Wharton State Forest, then east 10 miles to Evans Bridge on Route 563, and then southeast a final 10 miles to Lake Absegami in Bass River State Forest. The trail, started in 1961, was built by the Batona (Ba-to-na—back to nature) Hiking Club of Philadelphia.

In addition to enjoyment of wilderness along the entire trail, a few other highlights along the route are Pakim Pond in Lebanon State Forest, the fire tower at Apple Pie Hill, Carranza Memorial, Quaker Bridge, cedar swamps along the Batsto River, and historic Batsto Village. Because of the generally level and sandy terrain of the pine barrens, the trail is easy to hike as well as to follow its pink trail blazes painted on the trees along the way. Since the trail cuts across a number of roads, such as Routes 70, 72, 532, and 563, the trail can be easily reached and hikes of varying lengths can be started and ended at a number of points.

A folder and map of the Batona Trail is available at Bass River, Lebanon and Wharton State Forest offices.

Camping

There are five different types of camping opportunities in the pine barrens:

- Organized and supervised campsites with facilities suitable for tent and trailer camping are available year-round at Allaire, Bass River, Belleplain, and Wharton State Parks and Forests, and seasonally at Lebanon and Parvin.

- Wilderness type camping with primitive facilities is offered at seven different camping areas in Wharton State Forest. Most of these are accessible by car or by canoe, but few are suitable for other than tent camping.

- A limited number of cabins are available at Bass River, Lebanon, Parvin, and Wharton. (Complete information on campsites and reservations is outlined in a folder, "Campgrounds. New Jersey State Parks and Forests," available at any state park or forest office.)

- People on two or three day canoe or hiking trips who want to camp overnight must do so at state established campsites along the route. Only tent camping is recommended. Camping is not permitted at any other sites along any of the canoe routes or the Batona Trail.

- There are a number of private campgrounds, most open year-round and most offering all types of camping, from tents to R.V.'s.

Canoeing

In recent years, canoeing has become one of the most popular forms of recreation in the pine barrens. Almost year-round, nearly every weekend, canoeists set out on one or two day trips on the several streams in the barrens.

As mentioned earlier, the outer coastal plain, on which the pine barrens is located, is generally quite flat. Most pine barrens streams rise in headwater swamps and have a very shallow gradient as they move downstream from inland toward the Atlantic Ocean or the Delaware River. These are not white water streams, but, depending upon high or low water conditions, they nevertheless flow rapidly and require a reasonable amount of skill to navigate. Among the most popular one or two day canoe routes are:

BATSTO RIVER—Quaker Bridge to Batsto. Joins Mullica River below Batsto at Pleasant Mills. Swift flowing, but shallow. Through wilderness area.

UPPER MULLICA (ATSION) RIVER—From below Atsion Lake, Rte. 206, to Pleasant Mills. Narrow, swift, deep in places. Considerable snag carries.

LOWER MULLICA RIVER—Pleasant Mills to Green Bank. Clear, wide, and quiet.

OSWEGO (EAST BRANCH OF WADING) RIVER—From below Oswego Lake dam in Penn State Forest to Harrisville Lake on Rte. 679. Clear and swift.

WADING RIVER, WEST BRANCH—Speedwell on Rte. 563 or Hawkins Bridge on Tulpehocken Creek to Evans Bridge (Rte. 563) or Bodine Field (if camping) or Beaver Branch (if not camping) or to Wading River Bridge (Rte. 542). Wildest river in state. Should be seasoned canoeists.

RANCOCAS CREEK—Browns Mills (Rtes. 530 & 545) to Pemberton (Rte. 530). Clear and quiet. Wilderness to New Lisbon. Also: Pemberton to Mt. Holly (Rtes. 537 & 541). Clear and quiet.

GREAT EGG HARBOR RIVER—From old teakwood dam at Penneypot on Black Horse Pike to Weymouth just off Black Horse Pike. Wilderness area. Narrow and swift. Considerable snag carries. Also: Weymouth to Mays Landing (Rtes. 40 & 50). Upper half narrow, swift, with considerable snags. Lower half clear, wide, but still swift.

For those who do not have their own canoes, there are a number of canoe liveries throughout the region. For further details, consult any or all of the following:

Exploring the Little Rivers of New Jersey, J. & M. Cawley, Rutgers University Press, New Brunswick, NJ, 1961;
Canoe Trails of the Jersey Shore, J. & B. Meyer, Specialty Press, Ocean (Asbury Park), NJ, 1964;
Canoeing the Jersey Pine Barrens, R. Parnes, East Woods Press, Charlotte, NC, 1984.

Hunting

For generations, dating back to our native American Indians and our early colonists, the pine barrens has been known as a fine area for hunting wild game. In earlier years most of this was taken to provide food for the table, but in more recent times, hunting has become more of a sport. Today, there are numerous hunting lodges located throughout the pine barrens which serve as headquarters for members of hunting clubs to participate in this form of recreation, particularly during the hunting season for white-tail deer.

Present & Future Uses

Brief History of Preservation Efforts

Over the past 30 years, there have been several efforts to establish programs for the protection and preservation of the pine barrens. Possibly the first of these, in the early to mid-1960s, was a proposal for a pine barrens national monument. This proposition was co-advanced by an informal group of concerned citizens who had banded together in 1957 under the name of the Pine Barrens Conservationists, by the New Jersey Audubon Society, and by the Wildlife Preserves, Inc. Other early groups were the Federation of Conservationists United Societies (FOCUS) based in Ocean County, and the Citizen's Committee to Preserve State Lands.

Preservation efforts were greatly stimulated in the early 1960s by the announced search by the New York Port Authority for a new jetport site. This was followed by the 1965 proposal of the Pinelands Regional Planning Board, controlled by Ocean and Burlington County Freeholders, to locate that jetport and a surrounding "New City" of a quarter of a million persons in the pinelands. A part of that proposal was to fill in the bed of the present Lake Oswego in Penn State Forest and transform that into the main runway of the proposed jetport!

The first state-legislated body charged with the responsibility of protection of at least a part (southern and eastern Ocean and Burlington Counties) of the pine barrens was the Pinelands Environmental Council (PEC), established in January, 1972. Its report, "A Plan for the Pinelands," was produced in 1975. Although the purpose of this council was commendable, it was forced to rely on hoped for pressure of public opinion because it lacked authority to mandate and enforce.

Several state-wide organizations then began to collaborate to develop a more comprehensive and effective program. Among these were the New Jersey Audubon Society (see *New Jersey Audubon* magazine, June 1976), the New Jersey Conservation Foundation, and The Pine Barrens Coalition, organized in 1977 as an umbrella organization composed of representatives from a wide variety of environmentally oriented organizations and individuals.

A new proposal entitled, "A Plan for a Pinelands National Preserve" was released in July 1978. This plan was developed by the Center for Coastal and Environmental Studies, Rutgers University, and was funded by the North Atlantic Regional Office, National Park Service. This plan called for the division of the pinelands into two zones, an inner preservation zone and an outer protection zone, and recommended a regional commission be established to administer land management techniques such as acquisition of property, purchase of development rights, and restrictive zoning policies. In many ways this report was similar to one proposed by the Governor's Pinelands Review Committee that was released later in 1978.

Finally, the Congress of the United States passed, and President Carter signed, on November 10, 1978, a new law creating a one-million acre Pinelands National Reserve, contingent upon collateral action by the State of New Jersey. Under prodding by Governor Byrne, the State Senate and Assembly passed the Pinelands Protection Act, and the governor signed it on June 28, 1979.

Thus came into being the present Pinelands Commission, a state agency, which is charged with the responsibility to preserve and protect that portion of the pine barrens (934,000 acres) for which it is responsible. The commission then proceeded to develop a Comprehensive Management Plan (CMP), which provides for an inner preservation area and a surrounding protection area. This was adopted in November, 1980, and approved by the state and federal governments in January, 1981. It is this CMP and the members and staff of the Pinelands Commission that are the authorities for the future protection and preservation of the pinelands. Finally, in 1983, the United Nations recognized the importance of the pinelands by designating the region an International Biosphere Reserve!

Preservation and Future

What of the future for the pine barrens? What lies ahead for this unique wilderness area? Once the center for many small and diverse industries and villages, now long gone, the pine barrens today is a relatively undisturbed, wilderness area.

At this writing (1990), there does not appear to be any major reason why the present status quo in the pine barrens should not continue into the foreseeable future provided the members and staff of the Pinelands Commission adhere to the provisions of its CMP and do not allow waivers from its standards. The major agricultural crops of cranberry and blueberry cultivation should continue at present levels or even increase somewhat rather than decline. The present limited cutting of cordwood for pulpwood and as a domestic fuel probably will continue at about its present level. Present recreational uses of camping, canoeing, hiking, hunting, and other activities should continue unabated, largely under state control due to the extent of state owned and/or controlled land in the barrens.

A major portion of what the future holds for the pine barrens rests partly with the federal government but mostly with the State of New Jersey and its citizens. This is because the federal and state combined own or control about 280,000 acres or between 20 to 25% of the pine barrens area,[2] the state owning nearly half of this in the core or preservation area in just three state forests: Wharton, Penn State, and Lebanon. If at a future time any of the federal facilities, Ft. Dix Army Base, McGuire Air Force Base, or Lakehurst Naval Air Engineering Center, should close, it would be incumbent upon the state and its citizens to acquire these properties to provide additional protection to the pinelands. In fact, to reduce or eliminate constant pressure—present and future—on the Pinelands Commission to reduce standards, to grant waivers, and/or to reach weakening compromises, all of which result in developmental encroachments, it may become necessary that more federal and/or state acquisition take place to provide permanent protection.

One of the more serious threats to the future of the pine barrens is that the vast water reserves in the aquifers under this land may be tapped to such an extent to meet regional, state-wide, and/or metropolitan demands that there may not be sufficient reserves left to meet the needs of local residents and agriculture. Excessive use of these water resources may also have an adverse effect on the natural vegetation of the pine barrens and could so drastically alter the surface conditions as to invite possible desertation.

But for the moment, the machinery is in place to preserve the pine barrens for the foreseeable future. The Pinelands Commission is charged with the responsibility and it is incumbent upon both commission members and staff, as well as the politicians whose legislation allows the commission to exist, to enforce the preservation mandate of the citizens of New Jersey. One present weakness is that the commission lacks adequate enforcement powers, including assessment of fines. Hopefully, this will be rectified with the passage of enabling legislation. Another key to future preservation is the appointment by the governor (for members at large) and by the counties (for their representatives) of environmentally oriented and concerned commissioners committed to pinelands protection.

Finally, it is important for citizens and citizen groups interested in the future of the pine barrens, such as the recently formed Pinelands Preservation Alliance, to closely monitor the work of the commission to make certain that standards are maintained and not lowered, and that these standards, as delineated in the Comprehensive Management Plan, are properly administered. In the final analysis, it is the citizens who will decide the ultimate fate of the pine barrens. It is our responsibility to pass this wilderness heritage on, in its natural state, to our heirs.

The following are recommended for further reading on the preceding four sections. See References Cited at back of book for publisher details.

Beck, H.C. 1936 Forgotten Towns of Southern New Jersey

Beck, H.C. 1937 More Forgotten Towns of Southern New Jersey

Beck, H.C. 1945 Jersey Genesis. Story of the Mullica River

Beck, H.C. 1956 The Roads of Home. Lanes and Legends of New Jersey

Bisbee, H.H. 1971 Sign Posts. Place Names in History of Burlington County

Boyer, C.S. 1931 Early Forges and Furnaces in New Jersey

Boyer, C.S. 1962 Old Inns and Taverns in West Jersey

Collins, B.R. & E.W.B. Russell 1988 Protecting the New Jersey Pinelands

Forman, R.T.T., ed. 1979 Pine Barrens Ecosystem and Landscape
(Thirty-three chapters by different authorities on major facets of pine barrens ecology)

Harshberger, J.W. 1916 Vegetation of the New Jersey Pine Barrens

McCormick, J. 1970 The Pine Barrens. A Preliminary Ecological Inventory

McMahon, W. 1964 Historic South Jersey Towns

McMahon, W. 1973 South Jersey Towns. History and Legend

Pierce, A.D. 1957 Iron in the Pines

Robichaud, B. & M.F. Buell 1973 Vegetation of New Jersey

Species Descriptions

The descriptions of flora and fauna which follow, from non-flowering plants through mammals, are generally of those forms (species) which are most common, most conspicuous, and most likely to be observed by the average individual on a field trip into the pine barrens. Also included are a few of the more unique or particularly characteristic forms known to occur in the pine barrens of New Jersey.

Every effort has been made to write descriptions in a plain, concise manner so as to enable identification without the use of technical, scientific terms or keys. As far as possible, technical terms have been transposed into everyday, lay language through the use of simple words and short phrases so the descriptions can be more readily followed.

Common names are used as the main listing for each plant and animal because of the non-technical nature of this field guide. For species which do not have a generally recognized common name, a literal translation of the scientific (Latin) name may be given. Common names can be very useful, especially if they are accurately descriptive. On the other hand, some common names are descriptively misleading, and many plants and animals have more than one common name, sometimes even several, so there is no accepted "right" common name. For these reasons, scientific names are included with all descriptions as these names are constant and universal, are an exact means of identification and reference, are used by biologists and serious amateurs everywhere, and are the only ones used in many technical and scientific references. All scientific names are in Latin, the first name being the genus to which the species belongs, the second name being the species itself.

Perhaps the most difficult task in authoring this guide has been the necessary decision making with respect to which species to include. In order to keep this within a reasonable physical format and pagination, it was agreed with the publisher to describe between 300 and 500 species of flora and fauna. To realize how difficult this has been, one only needs to review the following table.

Species Descriptions

Taxon	No. known species in pine barrens (or N.J.) and reference	Number included	% included
Algae	360 (Moul & Buell, 1979)	3	1.0
Fungi	?	5	?
Lichens	275 (NJ) (Britton, 1889)	17	6.2
Liverworts	78 (NJ) (Britton, 1889)	3	3.8
Mosses	274 (NJ) (Britton, 1889)	16	5.8
Horsetails & Club-mosses	8 (Hand, 1965)	4	50.0
Ferns	21 (Hand, 1965)	10	47.6
Flowering plants	385 (Stone, 1911)		74.5
	800 (McCormick, 1970)		
	850 (Fairbrothers, 1979)	288	33.9
Total plants		346	
Anthropods other than insects	?	13	?
Insects	10,385 (NJ) (Smith, 1910)	280	0.027
Fishes	24 (McCormick, 1970)		
	36 (Hastings, 1979)	16	44.4
Amphibians	23 (McCormick, 1970)		
	24 (Conant, 1979)	13	54.2
Reptiles	30 (McCormick, 1970)		
	30 (Conant, 1979)	19	55.8
Birds	144 (McCormick, 1970)		
	73 (Leck, 1979)	90	62.5
Mammals	34 (McCormick, 1970)		
	34 (Wolgast, 1979)	27	79.4
Total animals		458	
Total species	13,000 (plus fungi and other anthropods)	804	

It is clear the 804 species finally included, of which 713 are illustrated, far exceed the original goal. For this I accept full responsibility, both for their inclusion and for convincing a very cooperative publisher their inclusion was essential in order to provide a reasonably comprehensive coverage of the subject.

A word needs to be included here on the status given for certain species characterized as rare, threatened, or endangered in the following descriptions:

rare - a general term denoting a species which occurs in such small numbers in the pine barrens that it may become threatened or endangered.

threatened - a species which may become endangered within the foreseeable future if the environment in the pine barrens deteriorates or other limiting factors are altered.

endangered - a species whose survival in the pine barrens is in imminent danger of extinction.

There are several authoritative listings of threatened and endangered species of plants and animals. Chief among these for the pine barrens are: (1) listings in the Comprehensive Management Plan (CMP) of the Pinelands Commission; (2) listings of Special Plants, Invertebrates, and Vertebrates of New Jersey by the N. J. Natural Heritage Program, Office of Natural Lands Management, N. J. Dept. of Environmental Protection (DEP); and (3) Endangered, Threatened, and Rare Vascular Plants of the Pine Barrens and their Biogeography, by David E. Fairbrothers, Chapter 22 in Pine Barrens Ecosystem and Landscape. The stata given in the following descriptions are based on these sources. Of the 54 species of plants listed as either threatened or endangered in the CMP, 24 are included in the following descriptions. Of the 33 species of animal life so listed in the CMP, only 6 are included.

"Kingdoms" of Living Organisms

Ever since man began to name and classify forms of life, the living world has been divided into two "kingdoms"—plants and animals. Plants are rooted in one location and feed by means of photosynthesis. Animals move about and ingest their food.

In recent years, more modern systems have been developed to better organize the world of living matter, especially regarding the inclusion of microorganisms. First developed was the "three kingdom" system but this has now been superseded by the "five kingdom" system. As presently recognized and propounded, this consists of five "kingdoms" as follows:[1]

Kingdom

Bacteria and cyanobacteria (blue-green alga). .... Monera

Simple, single cells with absorptive nutrition

Protozoa, unicellular algae, primitive fungi., ... Protista and slime molds

Simple, single or colonial or multi-cellular organisms

Living Organisms

	Kingdom
Fungi, including lichens.	Fungi

    Non-mobile, uni- or multi-(usually) cellular organisms with absorptive nutrition

Animals	Animalia

    Multicellular, usually mobile organisms with ingestive nutrition

Plants	Plantae

    Multicellular, usually non-mobile organisms with photosynthetic nutrition

Since this book is primarily concerned with the two "higher kingdoms" of plants and animals, and, with the exception of lichens, touches only briefly or not at all on the other three "kingdoms," the more traditional "two kingdom" system is retained and used in the following pages.

The Plant Kingdom

Plants are the basis of all forms of life, providing food, oxygen, and other products to the animal kingdom, including mankind. Briefly, there are four major divisions of plants, the first three of which, taken together, are called the non-flowering plants because none have any flowers, seeds, or fruits, having instead, spores or other non-seed mechanisms for reproduction.

NON-FLOWERING PLANTS
(Divisions I, II, and III, below)

I. ALGAE, FUNGI, and LICHENS* (Thallophyta) - the simplest and lowest forms of plant life. The plant body is called a thallus which has neither any conductive (vascular) tissue nor any roots, stems, or leaves.

II. MOSSES and LIVERWORTS (Bryophyta) - plants somewhat more developed than thallophytes, with simple leaves, but still without any conductive tissue.

VASCULAR PLANTS

III. FERNS and FERN ALLIES (Pteridophyta) - plants that have developed a connective tissue (vascular) system to conduct water and nutrients from roots up through stems to the leaves.

Ferns and fern allies may be called a "swing" division, being the most highly developed of the non-flowering plants while, at the same time, being the least developed of the higher, vascular plants.

IV. FLOWERING and SEED-PRODUCING PLANTS (Spermatophyta) - of which there are two subdivisions:

1. NAKED SEED PLANTS (Gymnospermae) - do not have flowers in the usual sense but produce seeds in cones. All are woody, perennial forms and most are evergreen with needle-like, linear, or scale-like leaves.

2. TRUE FLOWERING PLANTS (Angiospermae) - produce flow-

---

*These groups now spread among three other 'kingdoms,' the Monera, the Protista, and the Fungi. See page 73.

ers and bear seeds within a closed structure (ovary). These far exceed any other group in the plant kingdom in number of species and importance to man. The number of described species exceeds all other plant groups put together. There are two classes of true flowering plants:

A. MONOCOTYLEDONS - a) leaves generally parallel-veined; b) flower parts of any one type (ex: petals) in three's or some multiple thereof; c) seed embryoes with only one seed part or seed leaf (ex: single kernel of corn).

B. DICOTYLEDONS - a) leaves generally net-veined, frequently toothed, lobed, or compound; b) flower parts of any one type (ex: petals) in four's or five's or some multiple thereof; c) seed embryoes usually with two seed parts or seed leaves (ex: bean or peanut).

Simple Field Key to Groups of Non-Flowering Plants

1. Without leaves, stems or roots. Some may not even look like plants. Color varied 2
1. With leaves, definitely recognizable as plants 4
2. Hardly look like plants. Seem to have no body. All contain chlorophyll. Most are green. Some look like a green scum. Most are microscopic. Mostly aquatic, few in damp habitats .... ALGAE
   (Page 78)
2. More clearly plant looking. Have definite body. Usually terrestrial. 3
3. Without chlorophyll, lack green color. Fine threads serve as roots. Saprophytic or parasitic .... FUNGI
   (Page 78)
3. With chlorophyll, contain green color. Grow flat against surface or grow erect .... LICHENS
   (Page 79)
4. With leaves or leafy bracts and stems but not true roots 5
4. With definite leaves, stems, and roots. Vascular plants with connective tissue 6
5. Usually grow prostrate against surface, in moist habitats. Leaves in 2 or 3 rows and flattened, not in spirals .... LIVERWORTS
   (Page 88)
5. Usually grow more or less erect, with quite leafy, definite stems .... MOSSES
   (Page 88)
6. Stems jointed. Leaves small, simple, arranged in

whorls ....HORSETAILS
(Page 96)
6. Stems not jointed, creep along ground. Leaves small, simple,
scale-like, irregularly or spirally arranged in vertical rows,
not whorled .... CLUB MOSSES
(Page 96)
6. Leaves usually large, flat, complex, arise from
perennial root-stalks ....FERNS
(Page 98)

Flora of the New Jersey Pine Barrens

The pine barrens of New Jersey is noted world-wide for its unique flora. As stated earlier (see Geologic History, page 6), it is believed that present pine barrens flora and fauna developed only after the retreat of the Wisconsin Ice some 12,000 to 10,000 years ago. It is likely that the glaciers, having scraped the terrain of northern territories, brought with them a number of northern plants which, after the glaciers melted, flowed south with the glacial melt run-off and remained to vegetate the coastal plain. Following the retreat of the glaciers, the climate became warmer thus allowing plants of southern origin to begin to move north into this same area.

Thus the pine barrens is a meeting ground for several northern as well as many southern species of plants. McCormick, 1970, states that 14 basically northern species reach the southern limits of their distributional ranges in New Jersey's pinelands. Most notable of these are **curly-grass fern** and **broom-crowberry**, while several other pinelands plants such as **pine barrens heather** or **hudsonia**, and **bearberry** also are of northern origin.

Many more, at least 109[2], plants of southern origin have moved north to reach the New Jersey pine barrens as the northern limits of their ranges. Most of these, like **pyxie**, **bog asphodel**, and **pine barrens gentian**, moved up along the coast while a few, like **turkeybeard**, moved north along the Appalachian Mts. Two plant species, **sand-myrtle** and **Pickering's morning-glory**, are known to occur only in the New Jersey pinelands, while two other species, **Knieskern's beaked rush** and **blazing-star**, occur only in the New Jersey pine barrens and in similar nearby habitats in Delaware[3]

As McCormick states, "We do not know why the contemporary range of any pine barren plant or animal species ends where it does....But it is clear that if we do not preserve a large segment of the Pine Region, the answers to many presently enigmatic problems may be lost forever."

ALGAE

Algae, fungi, and lichens, together, are known as thallophytes, after their non-conductive (non-vascular) plant body called a thallus. None of these plants have either stems, roots, leaves, or flowers.

Algae are small, green plants that live in water (usually) and contain chlorophyll, thus are able to manufacture their own food. They are the lowest form of plant life and are important as a basis of many food chains, directly or indirectly serving as an important source of food for aquatic animals. They also help maintain the oxygen content of water by releasing oxygen as a by-product of photosynthesis.

Of the 360 known species of algae in the streams, ponds, and bogs of the pine barrens[4], the most common are several species of desmids (green algae), single-celled diatoms (yellow-green algae), and at least two species of red algae, all of which are characteristic of acid pine barrens waters. A number of green algae are filamentous and occur as long, stringy, green masses floating in ponds or strung out in slow-moving streams. The occurrence of blue-green algae in the pine barrens is limited and relatively unimportant.

None of these are described further or illustrated in this field guide.

FUNGI

Fungi are plants that do not contain any chlorophyll, thus cannot manufacture their own food. These exist either as parasites (living in or on and consuming the tissues of another living organism (host)), or saprophytes (consumers of dead tissues). In most fungi, the vegetative body consists of fine threads (hyphae) that, in a mass, form the root-like portion (mycelium) of the plant body (thallus). The mycelia of mushrooms grow underground or in decaying wood or humus. The mycelia of parasitic fungi invade the host.

No attempt can be made here to enumerate the almost countless species of bacteria, mildews, molds, rusts, mushrooms, and stinkhorns that occur in the pine barrens. Mention is made of only five representative specific forms.

Black Knot *Dibotryon morbosum*
Appears as heavy, roughened, black growth on twigs, limbs, and trunks of (principally) wild black cherry. Growth may be honeycombed and give off a gummy secretion. A destructive fungus.(Illus. on pg. 81)

Bracket or Artist's Fungus *Fomes applanatum*
Generally fan-shaped, shelf-like, woody fungus that grows on the sides of trunks of hardwood trees, notably oaks. The upper surface is hard and zoned. The undersurface is smooth and white (when fresh), on which pictures may be drawn. This fungus is a favorite habitat of the forked fungus beetle. (Illus. on pg. 81)

Earthstar *Geastrum triplex*
First appears as small puff-ball on ground. As development proceeds, the outer layer splits and curves back to form four to eight star-like rays.. This exposes an inner sack with a thin, papery coating. The spores in this sack are ejected through a central pore when pressure is applied. Star-like base may close again in wet weather and reopen again in dry weather. A common and characteristic pine barrens species, occurring on white, sandy substrate. (Illus. on pg. 81)

Puffballs *Lycoperdon* spp.
Several species. Round to pear-shaped fungi that grow up out of ground. Usually white in early growth stages. Become brown to dark brown or black upon maturity. (Illus. on pg. 81)

Fly Mushroom *Amanita muscaria*   height to 10", diameter to 7"
A gilled and stalked mushroom with cap, gills under cap, ring around stalk just below cap, heavy stalk, and enlarged, heavy, bulbous-like base (volva). Cap often brightly colored from yellowish to orange or even reddish, usually with whitish or pale yellow flecks or "warts" on surface, somewhat like particles of cottage cheese. Most often seen in open woods, especially under pines and other conifers. Quite poisonous.
(Illus. on pg. 81)

LICHENS

Lichens are small, low-growing, "dual" plants, being composed of both fungi and algae growing together in a mutually beneficial (symbiotic) relationship. The fungus, in the form of microscopic white strands (hyphae), forms the actual plant body (thallus) and is the more conspicuous part of the lichen. The alga, in the form of green cells which contain chlorophyll, carries on photosynthesis and thus manufactures food for the fungal partner. The fungus supplies the water and the minerals.

Lichens

Lichens are pioneer plants that can grow where other plants cannot, on soil, sand, rock, bark, and dead wood. They stabilize soil surfaces, break down rocks, and by their death and decay contribute to soil formation. All species provide habitat for small invertebrates, and some species are eaten by higher animals. Lichens grow best where the air is clean and unpolluted and thus are good indicators of atmospheric conditions.

There are three basic types of lichens: 1) **crustose** (crusty), 2) **foliose** (leaf-like), and 3) **fruticose** (leafy or shrubby). Some lichens are called squamulose (lobed) because they are considered to be intermediate between crustose and foliose. Plant bodies of crustose lichens grow flat on rocks, tree trunks, and other surfaces; those of foliose lichens are attached to the growing surface in spots by hair-like structures (rhizines), leaving the often lobed margins free. Fruticose lichens grow like small, branched plants, with either upright or hanging stems. The general pattern of squamulose lichens is similar to that of foliose lichens.

It is difficult to walk in the pine barrens without stepping on lichens. Virtually all terrestrial habitats contain these plants in abundance.[5]

Tuckerman and Eckfelt (ed.) in Britton, 1889, named 279 lichens in New Jersey, but the number of lichen species in the pine barrens is unknown. Forman, 1979, provides illustrations (6) and brief descriptions (13) of common pine barrens lichens.

A Byrological Society of America field trip in the pine barrens in 1982 collected 72 species of lichens at four locations. A lichen workshop and foray, sponsored by Rancocas Nature Center, New Jersey Audubon Society, in the vicinity of Deep Hollow Pond, Lebanon State Forest, on November 2, 1985, produced 29 different lichens in one afternoon (both statements per Karl Anderson (K.A.), pers. comm.).

In the identification of lichens, it is important to know the meaning of the following terms:

Apothecium(a) - disk or cup-shaped fruiting body containing fungal spores found on surface of plant body.
Isidium(a) - minute, usually cylindrical, outgrowths ("dots") from surface of plant body.
Hypha(ae) - elongate cells of a fungus which largely determine shape, appearance, and consistency of a lichen plant body.
Podetium(a) - upright fruiting stalk (thallus), usually tipped with apothecia, of some fruticose lichens, principally in the genus *Cladonia*.

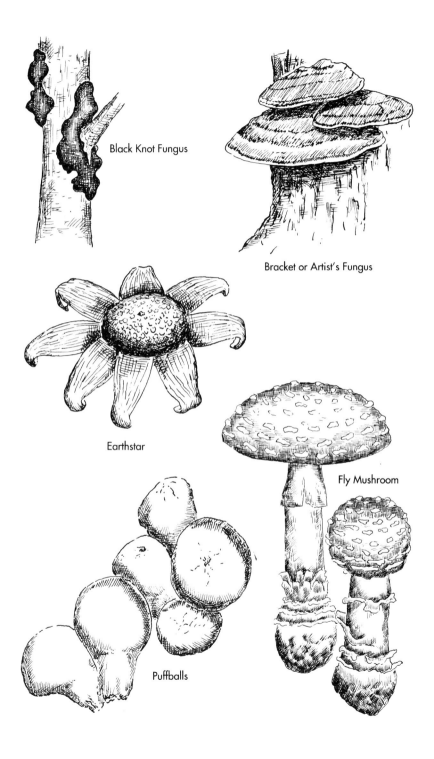

Rhizine(s) - root-like structures looking like dense, dark brown to black "strings" of hyphae extending down from underneath plant body to substrate, firmly anchoring lichen to support base.

Soredium(a) - microscopic, asexual, reproductive cells that appear as minute, powdery masses on the surface of a lichen plant body.

Squamule(s) - small, scale-like thallus or scale-like appendage (lobe) of a thallus.

Thallus - plant body of a lichen, composed of microscopic white strands of a fungus, and green algal cells. The fungal component dominates to form the actual plant body.

The following 17 lichens are among the most common, most characteristic, and most easily observable in the pine barrens of New Jersey.

CRUSTOSE LICHENS

Tar Lichen *Lecidea uliginosa*

Flat on sandy surface
Size of patches varied

Very dark- or olive-brown to, most often, black. Very minute particles make a felt-like surface, consolidating the sandy surface into large, blackish patches. Appears as though tar had been spilled on sand some time ago and had hardened. Covers sand and bits of rotten wood with a crusty cover. Common on sand. (No illus.)

FOLIOSE LICHENS

Rough Shield Lichen *Parmelia rudecta*

Flat against surface
6" - 1' across

Gray or blue-gray-green rosettes. Central (older) portion of upper surface appears rough (covered with insidia). Outer (younger) portion covered with tiny white pores. Lobes at edges of plant body relatively large, wide. Likely most conspicuous, most common foliose lichen on bases, trunks, and bark of oaks and other deciduous trees in upland areas.

Green Shield Lichen *Pseudoparmelia caperata*

Flat against surface
6" - 1' across

Large yellow-green, broad-lobed rosettes, with the older growth covered with coarse soredia. Lower surface black, edged with dark brown. Apothecia almost unknown. Common on bark of deciduous trees. The only large, yellow-green, bark-growing lichen in the pine barrens. (K.A.)

Shield Lichen *Parmelia squarrosa*
Furrowed Shield Lichen *Parmelia sulcata*

Both flat against surface
4" - 8" across

Two similar species. Both gray, with wide lobes having angular white markings at edges of plant body. *P. squarrosa* has insidia and lacks soredia

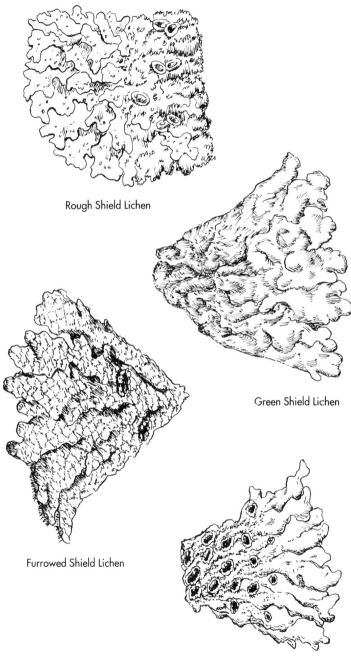

but *P. sulcata* has soredia erupting on surface and along angular white markings and ridges of its lobes and lacks insidia (K.A.).

Lace-tipped Lichen *Physcia millegrana*
Flat against surface
Up to 2" across

Gray-green. Small. Lobes finely divided, heavily sorediate on tips. Apothecia frequent, dark gray or black, thick-rimmed. Extremely common on bark of deciduous trees (K.A.). (No illus.)

Goat or Gray Star Lichen *Physcia aipolia*
Flat against surface
Up to 2" across

Dull silver to slate-gray, with pale tips. Small, dainty, but rather conspicuous. Central area a confused, crust-like, often warty mass. Apothecia frequent, dark gray or black, thick rimmed. Has neither isidea nor soredia. Occasional on bark of deciduous trees (K.A.). (Illus. on pg. 83)

FRUTICOSE LICHENS

By far the dominant lichens in the pine barrens are nearly two dozen fruticose species of the genus *Cladonia*. Under the usual hot, dry surface conditions in the pines, these plants tend to get quite dry and brittle and "crunch" under foot, but when recently dampened by rainfall, these tend to become rather soft and pliable.

False Reindeer Lichen *Cladonia subtenuis*
1" - 3" high
Patches up to 10" across

Stalks whitish to grayish-brown, fibrous surfaced, tips very slender, often hair-like, may droop or stand erect but tend to split into several drooping fingers. Stalks branch intricately by forking. Common to abundant on sandy soil, often covering large areas.

Thorn Lichen *Cladonia uncialis*
1" - 2" high
Patches up to 10" across

Stalks grayish, yellowish, or brownish-green, sometimes nearly white, usually smooth, shiny, or cracking into patches, sponge-like in shape with intertangled branches. Tips normally end rather abruptly in sharp, even spiny bristles. Common, growing in mats on sandy soils, especially in low lying areas.

Mealy Goblet or
Pyxie-cup Lichen *Cladonia chlorophaea*
Up to 1" high

Stalks greenish-gray, cup-shaped on top, covered with soredia as fine granules, growing up from well developed base of leafy squamules. Apothecia brown, growing from rims of cups. Common on soil or rotting wood, especially in dry, upland areas. Indistinguishable in field from several morphologically identical chemical species (K.A.).

False Reindeer Lichen

Thorn Lichen

Pyxie-cup Lichen

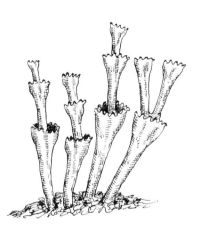

Slender Ladder Lichen

Awl Lichen

Lichens

Coastal Plain or
Slender Ladder Lichen *Cladonia calycantha*   2" plus high
Stalks greenish- to bluish-gray, cup-shaped, multi-tiered. Basal stalk grows out of sparse cluster of greenish-gray squamules. Each upper stalk grows out of center of lower cup. Often abundant in sand, old stumps, tussocks in marshy and swampy areas, including cedar swamps. (Illus. on pg. 85)

Broccoli Lichen *Cladonia atlantica*   3/8" - 1" high
Stalks squamulose, many-branched, variable, often vaguely cup-shaped, without soredia. Gray, but in sunny places it "sunburns" to brown. Basal squamules small or lacking. Common on sandy soil, especially in repeatedly burned areas in dry uplands (K.A.). (No illus.)

Awl Lichen *Cladonia coniocraea*   3/8" to 1/2" high
Stalks pale green or whitish, unbranched, usually all curved in same direction, shaped somewhat like cow's horns, covered with soredia, arise from cluster of basal, lobed, grayish-green to bluish-gray squamules. Common on sandy soil, tree bases, dead wood. (Illus. on pg. 85)

British Soldier Lichen *Cladonia cristatella*   3/8" - 7/8" high
Stalks yellow-green with bright red fruiting caps, usually branched, rough surfaced. Red fruited caps on tips of virtually all stalks. Basal squamules greenish-gray (straw-colored when dry), very small and lack soredia. Common on tree bases, soil, sand, dead wood. Possibly easiest of all lichens for beginners to recognize.

Swamp Lichen *Cladonia incrassata*   3/8" - 7/8"
Very similar to above but differs in two regards: 1) basal squamules usually somewhat larger and 2) distinguished by presence of dense growth of rather bright yellowish-white soredia on undersurfaces. On rotten stumps, tree bases, occasionally on ground, most frequently in bogs and swamps (K.A.). (No illus.)

Tunnel or Squamose Lichen *Cladonia squamosa*   3/8" - 2" high
Stalks variable, either smooth, slender rods under 1/2" or many-branched, tree-like, up to 2", thickly covered with small greenish-gray squamules. Tips often broadened to form a distinct tunnel with a toothed lip. Common in oak forests, especially in repeatedly burned areas, and in hardwood swamps and thickets, usually in somewhat moist and shady locations.

Santee Lichen *Cladonia santensis*   3/8" - 1"
This species, somewhat similar to the above, is often abundant as a bluish-gray-green "bloom" around the bases and up the trunks of cedar trees in deep cedar bogs and swamps. In general, the podetia of this lichen are shorter, less likely to branch, and have smaller and fewer squamules than squamose lichen. It fruits only sparingly (K.A.). (No illus.)

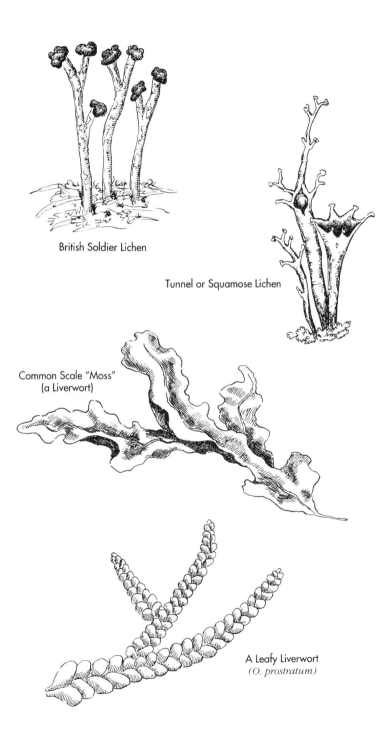

LIVERWORTS

Liverworts and mosses are closely related and, together, are called bryophytes. These are plants with more complex structures than the thallophytes (algae and fungi) but still do not have any conductive (vascular) plant tissue. However, all have filamentous structures (rhizoids) that serve as roots.

True liverworts lack leaves, having, instead, a flat, thin, ribbon-like, often divided stalk (thallus). Other liverworts are leafy but the leaves are in two or three rows, and flattened to have two distinct surfaces, the leaves never in spirals or radially symmetric and never have a mid-rib. In general, liverworts are small, prostrate, rather dark green plants, often found on moist soil and in damp woods on trees and old logs, sometimes growing along-side mosses.

Austin, in Britton, 1899, listed 78 species of liverworts in New Jersey but the number of forms in the pine barrens is unknown. Forman, 1979, provides illustrations (2) and brief descriptions of three pine barrens liverworts. Olsson, 1979, states that most liverworts in the pine barrens are characteristic of cedar swamps. Only three representative species are described here.

Common Scale "Moss"
(A Thallose Liverwort) *Pallavicinia lyellii*   Thallus up to 4" long
Large, wavy-margined, ribbon-like stalk (thallus), with conspicuous, thickened mid-rib and branching, rounded tips. Lies prostrate and forms mats on wet soil or tree bases. Common in lowlands, including cedar swamps. (Illus. on pg. 87)

A Leafy Liverwort *Odontoschisma prostratum*
Leaves nearly round, in two symmetrical rows on opposite sides of prostrate, branching stem, giving a horizontal, flattened appearance when moist, a vertical flattened appearance when dry. Common in lowlands in moist soil and in sphagnum and cedar bogs and swamps.(Illus. on pg. 87)

Frullania *Frullania inflata* and other species   Minute, thread-like
Small-leaved, leafy liverworts. Leaves nearly round with cupshaped ventral lobes that may retain water. These plants occur as dark green, purplish, or blackish patches on the bark of deciduous trees, particularly oaks. Common in many upland areas, but often overlooked. (K.A.) (No illus.)

MOSSES

Mosses are small, green, leafy plants that, in the pine barrens, grow in mats or clumps on sandy soil, on bark and the bases of trees, on stumps and old logs, in aquatic habitats as bogs, and even

Mosses

in water. Plants consist of stems, leaves, and root-like rhizoids, the stems either erect, prostrate, or ascending, and either branched or unbranched. Leaves usually grow all around both the stems and their branches.

The simple structure of mosses allows them to absorb large amounts of water, thus they are valuable by holding rain water, allowing it to soak gradually into the soil. When moist, leaves usually are well expanded and spreading. When dry they fold up, closely overlapping against the stem, or they become twisted and curled (crisped). Plant bodies usually are cylindrical in outline.

Mosses grow from spores formed in capsules which develop on top of slender stalks. These capsules often are quite conspicuous, growing slightly above the mats of plant bodies. Identification of mosses is based to a large extent on position and structure of the leaves and on characters of the spore-bearing capsules, including the capsule itself, its lid which falls off when the spores mature, its "teeth" (peristome) around the capsule opening, and the length of its stalk (seta). Because much of this identification is best done with both a hand lens and a good microscope, no attempt is made here, in what is intended only as a field guide, to go into these details in the following descriptions.

Austin, *in* Britton, 1889, listed 274 species of mosses in New Jersey but the number of forms in the pine barrens is unknown. Forman, 1979, illustrates 10 and briefly describes 24 pine barren species. Of these, the following 16 mosses may be among the most common, most characteristic, and most easily observable.

Sphagnum or Peat Moss *Sphagnum* spp.

Large for a moss. Plants crowded together, often forming dense, extensive mats. Leaves vary from pale gray-green to bright green, but in some species pink to deep red, usually depending on the amount of sunlight. Stems somewhat floppy, much branched, sometimes growing a foot or more in length in thick, compact masses close to the ground. Clusters of branches at tips of stems give a "star-shaped" appearance. Leaves short, sharp-pointed, or spoon-shaped, with large cells, especially adapted for holding water. Spore bearing capsules not common. Abundant along edges of, and in, bogs, small ponds, and cedar swamps.

It is almost impossible to walk in the cedar swamps and other boggy areas of the pinelands without stepping on extensive mats of *Sphagnum* spp. moss. Rau, *in* Britton, 1889, listed 19 species of sphagnum mosses in New Jersey. Forman, 1979, illustrates and briefly describes one of these and states there are more than 20 species present in the pine barrens. Of the species of sphagnum that occur in the pinelands, perhaps the most common is *S. magellanicum,* closely followed by *S. fallax, S. papillosum, S.*

cuspidatum, and, less frequently, *S. pulchrum* and *S. flavicomans*.

Sphagnum mosses are valuable in many ways. Their ability to hold large quantities of water on the surface of the soil is of great importance in water conservation, soil erosion, and even flood control. Authorities differ on the amount of water these plants can absorb and hold, but it certainly is many times their own weight!

Over a long period of time, as older sphagnum plant portions become decayed, compressed masses at the bottom of bogs become peat which, dug, cut into bricks, and dried, can be used as fuel. Peat bogs sometimes contain preserved plant and animal forms and thus may be valuable depositories of former life forms and indicators of past ecological conditions. However, the quantity and quality of peat in the New Jersey pine barrens has never been good enough to justify exploitation. Sphagnum also is valuable as a packing material for insulation and, by gardeners and florists, for packing vegetables and nursery plants. Gardeners use peat moss to loosen and improve their soils. In both world wars, sphagnum was used as an absorbent surgical dressing. For these purposes, over many years, the gathering of sphagnum or bog mosses was a major "piney" industry in the barrens. Finally, it is said that native American Indians used sphagnum as absorbent diaper material for their infants!

Purple Horntooth Moss *Ceratodon purpureus* 1/2" - 1" high
Small plants crowded together. Dark green. Stems short, erect, close together, slightly branched. Leaves short, hair-like, margins slightly curved back, wide spreading when moist, tips curled and twisted when dry. Mature capsules and stalks dark red, conspicuous, abundant. Capsules cylindrical, with lengthwise grooves, inclined toward horizontal. An early moss, fruit-stalks at full height and beginning to turn from green to red by late March. Capsules mature May-June. Common in upland, sandy areas.

Little Broom Moss *Dicranella heteromalla* Up to 1" high
Small plants close together. Bright to dark green. Stems erect, slender. Leaves long, narrow, hair-like, usually turned to one side so plants often appear "wind swept." Capsules common, on yellow stalks, relatively short and fat with threadlike caps, reddish-brown when mature in fall. Common in uplands and around bases of fallen trees.

Broom Moss *Dicranum scoparium* Up to 4" high
Large plants. Yellow-green. Variable. Stems erect, close together. Leaves like long, green hairs, slightly curved, pointing in same direction so plants appear "windswept." Capsules common, cylindrical, on tall stalks, mature August-September. Common, forming rounded clumps or extensive mats. Dominant in oak forests near tree bases.

A Broom Moss *Dicranum flagellare* Up to 1" high
Small. Bright or yellowish-green. Stems erect, close together. Fine branchlets (flagella) often at tips of stems. Leaves slightly curved, turned to one side,

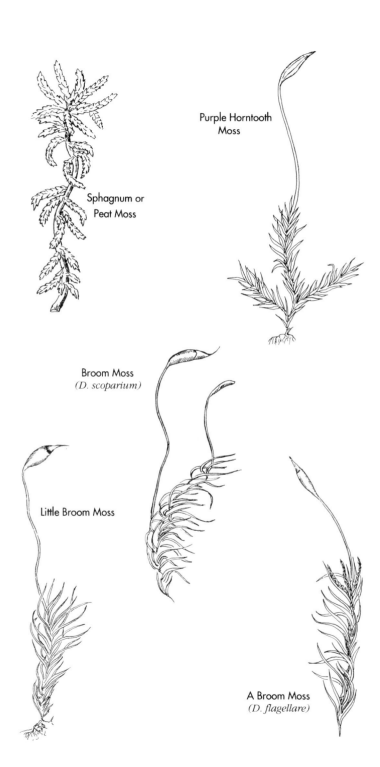

hair-like at tips, crisped when dry. Capsules common, erect, thin, cylindrical, yellowish or reddish, mature late summer. Common in uplands as mats on soil and as clumps on moist, decaying logs.

A Broom Moss *Dicranum condensatum* 1" - 2" high
Medium size. Stems somewhat apart. Leaves relatively wide with long points, curled and twisted when dry. Capsules similar to *D. scoparium*. Most common as mats on sand in repeatedly burned areas, especially the pygmy forests of the plains.

Pin Cushion Moss *Leucobryum glaucum* 1" - 3" high
Large. Plants grow in cushions. Pale yellowish- or whitish-green above when dry, darker green when moist. Stems erect, thick but easily broken. Leaves long, narrow, tubular. Capsules cylindrical, curved, inclined, dark brown, mature in fall, but rare. Common as clumps in oak, maple, and black gum lowlands.

A Pin Cushion Moss *Leucobryum albidum* 1" - 2" high
Medium size. Similar to above but stems closer together. Leaves whitish, greener when wet or young, many detached and lying across plant surface. Capsules uncommon, relatively short and fat, inclined but not horizontal, dark when dry. Common in uplands as rounded "pin-cushions" on soil. (No illus.)

Common Pohlia *Pohlia nutans* Up to 1" high
Small plants thickly crowded together. Usually dark green. Stems erect, slender, sometimes branched, often reddish. Leaves long, narrow, erect, forming tufts at tips of stems. Capsules inclined, horizontal or drooping, pear- or club-shaped, relatively uncommon, reddish-brown when mature in early summer. Common in uplands on better soils and around tree bases.

Hairy Bryum *Bryum capillare* Up to 1/2" high
Small plants, similar to above but with very short, erect, somewhat branched, often reddish stems. Plants grow close together. Yellowish- to dark green. Leaves wide but longer than wide, sometimes too small to be seen easily. Capsules similar to *P. nutans*. Occasionally common on upland sites.

Awned Hair-cap Moss *Polytrichum piliferum* Up to 1" high
Smallest of hair-cap mosses, all of which are so-named for fine, light-colored hairs covering young cap (calyptra) of capsule. Plants grow in patches. Leaves dark green, crowded at ends of erect stems, end in long, whitish hair-like tips giving plants hoary appearance. Plants conspicuous in spring with abundant, red fruit stalks and pale yellow hoods over capsules. Widespread in uplands. (No illus.)

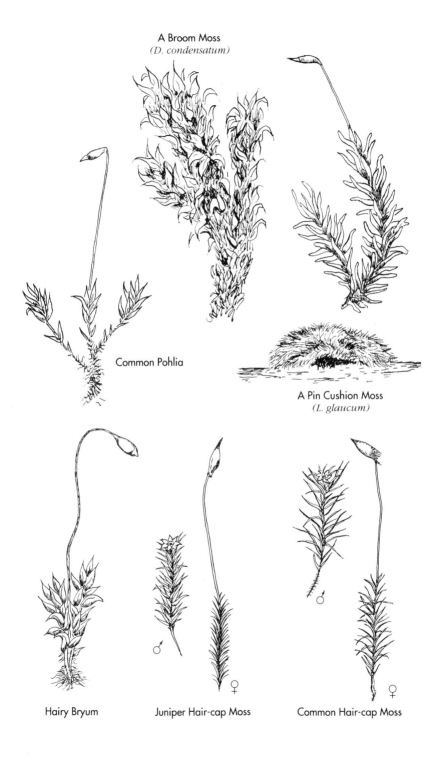

Vascular Plants

Juniper Hair-cap Moss *Polytrichum juniperinum* 1" - 4" high
Large plants. Erect, rarely branched stems relatively far apart. Bluish-green. Leaves long, wide margins folded in, giving appearance of being tubular. Capsules oblong, sharply 4-angled on tall, reddish stalks, with whitish, hairy caps. Widespread on upland pine-oak forest floor.(Illus. on pg. 93)

Common Hair-cap Moss *Polytrichum commune* Up to 12" high
but generally nearer 6"
Our largest moss. Very similar to above except for larger size, color more yellowish-green to olive or dark green, leaf margins not folded in, and capsule nearly cubical. A widespread, cosmopolitan species. (Illus. on pg. 93)

Ohio Hair-cap Moss *Polytrichum ohioense* Up to 3" high
Similar to both above but leaves more olive or dark green. Leaf margins toothed but not folded in (non-tubular). Capsules angled, gradually tapering to stalks which are reddish below, yellowish above. Widespread in uplands. (No illus.)

Common Thelia *Thelia hirtella* and No height; flat against surface
Thelia asprella
Both small. *T. asprella* whitish to bluish. *T. hirtella* more whitish to greenish. Both grow as low mats (not erect) close to growing surface, *T. hirtella* possibly flatter. Both forms fruit abundantly. Both grow as dense, thick mats on tree bases, mainly in upland oak forest areas, *T. hirtella* more likely in drier woods. These two cannot be separately identified without the use of a good microscope. (Illus. on pg. 97)

VASCULAR PLANTS

Vascular plants have developed a connective tissue system to conduct water and nutrients from their roots up through their stems to their leaves and other plant parts. These are considered to be the most highly developed forms of plant life.

In developing the following descriptions of vascular plants, including ferns and fern allies (pteridophytes) and all flowering plants (spermatophytes), the following references have been consulted: Fernald, 1950; keys in Ferren *et al,* 1979 (see below); Gleason, 1962; Gleason and Cronquist, 1963, and Stone, 1911.

Nomenclature follows Fernald, 1950, with updated synonymy (=) from Kartesz and Kartesz, 1980. Flowering and fruiting times are from Stone, 1911, supplemented by personal records.

For those who are experienced in the use of keys, the most recent keys to the identification of selected vascular plant species of the New Jersey pine barrens are by Ferren, W. R., Jr., J. W. Braxton, and L. Hand in their chapter, Common Vascular Plants of the New Jersey Pine Barrens *in* Forman, R. T. T., ed., Pine Barrens Ecosystem and Landscape. 1979. Chap. 21, pp. 374-382.

Stone, 1911, listed 175 species of vascular plants as being "Characteristic Pine Barrens Species" (CPB). 52 of these are grasses, sedges, or rushes. 108 of the 175 CPB species are included in the following descriptions and are indicated by two asterisks (**) preceding their names. Stone also listed another 197 species as being "common to both the pine barrens and the middle district" (PBM). Again, 52 of these are grasses, sedges, or rushes. Stone defined his middle district as "that portion of the (Atlantic) coastal plain which lies west and north of the pine barrens, reaching around the bay shore to Dennisville." 144 of the 197 PBM species which Stone included in this second category are included in the following descriptions and are indicated by one asterisk (*) preceding their names.

FERNS AND FERN ALLIES

Ferns and fern allies (pteridophytes) are the earliest and lowest forms of vascular plants. These plants have true roots, stems, and leaves and are "vascular" because their plant tissues have the ability to conduct water and food from their roots, up through their stems to their leaves. However, these plants do not have flowers or seeds, and so depend on the dissemination of spores, as in the lower mosses, for their reproduction.

Today, pteridophytes are a small group of relatively less important plants, but there was a time, back in the Carboniferous Period of the later Paleozoic Era, around 300 million years ago, when these were the dominant plants on earth. In the great swamp forests of those times, many towering horsetails (principally) and giant ferns and seed ferns attained sizes comparable to our present forest trees. Some ancient horsetails reached heights of 100 feet or more, with great branching, leafy crowns. The preserved remains of these plants form the basis of one of today's best known fossil fuels, coal, for it was during the Carboniferous Period that today's great coal beds were deposited.

Fossilized remains indicate that even though pteridophytes today are much smaller than formerly, the form of these plants has changed only slightly since those times, so these are among the most ancient of plants. Fossilized remains also indicate these spore-bearing plants evolved earlier than the more highly developed flowering and seed producing (spermatophyte) plants.

FERN ALLIES

Fern allies in the pine barrens consist of only two families of plants: horsetails and club-mosses. Of these two families, only the club-mosses are truly characteristic of the pinelands.

HORSETAILS

Horsetails are considered to be the oldest and most primitive of all pteridophytes. Only one species occurs with any frequency in the pine barrens, due largely to its introduction into disturbed areas along roadsides and railroad embankments.

Horsetail Family (Equisetaceae)

Common or Field Horsetail *Equisetum arvense* — Fertile stalks: 6" - 12"
Sterile plants: 16" - 24"

Two types of plant growth. First, from early April to mid-May, are the whitish to brownish or flesh-colored fertile stalks, without leaves. These are relatively stout, succulent, short-lived. Each stalk ends in a distinct, terminal cone about 1" high, from which clouds of spores are released, after which these stalks wither and die.

Next to appear are the green, "leafy," sterile plants which have cylindrical, bamboo-like stems with branches occurring at regular whorls. Stems are mostly hollow tubes joined at nodes. Outer surfaces of both stems and branches alternately grooved and ridged and contain gritty particles. Surrounding each node is a thin, parchment-like skin (leaf-sheath) shaped around the node like a tight funnel, with several sharp, upward-pointed teeth. Both stems and branches perform photosynthesis.

CLUB-MOSSES

Club-mosses are small, green plants consisting of stems, with a few coarse roots, that creep along the ground, growing from a running tip, the growth of former years dying behind. Stems bear small, simple, overlapping, scale-like leaves, crowded together in rows or whorls. Reproduction is by spores produced near the bases of modified leaves. The three most characteristic pine barrens species are:

Club-moss Family (Lycopodiaceae)

**Carolina Club-moss *Lycopodium carolinianum* — Fertile stems up to 6" high

Firmly rooted main stem creeps prostrate along ground, sends up erect, slender, fertile stems, nearly naked of leaves, each with small, tight, yellowish, terminal cone. Almost entirely restricted to damp, sandy, acid, open, pine barrens bogs.

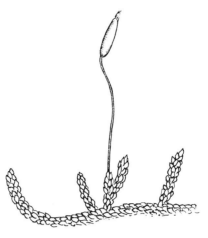

Common Thelia (Moss)

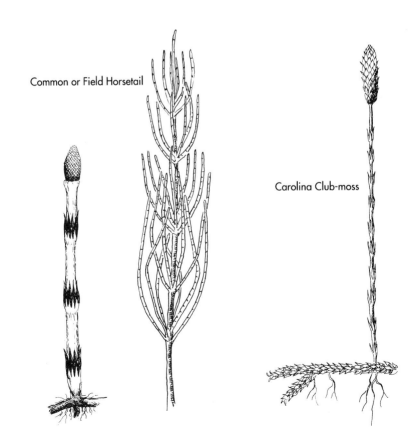

Common or Field Horsetail

Carolina Club-moss

Ferns

Fox-tail Club-moss *Lycopodium alopecuroides*
Fertile stems up to 10" high

Main stem creeps along ground, rooting at only a few spots, the stem rising above ground in low arches between rooting points. Usually turns yellow in winter. Upright, fertile stems semi-evergreen, densely or bushy-leaved, ending in enlarged, open, busy cones, like "foxes tails." Cones thicker than ordinary lead pencil. Leaves with distinctly toothed edges (use hand lens). In damp, sandy, acid, sphagnum bogs and edges of ponds.

Bog Club-moss *Lycopodium inundatum*
var. *bigelovii* = *L. appressum*

Fertile stems up to 8" high

Main stem creeps along ground, rooting at intervals, slightly arched between rooting points. Deciduous except for evergreen tips. Somewhat similar to fox-tail, above, but stem thinner, only slightly arched, with rooting points more frequent and closer together. Upright, fertile stems with few leaves, ending in bushy cones with evergreen tips. Cones thinner than ordinary lead pencil. Leaves slightly toothed at bases (use hand lens). In damp, sandy, acid bogs, and pond edges.

FERNS

Ferns are non-flowering plants that grow from perennial rootstalks. Usually, leaves first appear rolled up in "fiddleheads" which uncoil and rise on stalks (fronds) to become erect leaves. Leaves or their leaflets bear spores in tiny spore cases (sori), or plants send up separate, fertile, spore-bearing fronds. Of 21 different ferns in the Wharton State Forest[6] the following 10 are the most common or characteristic throughout the pinelands, although three (bog and curly-grass ferns and ebony spleenwort) are more localized and restricted in distribution.

Flowering Fern Family (Osmundaceae)

Royal Fern *Osmunda regalis*
Up to 2 1/2' - 3' in p.b.

Large, graceful fern growing from heavy rootstalk. Fronds bear opposite branches, each of which bears pairs of oblong, green leaflets, looking somewhat similar to leaflets on a locust tree. Spores borne on light brown, fertile leaflets at tip ends of fronds. In bogs, swamps, wet woods, often growing in shallow water.

Cinnamon Fern *Osmunda cinnamomea*
Up to 2 1/2' - 3'

Large, vigorous fern, fronds first appearing from heavy rootstalks as stout, conspicuous "fiddleheads," then growing into arching, circular clusters of tall leaves. Sterile fronds bear opposite, elongate, deeply-cut, green leaflets. Bases of fronds woolly. Spores borne on separate spore-bearing fronds which turn cinnamon-brown when mature and are the first to wither. Widely distributed in nearly all low, swampy, partially shaded locations.

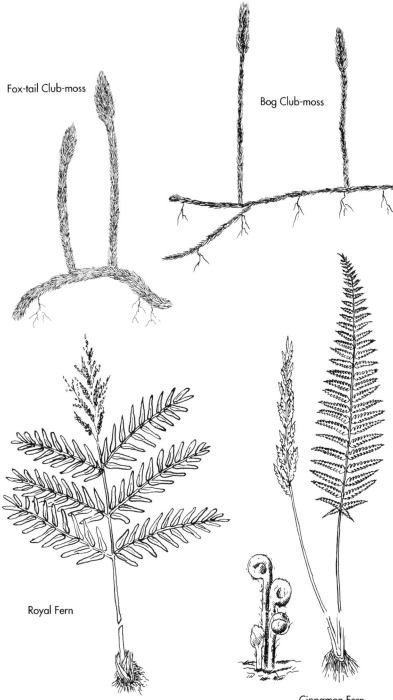

Ferns

Curly-grass Family (Schizaeaceae)

**Curly-grass Fern* *Schizaea pusilla*

Sterile fronds 2" high
Fertile fronds up to 4" - 5"

Very small, quite unfern-like. Sterile fronds look exactly like its name: tiny, evergreen blades of very curly or twisted grass. Spores borne on separate, fertile fronds topped by one-sided, segmented, fruiting "heads." Local in moist, sandy spots in and around edges of acid, sphagnum and cedar bogs and swamps. Rare. Classified as threatened[7] or endangered (CMP) but may be more common as may often be overlooked due to its tiny size.

Fern Family (Polypodiaceae=Aspleniaceae)

Sensitive or Bead Fern *Onoclea sensibilis*

Sterile fronds up to 2'
Fertile fronds 1' - 1 1/2'

Sturdy, broad, sterile fronds have opposite, wavy-edged, light green, veined leaflets which spread at base into narrow wings along each side of main stem. Spores borne on separate fronds, spore cases enclosed in small, hardened, bead-like leaflets which become dark brown when mature and may persist for 2-3 years. Leaves wilt quickly when picked, die soon after first frost. In pine barrens only as introductions into disturbed areas.

Marsh Fern *Dryopteris thelypteris* = *Thelypteris palustris*   1' - 2'

Fronds thin, delicate, green or yellow-green, on tall, thin, green stalks which are blackish at bases. Sterile fronds appear first, followed by taller and narrower fertile fronds. Leaflets paired, opposite, stemless and lance-shaped. Sub-leaflets broad based, rounded or blunt-tipped with smooth margins. Spore cases in two rows, parallel to and near midveins, often partially hidden by reflexed margins, on upper leaflets of fertile fronds. Common in sunny, moist meadows, marshes, damp depressions, and wet, open woods.

**Bog or Massachusetts Fern* *Dryopteris simulata*

Fronds 1' - 2'
= *Thelypteris simulata*

Sterile fronds yellowish-green, lance-shaped, pointed at tips, slightly narrowed at bases. Leaflets oblong, cut almost to mid-vein into pairs of blunt lobes, margins not toothed. Lowest pair of leaflet branches usually tipped downward. Veins simple, not forked. Spore cases appear as rounded dots on underside of leaflets on separate, fertile fronds. Restricted to acid, sphagnum and cedar bogs and swamps, especially along edges of pine barrens.

Ebony Spleenwort *Asplenium platyneuron*   Fertile fronds 6" - 18"

Fertile fronds erect. Stalks, especially lower, smooth, shiny, dark brown. Leaflets arranged pinnately or ladder-like. Each leaflet "winged" at base, tardily deciduous. Spore cases (fruit dots) short, straight, on undersides of fertile leaflets, nearer mid-veins than margins. Sterile fronds short, sometimes curled, spreading, lie close to ground, may form small evergreen tufts.

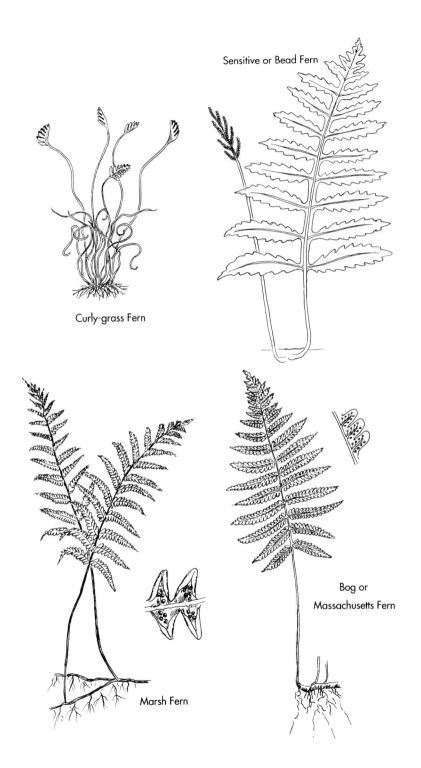

Ferns

Not common in pine barrens. Most frequently found growing out of old masonry foundations or in other shaded, well-drained, sub-acid soils.

Fern Family = Chain-fern Family (Polypodiaceae = Blechnaceae)

*Virginia Chain-fern *Woodwardia virginica*   Up to 2' - 3' in p.b.
Fronds erect, with dark purplish stalks. Leaves somewhat leathery with narrow, tapered, deeply cleft, almost opposite leaflets. Spore cases rusty to dark brown in double, chain-like rows on undersides of leaflets of separate fertile fronds. Abundant in bogs, swamps, low woods, and other damp to wet localities. A pernicious weed in cranberry bogs.

*Netted or Narrow-leaved Chain-fern   *Woodwardia areolata*   Fronds 1' - 2'
Superficially resembles a small sensitive fern. Fronds with pairs of alternate but nearly opposite, narrow, glossy green leaflets which are sharp-pointed with wavy margins and winged along stalk. Veins raised, prominent, netted in form. Leaflets of separate fertile fronds very narrow, covered on undersides with double rows (chains) of oblong spore cases. Plentiful in acid bogs, swamps, wet woods.

Fern Family = Hay-scented Fern Family (Polypodiaceae=Dennstaedtiaceae)

*Bracken Fern or Brake *Pteridium aquilinum*   Fronds 1 1/2' - 3'
Strong, stout fronds usually bear three nearly equal sections of roughly triangular, dark green leaves that often flatten out into a nearly horizontal position. Spores borne in almost continuous narrow lines along under edges of leaflets which are rolled back over spore cases as a protective covering. Widely distributed. Often grows in large colonies in poor, barren, acid sands and dry woods. The most common upland fern in pine barrens and a major part of the undergrowth in open pine woods.

The following are recommended for further identification and reference:

NON-FLOWERING PLANTS
Shuttleworth, F.S. and H.S. Zim. 1967. Non-flowering Plants. Golden Press, NY

ALGAE
Moul, E. T. and H. F. Buell. 1979. Algae of the Pine Barrens *in* Forman, R. T. T., Jr. Pine Barrens Ecosystem and Landscape. Academic Press, NY

FUNGI
Bigelow, H. E. 1974. Mushroom Pocket Field Guide. Macmillan Pub. Co.
Lincoff, G.H. 1981. The Audubon Society Field Guide to North American Mushrooms. A.A. Knopf, NY

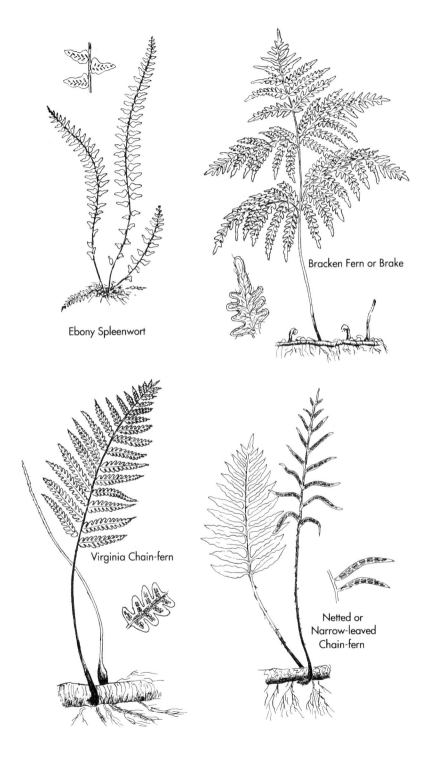

McKnight, K.H. and V.B. McKnight. 1987. A Field Guide to Mushrooms. Houghton Mifflin Co., Boston, MA

Smith, A. H. 1964. Mushroom Hunter's Field Guide. Univ. of Michigan Press.

LICHENS

Brodo, I. 1968. Lichens of Long Island, NY. Bull. #410. NY State Museum and Science Service, Albany, NY

Forman, R. T. T., Jr. 1979. Common Bryophytes (Mosses & Liverworts) and Lichens of the New Jersey Pine Barrens *in* Pine Barrens Ecosystem and Landscape. Academic Press.

Hale, M. E., Jr. 1961. Lichen Handbook. A Guide to the Lichens of Eastern North America. Publ. #4434. Smithsonian Inst. Washington, DC

Hale, M.E., Jr. 1979. How to Know the Lichens. W.C. Brown Co., Dubuque, IA

Nearing, G. G. 1947. The Lichen Book. Handbook of Northeastern United States. Ridgewood, NJ. (Published privately. Out of print).

MOSSES

Conard, H.W. 1944. How to Know the Mosses. W. C. Brown Co., Dubuque, IA

Dunham, E. M. 1916. How to Know the Mosses. A Popular Guide to the Mosses of the Northeastern United States. Houghton Mifflin Co., Boston, MA

Grout, A. J. 1903. Mosses with Hand-lens and Microscope. A non-technical Handbook of the more common Mosses of the Northeastern United States. Brooklyn, NY

FERNS

Cobb, B. 1956. A Field Guide to the Ferns and their related Families of Northeastern and Central North America. Houghton Mifflin Co., Boston, MA

Chrysler, M.A. and J.L. Edwards. 1947. The Ferns of New Jersey, Including the Fern Allies. Rutgers University Press.

Durand, H. 1928. Field Book of Common Ferns. G.P. Putnam's Sons, NY

Parsons, F.T. 1899. How to Know the Ferns. A Guide to the Names, Haunts, and Habits of our Common Ferns. 1961 Edition, Dover Publ., NY

Wherry, E. T. 1961. The Fern Guide. Northeastern and Midland United States and adjacent Canada. Doubleday & Co., Garden City, NY

FLOWERING and SEED PRODUCING PLANTS

To enable easy and quick identification of pinelands flora, the descriptions of all flowering and seed-producing plants which follow are grouped by similar types of plant growth rather than by following a strict "natural" (phylogenetic) sequence.

These groupings are under the headings of:

	Page
A. TREES	106
B. SHRUBS	118
C. SUB-SHRUBS	134
D. VINES and VINE-LIKE PLANTS	138

Flowering Plants

   E. HERBACEOUS PLANTS 144
      1. AQUATIC PLANTS 144
      2. PLANTS WITH UNUSUAL STRUCTURES or
         HABITS (INSECTIVOROUS and CACTUS) 152
      3. GRASSES, SEDGES, and RUSHES 156
      4. with PARALLEL-VEINED LEAVES 172
      5. with NETTED-VEINED LEAVES 182
      6. with COMPOSITE FLOWER HEADS 206

A simple field key to the identification of these groups of flowering and seed-producing plants follows.

Simple Field Key to Groups of Flowering and Seed-Producing Plants

1. Woody plants, including trees, shrubs, and sub-shrubs. Have stems which live over winter above ground and resume growth next season 2
1. Vines, both woody and herbaceous. Stems lack necessary strength and rigidity to grow erect without support. Either creep along ground or climb by twining or by means of tendrils ....VINES
(Page 138)
1. Herbaceous plants (herbs) with stems that die down to the ground or to underwater roots each year 3
2. Main stem (trunk) single, tall, usually 20' or more, with no or only few branches on lowest part..TREES
(Page 106)
2. Several stems or trunks, all usually under 20' .... SHRUBS
(Page 118)
2. Grow very close to ground, seldom higher than a few inches, almost always under 12" ....SUB-SHRUBS
(Page 134)
3. Primarily aquatic or semi-aquatic. Grow in water. Plants free-floating or rooted. Leaves submerged or floating ....AQUATIC PLANTS
(Page 144)
3. Unusual leaves, either hollow tubes, or with small, sticky hairs, or armed with tiny bladders, or stem divided into broad, flat, oval, fleshy joints ..
....INSECTIVOROUS PLANTS and CACTI
(Page 152)
3. All other herbaceous plants, producing flowers in season followed by seeds 4
4. Leaves narrow, linear, grass-like with parallel veins and small, inconspicuous flowers .... GRASSES, SEDGES, and RUSHES
(Page 156)

4. All others with parallel-veined leaves, flowers in 3's or multiples thereof .... remaining MONOCOTYLEDONS
(Page 172)
4. Leaves net-veined, frequently toothed, lobed or compound, not parallel-veined. Flowers usually in 4's or 5's or multiples thereof 5
5. Flower heads composed of many closely arranged, small florets, either tubular or ray type, or both, the whole resembling a single flower like asters, daisies, and goldenrod ....COMPOSITES
(Page 206)
5. All others with leaves net-veined, frequently toothed, lobed, or compound, not parallel-veined ....remaining DICOTYLEDONS
(Page 182)

TREES

Trees are woody, perennial plants which usually have only a single, tall, usually 20' or more, main stem or trunk, with no or only a few branches on its lowest part. The two types of trees in the pine barrens are evergreen and deciduous. Evergreen species include the pines, cedars, and American holly. Trees in the pine family, below, are gymnosperms. The balance of the trees described below are deciduous, are true flowering plants (angiosperms) and are dicotyledons (see pages 75 and 76).

Pine Family (Pinaceae)

Pines and Cedars

Naked seed plants (gymnosperms) that produce seeds in cones, thus often called conifers. All our pinelands forms are woody, perennial, evergreen, with needle-like or scale-like foliage. Although considered evergreen, these trees actually shed some of their needles and scale-like leaves throughout the year, gradually replacing them with new growth.

Male flowers of pines are small, pollen producing organs near the tips of branches that are obvious only during the spring period, usually in May, when their yellow pollen grains spread like dust through the air and carpet the ground. These are especially noticeable as a cloud of dust from a shaken branch, or after a rain when puddles form and rings of yellow pollen remain as the waters soak into the sand. The spent male catkins fall to the ground and resemble small brown caterpillars. Female flowers are tiny cones that take at least two years to develop into mature cones. Seeds of pines are extremely small, light, winged, and windblown, and are formed deep down between the scales (bracts) next to the stalk (stem) of the cone.

*Short-leaf or Yellow Pine *Pinus echinata*   80' - 100'
Large, tall tree. Bark dark reddish-brown in irregular, scaly plates. Needles

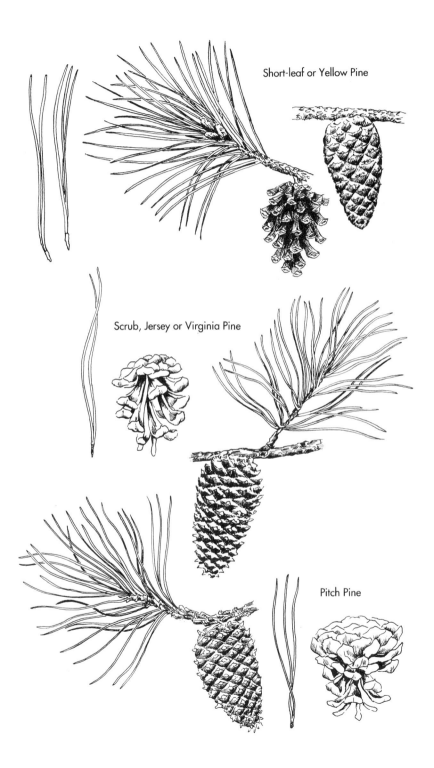

Trees

dark green, slender, 2"-5", usually in bundles of 2, sometimes 3. Cones small, oval, about 1 1/2"-2 1/2" with weak prickle on tip of each scale. Dry, sandy soils.

Scrub, Jersey or Virginia Pine *Pinus virginiana* 30' - 40'
Small, scrubby, irregular tree with spreading branches. Bark reddish-to-grayish brown with thin, scaly plates. Twigs slightly purplish. Buds resinous. Needles deep yellowish-green, short, 1"-3", twisted, widely spread, in bundles of 2. Cones small, ovoid, sharp-pointed, 1 1/2"-2 1/2", with short, straight or recurved spine on slightly thickened tip of each scale. In poor, sandy soils around edges of but rarely in the interior of pine barrens.
(Illus. on pg. 107)

**Pitch Pine *Pinus rigida* 40' - 70'
Medium size, ragged, irregular tree with heavy, gnarled branches. Bark dark reddish-brown to black, broken into thick, irregular plates which peal off in thin scales. Thick bark makes tree very resistant to fire. Often has sprouts from base of tree, especially after fires. Often has dense mats of needles growing from sides of main trunk. Needles dark olive-green, coarse, stiff, 2"-5", in bundles of 3. Cones small, 1 1/2"-3", oval, stemless, with a stout recurved spine or prickle on thickened tip of each scale. In the dwarf (pygmy) forests of the pines plains, cones remain unopened (serotinous) on branches for many years, until opened by the heat of fire. Abundant. The most common pine in poor, sandy soils throughout pinelands.
(Illus. on pg. 107)

**Atlantic White Cedar *Chamaecyparis thyoides* 50' - 80'
Tall, straight tree with dark green, spire-like crown. Often grows in stands so dense that all branches have died except those near top of canopy. Bark thin, light, reddish-brown, with loose, easily shredded, sometimes spirally twisted surfaces. Leaves bluish-green, small, compressed, overlapping, scale-like, entirely covering the slender twigs. Cones small, 1/4", nearly round, somewhat fleshy and wrinkled, bluish, becoming dark red-brown and persisting when mature. Most characteristic tree along courses of pine barrens streams, spreading out into large sphagnum and cedar swamps.

Red Cedar *Juniperus virginiana* 25' - 60'
Small to medium tree with close, spire-like crown. Bark reddish-brown, fibrous, easily shredded. Leaves dark green, needle-shaped and prickly on young branches, scale-like and overlapping on older branches. Cones small, less than 1/4", round, berry-like, fleshy, bluish, covered with a grayish bloom, deciduous. Mainly peripheral to the pine barrens but introduced and locally established in dry, open, sterile, pine barrens sands and disturbed areas.

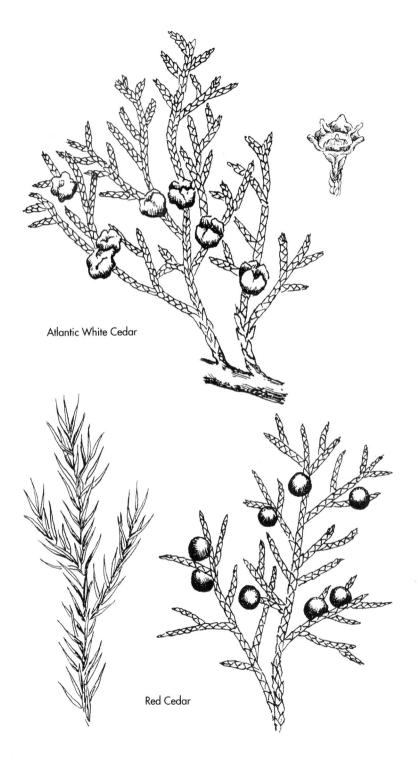

Trees

Hazel Family = Birch Family (Corylaceae = Betulaceae)

**Gray Birch *Betula populifolia* 20' - 30'
Small, slender tree with smooth, grayish-white, non-peeling bark on mature trees. Bark brownish on trunks of young trees and on branches. Trunks marked with dark, triangular patches below bases of branches. Commonly grows in groups. Leaves alternate, simple, triangular, sharp-pointed, fine-toothed, smooth, light green. Flowers small, inconspicuous, in hanging catkins from mid-April to early May. Common in poor, dry to moist soils.

Beech Family (Fagaceae)

Oaks

In combination with pitch pine, the many species of oaks are clearly the other dominant trees in the New Jersey pinelands. Oaks usually are tall trees with their end buds clustered at the tips of twigs. Male flowers appear in May as long, slender, hanging catkins which pollinate the inconspicuous female flowers, leading to the development of their seeds, acorns.

Oaks are divided into two main groups: white oaks, including the post and chestnut oaks; and red/black oaks, including the scarlet, scrub, and blackjack oaks. In the white oak group, bark on mature trees is light colored, often scaly, leaves have rounded lobes without tiny bristles at the tips, and their acorns mature in one season. In the red/black oak group, bark on mature trees is dark and furrowed, leaf lobes have tiny, pointed, hair-like, bristle tips, and their acorns require two seasons to develop.

*White Oak *Quercus alba* 60' - 80'
Tall tree with wide, spreading crown and whitish-gray, scaly bark. Leaves smooth, whitish underneath when mature, 4"-9", with 4-10 rounded lobes, without bristle tips, usually evenly divided or paired side to side. Acorn 3/4" long, ovoid, light brown, enclosed for one quarter its length in bowl-like cup with thickened, knobby scales. Locally common in dry, upland, pine woods.

Post-oak *Quercus stellata* 35' - 50'
Small tree with crown of gnarled, twisted branches. Bark thick, rough, ridged, gray-brownish. Leaves 3"-7", somewhat leathery, brownish-downy underneath, margins somewhat curled under, usually deeply five lobed, middle pair of lobes longer than others, at right angles to mid-rib, somewhat resembling a cross. Acorn 1/2"-3/4" long, oval, brown, enclosed for 1/3-1/2 its length in thin, bowl-like cup, with thin, almost smooth scales. Occasional to frequent in dry, sterile, sandy pinelands.

*Chestnut-oak *Quercus prinus* = *Q. montana* 50' - 80'
Large tree with wide, spreading crown. Bark thick, deeply ridged and furrowed, dark brown to black. Leaves oblong, 4"-7", somewhat leathery, undersurface smooth or very finely hairy, with 8-16 pairs of rounded teeth.

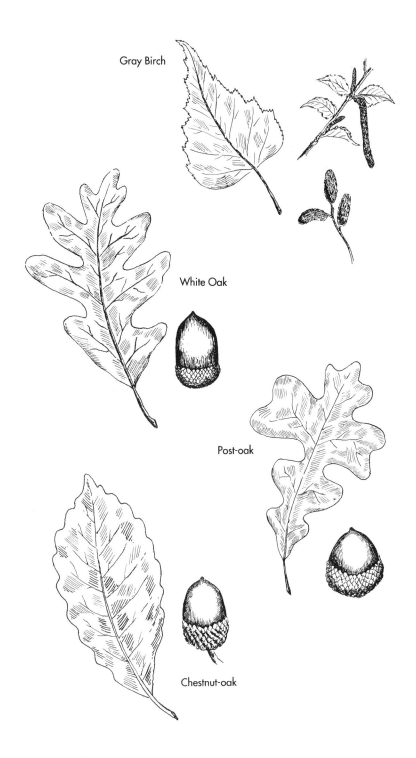

Trees

Acorn 7/8"-1 1/2" long-ovoid, slender, smooth, glossy, chestnut red-brown, enclosed up to one-half its length in thin cup with knobby scales. Locally common in dry, upland, sandy woods.

Dwarf Chestnut-oak *Quercus prinoides*
A shrub. See page 120 for description.

Black Oak *Quercus velutina* 50' - 80'
Large tree with rounded crown. Bark rough, ridged, almost black. Inner bark orange. Leaves 4"-9", moderately cut into 5-7 toothed, bristle-tipped lobes, glossy dark green above, slightly hairy under. Acorn 1/2"-3/4" long, ovoid, light red-brown, enclosed for 1/3-1/2 its length in thin, bowl-shaped cup with sharp, pointed scales that form a loose fringe at rim. Locally common in dry pineland woods.

Scarlet Oak *Quercus coccinea* 40' - 50'
Medium size tree with spreading crown. Bark brown, finely grooved. Inner bark reddish. Leaves 3"-6", shiny, green, smooth underneath, deeply cut nearly to mid-rib making 5-9 narrow, toothed, bristle-tipped lobes. Leaves turn scarlet in fall. Acorn 1/2"-1" long, ovoid, reddish-brown, enclosed for one-half its length in thin, bowl-shaped cup with sharp, pointed scales that form a fringe at cup margin. Occasional to frequent in dry, sandy, pineland woods.

Southern Red or Spanish Oak *Quercus falcata* 50' - 60'
Medium size tree with rounded crown. Bark thick, furrowed with scaly ridges, dark reddish-brown to nearly black. Inner bark yellowish. Leaves 5"-7", shiny, dark green above, downy underneath, variable in outline, usually deeply cut to form 3 to 7 bristle-tipped lobes. Terminal lobe often longer and narrower than others, itself often shallowly 3-lobed. Acorn small, 1/2", egg-shaped to round, in thin, shallow, short-stalked cup covered with reddish scales. In low, oak-pine woods, more frequent south of Mullica River.[8]

**Black-jack Oak *Quercus marilandica* 20' - 40'
Small, low, ragged tree, often no more than a shrub, especially in plains areas, but in the absence of fire it sometimes grows into a large tree. Bark black, broken into squarish blocks on mature trees. Leaves 4"-8", thick, leathery, roughly triangular, widest and shallowly three-lobed near apex, often rusty-downy underneath. Acorn 3/4" long, globular-ovoid, light brown, enclosed for one-half or more of its length in deep cup with large, red-brown, loosely overlapping scales. Common in dry, poor, sandy barrens, especially pine plains.

**Bear- or Scrub-oak *Quercus ilicifolia*
A shrub. See page 120 for description.

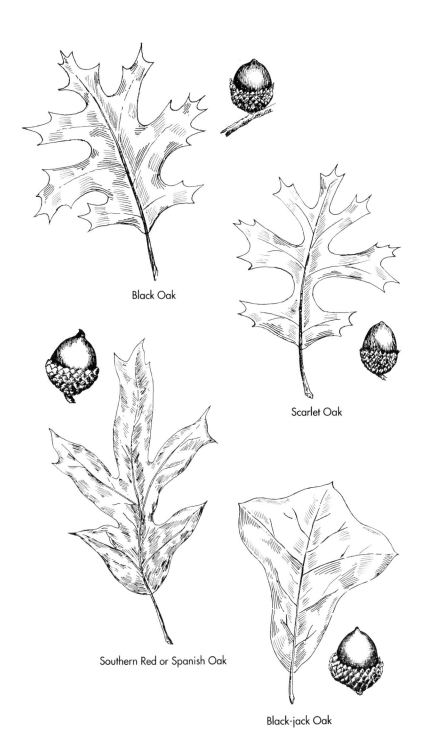

Trees

Magnolia Family (Magnoliaceae)

*Sweet Bay or Swamp Magnolia *Magnolia virginiana*   10' - 30'
Small tree with thin, light brown or grayish bark. Leaves large, 3"-6", oval to broadly lance-shaped, smooth, untoothed, thick, leathery, partially evergreen, shiny green above, distinctly whitish beneath, spicy fragrant when crushed. Flowers large, 2"-4" across, cup-shaped, creamy-white, showy, very fragrant. Fruit bright dark red. Common in swamps, thickets, low woods. Blooms late May-early July. Fruits August into early October.

Laurel Family (Lauraceae)

*Sassafras *Sassafras albidum*   10' - 40'
Rarely over 20' in p. b.
Small tree with red-brown, furrowed bark on mature trees. Often grows in small groups. Leaves smooth, thin, oval-elliptical, 3"-5", in 4 patterns: leaves with 2 lobes, with left lobes, with right lobes, without lobes, all on same tree. Leaves turn yellowish to orange-reddish in fall. Roots, twigs, bark, and leaves spicy-aromatic. Flowers greenish-yellow, in loose terminal clusters. Fruits on female trees blue, on bright red-cupped stalks. Frequent in woods, thickets throughout pines. Flowers mid-April into early May. Fruits late July into early September.

    Sassafras was one of the most sought after new plants found in the New World colonies. It was in great demand to be sent back to Europe where it was regarded as a cure-all for almost all kinds of ills.[9]

Witch-hazel Family (Hamamelidaceae)

Sweet Gum *Liquidambar styraciflua*   45' - 70'
Trunk rises to wide spreading crown. Bark grayish-brown, grooved, furrowed. Twigs often have cork-like ridges or wings. Leaves smooth, deep glossy green, 5"-7", star-shaped with 5-7 pointed, toothed lobes, fragrant when crushed. Flowers inconspicuous, greenish, in dense heads. Fruit long-stalked, woody, round, prickly, ball-like head of capsules. Not found naturally within pine barrens but occasional due to introductions and intrusions from adjacent, surrounding regions where it is native and abundant in moist soils near streams.

Rose Family (Rosaceae)

Black Cherry *Prunus serotina*   40' - 70'
Medium to large tree with irregular, rugged contour and drooping foliage. Bark on mature trees rough, dark reddish-brown; on young trees smooth, with horizontal markings. Leaves 2"-5", oblong-lanceolate, sharp-pointed, finely toothed, dark green. Flowers white in drooping clusters, mid-May to early June. Fruit dark red when immature, purple-black when ripe in September. Native peripherally and intrusive into disturbed areas in pinelands.

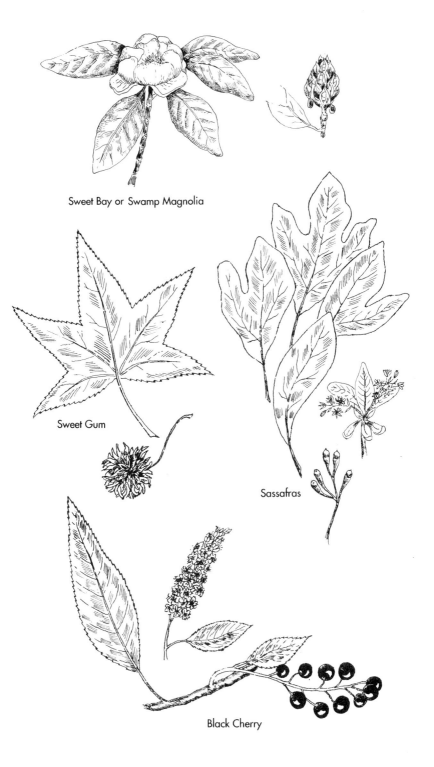

Trees

Holly Family (Aquifoliaceae)

American Holly *Ilex opaca* 15' - 50'

Bark thin, light gray to brownish-gray. Leaves oval-elliptical, leathery, evergreen, with 5-9 spine-tipped teeth along margins and a sharp spine at tip. Flowers small, white to greenish-white, male and female flowers on separate trees. Fruits bright red berries on female trees. Primarily a coastal species but frequent in moist woodlands along stream margins as it follows streams for some distance up into pines. Flowers late May through June. Fruits ripen bright red by December, persist over winter.

Maple Family (Aceraceae)

Red or Swamp Maple *Acer rubrum* 20' - 80'

Large tree. Bark dark gray-brown with furrows and ridges on mature trees; smooth, light gray on young trees. Twigs and buds reddish. Leaves opposite with 3-5 coarsely toothed lobes, lighter underneath than above, turn shades of yellow-orange-red in fall. Flowers small, bright reddish, open before leaves. Seeds paired, with wings, often bright reddish. May be most common deciduous tree in low woods, swamps, and stream edges. Flowers late March to mid-April. Seeds by May.

Some authorities recognize a variety with very small, three-lobed leaves, known as **trident red maple**, *Acer rubrum* var. *trilobum*, which is common in the pine barrens.

Sour Gum Family (Nyssaceae)

*Sour or Black Gum or Tupelo *Nyssa sylvatica* 20' - 60'

Medium size tree with horizontal, contorted branches. Bark dark gray-brown, grooved, ridged. Leaves alternate, ovate, entire or with vestigial teeth, 2"-5", shiny, smooth, dark green, lighter beneath, turn maroon to deep red early in fall. Flowers inconspicuous, greenish, in small clusters. Fruits blue-purple berries on long stalks, usually in pairs or clusters. Common in moist woods, swamps. Flowers mid-May to mid-June. Fruits September into October.

A number of other trees have been introduced locally into the pine barrens, often brought in by early settlers for shade or ornamentation. Some are still maintained around functioning historic centers as Allaire, Atsion, Batsto, Double Trouble, and Whitesbog. Some still struggle for survival in the wild as indicators of past human activities at now abandoned sites as Buckingham, Harrisville, Martha, and Mount. Few, if any of these in the wild are reproducing themselves and, in time, most likely will die out. Among these are:

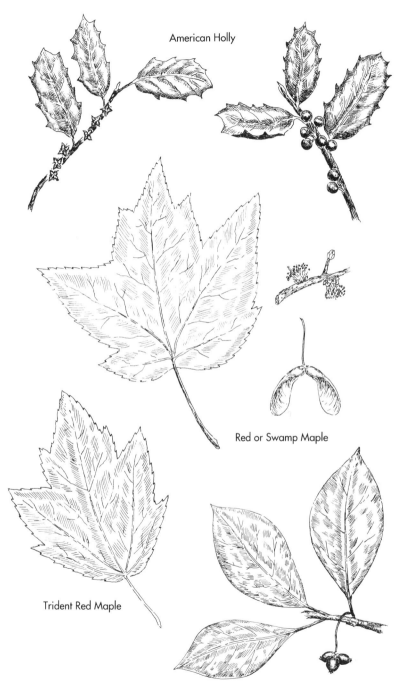

Shrubs

Norway spruce *(Picea abies)*, **white pine** *(Pinus strobus)*, **black willow** *(Salix nigra)*, **large-toothed aspen** *(Populus grandidentata)*, **black walnut** *(Juglans nigra)*, **tulip** or **yellow poplar** *(Liriodendron tulipifera)*, **buttonwood** or **sycamore** *(Platanus occidentalis)*, **black locust** *(Robinia pseudo-acacia)*, **flowering dogwood** *(Cornus florida)*, and **catalpa** *(Catalpa bignonioides)*.

SHRUBS

Shrubs are woody and perennial, like trees, but are lower, usually under 20', usually with several stems or trunks, often densely branched. The following are all true flowering plants (angiosperms) and are dicotyledons (see page 76).

Wax-myrtle Family (Myricaceae)
Shrubs with alternate, simple, resinous-dotted (use hand lens) leaves. Aromatic when crushed. Flowers in catkins.

Bayberry or Candleberry *Myrica pensylvanica* 2' - 6'
Compact, bushy shrub with smooth, light brown stems. Twigs covered with gray hairs. Buds white. Leaves thin, wedge-shaped, pointed at tips, smooth above, somewhat hairy beneath, green, resinous-dotted, fragrant when crushed, deciduous though sometimes persistent. Frequent on damp, sterile sands, borders of swamps. Catkins in May through June. Fruits clusters of small, round nuts covered with whitish-gray wax, used in making candles.

Evergreen Bayberry or Wax-myrtle *Myrica heterophylla* 2' - 6'
Similar to above but twigs covered with black hairs. Leaves oblong, shiny, leathery, evergreen, with minute, yellowish resin dots on under surface, fragrant when crushed. Dry to wet thickets and woods. Catkins in May. (No illus.)

*Sweet-fern *Comptonia peregrina* var. *asplenifolia* 1' - 3'
Low, bushy shrub with many branches. Leaves long, narrow, deeply cut, with rounded lobes on sides, appear fern-like, resinous dotted (use hand lens), fragrant when crushed. Common in dry, sandy barrens and pinelands. Catkins late April - early May.

Hazel Family = Birch Family (Corylaceae = Betulaceae)

*Common or Smooth Alder *Alnus serrulata* 6' - 12'

Shrub that forms thickets along water courses. Leaves eggshaped, broadest at or above middle, narrowed at base, rounded at tip, finely toothed on

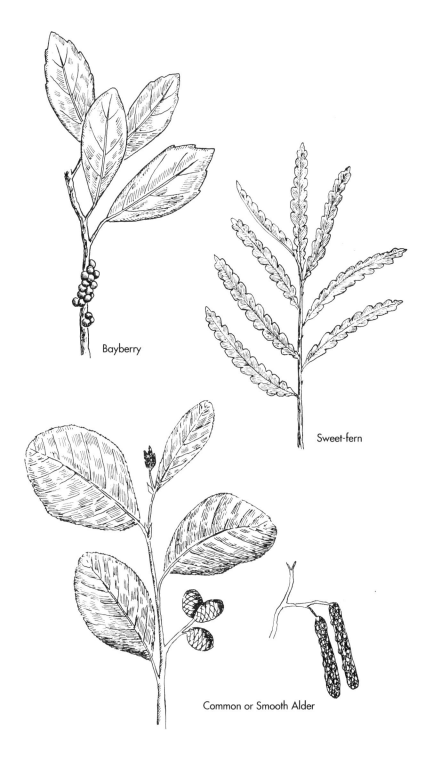

Shrubs

outer edges. Male catkins long, pendant, mid-March to early April. Fruits woody, cone-like, persist through winter. Occasional along stream edges, thickets, cedar swamps.

Beech Family (Fagaceae)

*Dwarf Chestnut-oak *Quercus prinoides* 2' - 10'
Often only 2' - 4' in plains

Shrubby oak to, rarely, small tree. Bark light brown, scaly. Leaves small, 2"-4", oblong-elongate, narrow, twice longer than wide, sharply toothed with 3-7 pairs of teeth, bright green above, grayish-white-hairy beneath. Acorn 1/2"-3/4" long, oval, light chestnut-brown, enclosed in deep, thin, knobby cup that covers about one-half of nut. Acorns usually borne in pairs. Locally frequent in dry thickets, edges of woods. Flowers mid- to late-May.

**Bear or Scrub-oak *Quercus ilicifolia* 3' - 12'
Often only 2' - 6' in plains

A straggling, thicket-forming, multi-branched shrub in plains areas to, rarely, a small tree elsewhere. Bark dark brown. Leaves shiny, hairy, whitish beneath, with 4-8 toothed lobes. Acorn 1/2", broadly ovoid, light brown, enclosed in top-shaped cup with overlapping, red-brown scales that form a fringe at rim and cover about one half of nut. Acorns usually borne in pairs and clustered on branches. Common in dry, poor, gravelly, sandy barrens, especially the pine plains. Flowers early to mid-May.

Saxifrage Family (Saxifragaceae)

**Virginia-willow *Itea virginica* 3' - 7'
Shrub with green twigs which show partitions in pith when split lengthwise. Leaves alternate, oblong, pointed, finely toothed. Flowers white with 5 petals and 5 stamens, in loose, open, terminal spikes. Superficially resembles sweet pepperbush (see page 126) but flowers much earlier. Frequent in swamps, wet thickets. Blooms June. Fruits September.

Rose Family (Rosaceae)

Large family of mostly (in pine barrens) shrubs with alternate, toothed leaves. Flowers usually with 5 petals, 5 sepals, and numerous stamens surrounding center.

*Red Chokeberry *Pyrus arbutifolia* = *Aronia arbutifolia* 3' - 10'
Shrub that grows in small clumps, with slender, dark brown stems. Twigs, buds woolly. Leaves oval, pointed, fine-toothed, smooth above, woolly-hairy beneath. Flowers white with rounded petals, usually tinged with magenta-pink, in terminal clusters. Common in low thickets, swamps, woods. Flowers late April through May. Fruits bright red in September - October.

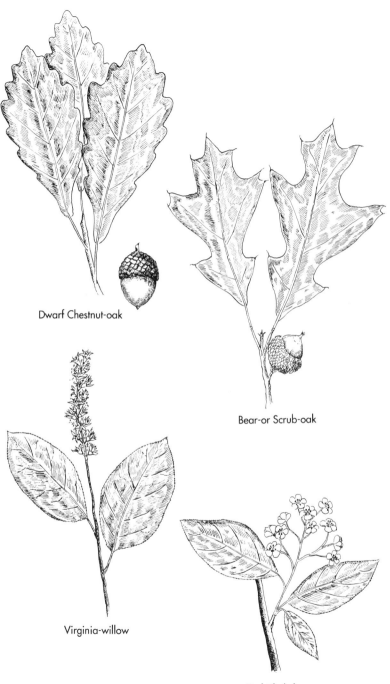

Shrubs

*Black Chokeberry *Pyrus melanocarpa = Aronia melanocarpa*   3' - 10'
Nearly identical to above except twigs and leaves entirely smooth, without any hairs or woolliness. May hybridize with above. Common in swamps, damp thickets. Flowers late April through May. Fruits black in July - August.' (No illus.)

*Shadbush or Serviceberry *Amelanchier canadensis*   2' - 15'
Shrub that grows in clumps, or, rarely, a small tree. Bark gray with brown striping. Twigs brownish. Leaf stalks woolly. Leaves oval-oblong, toothed, conspicuously white-woolly when developing, but not fully expanded at flowering time, remain somewhat hairy. Flowers white, petals longer than wide. Occasional in swamps, thickets, low woods, roadside ditches. Blooms April - early May. Fruits blackish, late June - early July.

*Dwarf Hawthorn *Crataegus uniflora*   3' - 5'
Small shrub with grayish bark and slender, spreading branches, with thorns. Leaves oval or broadly elliptical, coarsely round-toothed, deep green above, paler beneath. Flowers small, solitary, white. Fruits round, dull orange-red. Occasional on sandy banks, upland woods. Blooms mid-May to early June. Fruits September - October.

*Dewberry *Rubus flagellaris*
*Swamp Blackberry *Rubus hispidus*
Both vines in rose family. See page 140 for descriptions.

*Sand Blackberry *Rubus cuneifolius*   1' - 3'
Stems erect or arching. Often grows in large patches. Stems and leaf stalks densely prickly. Leaflets 3-5, woolly beneath. Flowers white, showy. Fruits compact clusters of fleshy, black seeds. Common in upland, sandy soils. Flowers late May - early July. Fruits late July - August.

Beach-plum *Prunus maritima*   2' - 6'
Low, straggly, much branched shrub with crooked, black trunks. Leaves broadly elliptical, finely toothed, hairy underneath. Flowers white, 5-petaled, in small clusters, appear before leaves. Fruit a bluish-purple to purplish-black plum. Occasional and local on sandy gravels in barrens. Flowers late April - early May. Fruits late August - September.

Black Cherry *Prunus serotina*
A tree. See page 114 for description.

Cashew Family (Anacardiaceae)
Trees, shrubs, vines with alternate, compound leaves divided into leaflets. Flowers small, inconspicuous, clustered, 5-parted, greenish- to yellowish-white. Fruits small, pea size, clustered, berry-like. Some of these plants have a resinous, acrid, caustic juice which is poisonous to human touch.

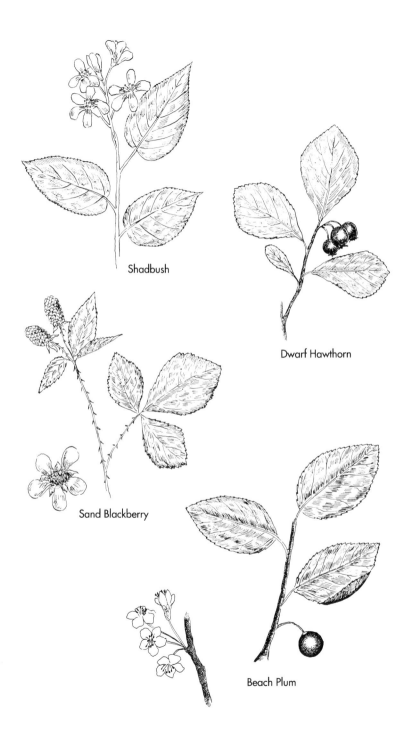

Shrubs

Dwarf or Winged Sumac *Rhus copallina* 2' - 15'
Branched shrub, often grows in small groups. Stems downy with large, soft centers (pith). Leaves alternate, compound, divided into 7-21 dark green, shiny, opposite leaflets. Margins of leaf stalk winged between leaflets. Flowers small, numerous, greenish-white to greenish-yellow, in terminal clusters. Fruits small, maroon-red, hairy, in conspicuous terminal clusters, persist well into winter. Introduced into disturbed openings, waste places, dry woods. Blossoms late July into September. Fruits late August into early October. Non-poisonous.

*Poison Sumac *Rhus vernix* = *Toxicodendron vernix* 6' - 18'
Tall shrub, occasional small tree. Bark gray-brown, smooth, or rough with age. Leaves compound, divided into 7-13 smooth, light green leaflets. Flowers dull white-greenish or greenish-yellow, borne in loose clusters from leaf axils. Fruits in hanging clusters of small greenish, grayish, or dull white berries which persist most all winter. Local and occasional in wet ground, swamps. Blossoms late May - June. Fruits late August - early September. Every part of entire plant highly poisonous.

Poison Ivy *Rhus radicans* = *Toxicodendron radicans*
A vine. See page 140 for description.

Holly Family (Aquifoliaceae)
Trees and shrubs with alternate, simple, often leathery leaves and small, greenish-white flowers in leaf axils. Four pine barrens hollies, one a tree, three shrubs, are described.

American Holly *Ilex opaca*
A tree. See page 116 for descriptions.

Black Alder or Winterberry *Ilex verticillata* 4' - 15'
Shrub with smooth, grayish stems and branches. Leaves oval-lance-shaped, often variable in width with coarse but shallow teeth, dull above, sparsely hairy beneath. Flowers small, inconspicuous, white. Male and female flowers on separate plants. Fruit orange-red berry which persists into winter. Wooded swamps, damp thickets, stream margins. Flowers mid-June to early July. Fruits mid-September - October.

*Smooth Winterberry *Ilex laevigata* 4' - 15'
Similar to above except leaves shiny above, finely toothed. Twigs and leaves smooth. Flowers and fruits on short, 1/4", stalks. Same habitats as above. Flowers late May - mid-June. Fruits early September - October, persisting. (No illus.)

**Inkberry *Ilex glabra* 2' - 10'
Erect, evergreen shrub. Stems and twigs yellow when young, becoming

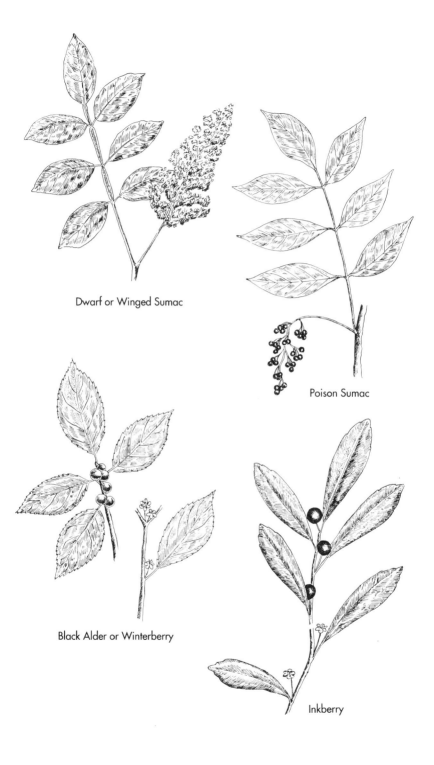

Shrubs

light gray, finely downy. Leaves leathery, evergreen, shiny, smooth, narrowly elliptical, dark green above, paler beneath, 1 or 2 notches each side near tips. Flowers small, inconspicuous, greenish-white, in leaf axils, male and female on separate plants. Fruits black, fleshy berries. Common in low, damp, sandy soils, wooded swamps, bog edges, pond and stream sides. Flowers mid-June to early July. Fruits mid-September - October, persisting into winter.

St. John's-wort Family (Guttiferae = Clusiaceae)

**Shrubby St. John's-wort *Hypericum densiflorum* 2' - 6'
Much branched small shrub. Stems ridged. Leaves narrow, almost linear, toward elliptical, with tufts of smaller leaves growing in leaf axils. Flowers in small clusters on sides and tops of stems and branches, forming dense flower cluster. Flowers bright, golden yellow, 5-petaled, with conspicuous stamens. Common in damp, sandy soils. Flowers early July- early September.

White Alder Family (Clethraceae)

*Sweet Pepperbush *Clethra alnifolia* 3' - 10'
Shrub with grayish stems, grayish-brown bark. Leaves alternate, widely ovate, short-pointed, sharply toothed beyond middle, deep green above, lighter beneath. Flowers small, 5-petaled, with 10 stamens, white, in slender, round, dense terminal clusters, very fragrant. Common in swamps, thickets, low sandy woods. Flowers late July - early September.

Heath Family (Ericaceae)
As the pines and oaks dominate the tree canopy of the pine barrens, the heaths, particularly the blueberries, huckleberries, and laurels, dominate and are the most characteristic plants in the understory layer of pine barrens vegetation.

Most pineland members of this family are shrubs with simple, usually alternate, leaves, and many have bell- or tube-shaped flowers with five lobe-like petals and five stamens.

**Swamp or Clammy Azalea *Rhododendron viscosum* 3' - 8'
Stem much branched. Branches smooth, gray-brown. Twigs hairy. Leaves blunt, lance-shaped, shiny light green above, lighter and mid-rib hairy below, sometimes waxy, somewhat clustered near tips of branches. Leaves may be strongly whitened beneath and sometimes above in form *glaucum*. Flowers in small, terminal clusters, white or pink-tinged, the sticky, hairy tubes longer than the spreading lobes, stamens prominent, very fragrant. Occasional to common in swamps, thickets. Flowers early June - early July.

**Sand-myrtle *Leiophyllum buxifolium* 6" - 3'
Low, spreading, with many scraggly stems and branches. Bark rough, shreddy, brown. Leaves almost opposite, crowded, ovate, shiny dark green,

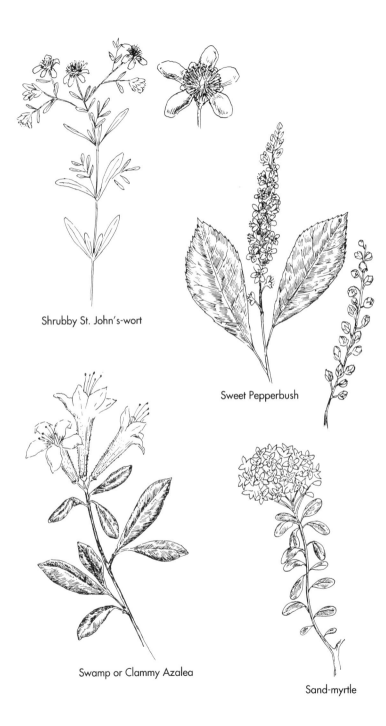

Shrubs

leathery, evergreen, like miniature boxwood. Flowers numerous in small, terminal clusters, white or pinkish, 5-petalled, with conspicuous lavender stamens. Type variety known only from damp open, sandy areas in heart of pine barrens of New Jersey.[10] Frequent to common in pitch pine lowlands. Flowers late April - early June.

*Mountain-laurel *Kalmia latifolia* 3' - 10'
Large, sturdy, often grows in thick clusters. Trunks irregular, gnarled, spreading. Bark ruddy brown. Leaves elliptical, 2"-5", thick, leathery, evergreen, smooth, shiny, dark green. Flowers waxy-white, pink-tinged, saucer-shaped, nearly 1" across, seemingly held open by the 10 stamens, in large, showy, dome-shaped, terminal clusters. Common in damp, sandy woods and swamps. Flowers late May - June.

**Sheep-laurel or Lambkill *Kalmia angustifolia* 1' - 3'
Stems irregular, branched. Bark brown. Leaves thin, narrow, elliptical, evergreen, dull olive-green, sometimes rust-spotted, lighter under. Newer leaves at tops of stems erect, leaves below flower clusters droop, often in whorls of 3 or more. Foliage said to be poisonous to livestock. Flowers small, saucer-shaped, with stamens as in *K. latifolia,* pale to deep crimson-pink, clustered around stem below top. Common in dry to mostly moist, sterile sands, especially near edges of bogs. Flowers late May - June.

**Stagger-bush *Lyonia mariana* 1' - 6'
Slender, smooth stems, branches. Leaves oblong-elliptical, olive-green, smooth above, finely hairy on veins under. Flowers large, showy, pendant, globular or urn-shaped, in open clusters on older, leafless branches, white to slightly pinkish. Common in sandy, peaty swamps, thickets, woods. Flowers late May - June.

*Maleberry or Privet Andromeda *Lyonia ligustrina* 2' - 10'
Ascending branches minutely hairy. Twigs hairy near tips. Leaves thin, variable, oblong-elliptical, minutely toothed or not. Flowers small, spherical, in small terminal clusters on leafless twigs, white. Frequent in swamps, wet thickets. Flowers mid-June to early July.

*Fetter-bush *Leucothoe racemosa* 3' - 12'
Erect branches. Leaves narrow-oblong or lance-shaped, pointed, finely toothed. Flowers hanging, bell-shaped, white, in long, single 1-sided rows from leafless tips of branches. Frequent to common in swamps, moist thickets, bogs. Flowers mid-May to June.

**Leather-leaf or Cassandra *Chamaedaphne calyculata* 2' - 4'
Profusely branched. Leaves small, narrow, elliptical, leathery, partially evergreen, dull green, underneath with rusty spots. Flowers bell-shaped, white, pendant in 1-sided, single rows from leafy tips of branches. Com-

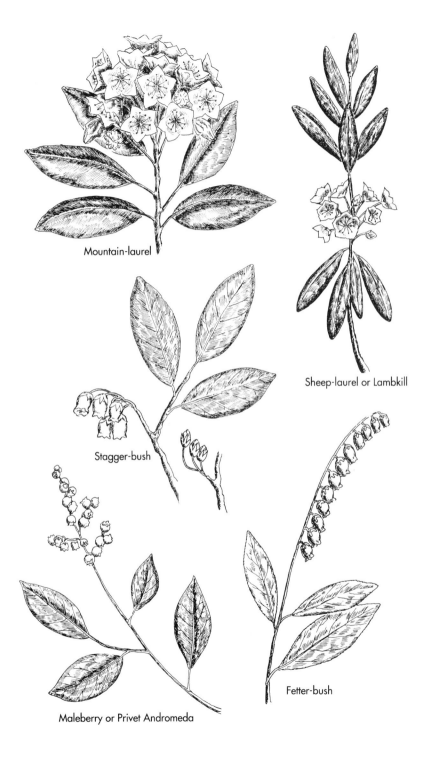

mon along edges of bogs, moist thickets, swamps, borders of cranberry bogs. Flowers early April - early May.

*Trailing Arbutus *Epigaea repens*
*Teaberry or Checkerberry *Gaultheria procumbens*
**Bearberry *Arctostaphylos uva-ursi*
Three subshrubs. See page 136 for descriptions.

Huckleberries *Gaylussacia* spp.
Blueberries *Vaccinium* spp.
Huckleberries and blueberries are closely related members of the heath family and, superficially, appear quite alike. However, two easily identifiable characters by which these can be separated are in the leaves and fruits. Leaves of huckleberries have tiny, yellowish resin dots (use hand lens) underneath; blueberries do not. Fruits of huckleberries contain 10 hard seed-like nutlets enclosed within a limited amount of pulp. Blueberries have many small, fine seeds, neither large nor hard, surrounded by considerable fleshy and very palatable pulp.

**Dwarf Huckleberry *Gaylussacia dumosa* 1' - 2'
Stems, branches upright, mostly leafless below. Young branches, twigs finely hairy. Leaves variable, thick, oblong, not toothed but bristle-tipped, dark green, shiny above, slightly hairy, covered with resin dots underneath. Flowers bell-shaped, pendant on short stalks in loose clusters, white to pinkish. Fruits black, hairy, resin-dotted. Occasional in moist, open sandy areas, more frequent in cedar swamps. Flowers late May - June. Fruits late July - early September.

*Dangleberry or Blue Huckleberry *Gaylussacia frondosa* 2' - 6'
Slender stemmed. Branches gray. Twigs smooth. Leaves oval-oblong, variable, thin, light green, whitish with resin dots under. Flowers bell-shaped, light green tinged with pink, in loose clusters. Fruits dark blue with a bloom. Common in dry woods, clearings. Also in bogs and moist thickets.

*Black Huckleberry *Gaylussacia baccata* 1' - 3'
Erect shrub with stiff, brown branches. Twigs hairy young, to smooth later. Leaves oval-oblong, pointed, covered on both surfaces with shiny resin dots, very resinous when young. Flowers bell-shaped, green, strongly tinged with brick-red, in 1-sided clusters. Fruits black. Abundant in dry or moist thickets, woods. Our most common species. Flowers early May - early June. Fruits early July - August.

*Low Blueberry *Vaccinium vacillans* = *V. pallidum* 1' - 2 1/2'
Low branching shrub, smooth throughout, often in patches. Leaves oval-elliptical, green above, whitish beneath, without resin dots. Flowers bell-

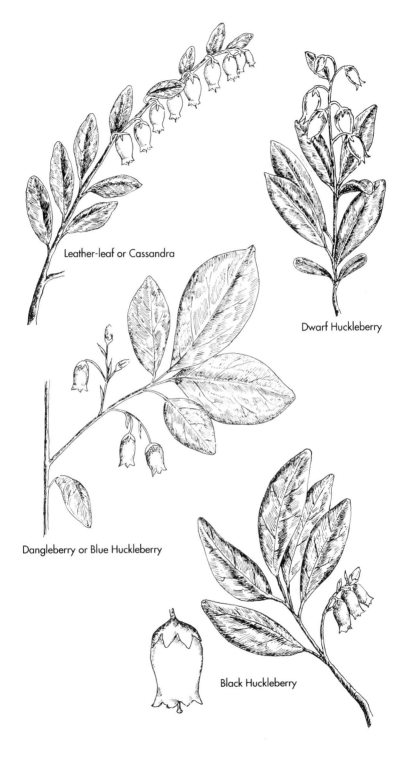

Shrubs

shaped, greenish-white tinged with pink. Fruit a blue berry with a bloom, sweet. Common in dry woods, clearings. Flowers May. Fruits late June - late July.

**Highbush Blueberry *Vaccinium corymbosum* 3' - 12'
Tall, branching. Leaves elliptical, toothed or not, pointed, green and smooth above, slightly paler green and somewhat hairy beneath, without resin dots. Flowers bell-shaped, white or pinkish. Fruits blue to blue-black with a bloom. This is the base stock for the cultivated highbush blueberry. Frequent in wet thickets, swamps. Flowers May. Fruits early July - early August. This and the following two are so similar they may all be only one species, *V. corymbosum,* with the following being only forms

New Jersey Blueberry *Vaccinium caesariense* = *V. corymbosum* 3' - 10'
Similar to highbush blueberry but bushes lower, leaves smaller, untoothed, entirely smooth, waxy and white beneath. Frequent in moist thickets. Flowers dull white, in May. Fruits dark blue with a bloom, July - early August. (No illus.)

*Black Highbush Blueberry *Vaccinium atrococcum* 3' - 12'
= *V. corymbosum*
Similar to highbush blueberry but leaves untoothed, deep green, woolly beneath. Flowers greenish-white, often pink-tinged, open before leaves unfold, late April to early May, earlier than highbush. Moist thickets. Fruits violet-black, without bloom, late June - July. (No illus.)

**American Cranberry *Vaccinium macrocarpon*
A vine-like shrub. See page 142 for description.

Madder Family (Rubiaceae)

*Buttonbush *Cephalanthus occidentalis* 3' - 10'
Stems widely branched. Bark smooth, brown-gray. Leaves large, over 3", ovate, rounded or narrowed at base, short-pointed at tip, entire, opposite or often in whorls of 3's or 4's, smooth, deep green above, strongly veined, lighter under. Leaf stalks often reddish. Flowers small, dull white, tubular, densely clustered, as are later fruits, in tight, round, long-stalked, ball-like heads. Occasional and peripheral to pine barrens along edges of streams, swamps. Flowers July - early August.

Honeysuckle Family (Caprifoliaceae)

*Witherod *Viburnum cassinoides* 2' - 12'
Stem, branches slender, brown-gray. Twigs dull light brown. Leaves opposite, thick, oval-lance-shaped, pointed, shallowly-toothed or wavy-edged, dull. Flowers small, in stalked, broad, flat topped, branching clusters. Ma-

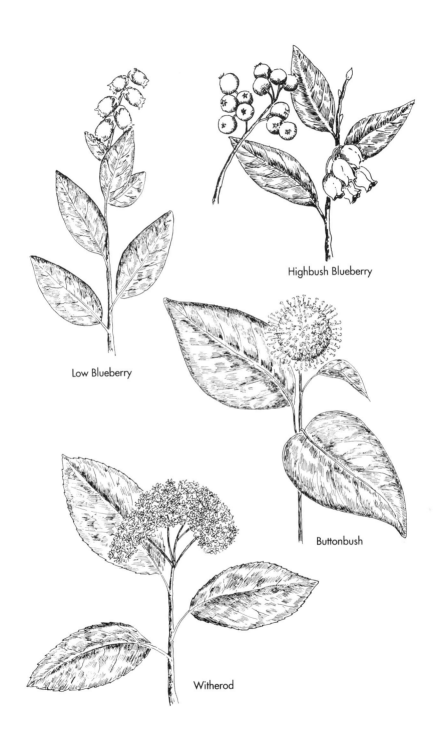

ture fruit bluish-black. Occasional in wet thickets, swamps. Flowers late May - mid-June.

*Naked Witherod *Viburnum nudum* 6' - 18'
Similar to above but leaves shiny, not dull, somewhat leathery, nearly entire, not toothed. Same habitats. Flowers later: mid-June to early July. More frequent than above. (No illus.)

SUB-SHRUBS

Sub-shrubs, usually woody and perennial, grow very close to the ground, seldom higher than a few inches, almost always under 12 inches. These are all true flowering plants (angiosperms) and are dicotyledons (see page 76).

Crowberry Family (Empetraceae)

**Broom-Crowberry *Corema conradii* 4" - 12"
Very low, multiple branching shrub, grows nearly prostrate, in roundish patches on bare sands. Stems woody, gnarled. Leaves tiny, 1/4", narrow, evergreen. Flowers tiny, inconspicuous, terminal, appear while plant still dingy-looking from winter, before new leaves. Fruits small, berry-like, dark brown-black. A northern plant which, in New Jersey, is almost restricted to east and west plains areas. Classified as an endangered species. Blossoms mid-March or earlier, into April, depending on season.

Rockrose Family (Cistaceae)

****Golden- (Pine-barren) heather
or Heath-like Hudsonia** *Hudsonia ericoides* 3" - 8"
Low shrub with bushy, branching stems. Leaves tiny, crowded, bristly, needle-like, spreading outward slightly from stem, greenish. Flowers small, 1/3", bright yellow, on hairy stalks at tips of stems and branches. Common on dry, open sands. Blooms late May - early June. Not a true heather.

Beach-heather or Woolly Hudsonia *Hudsonia tomentosa* 3" - 8"
Similar to above but leaves are scale-like, soft-woolly, close to stem, not spreading, grayish-white. Flowers without stalks. Occasional in same habitat as, and sometimes side by side with, golden heather. May hybridize with above. Same flowering times as above.

Wintergreen Family = Heath Family (Pyrolaceae = Ericaceae)

Spotted Wintergreen *Chimaphila maculata* 6" - 10"
Low, smooth, evergreen. Stem erect, smooth. Leaves lance-shaped, tapering to sharp point, strongly but sparsely toothed, arranged in whorls around

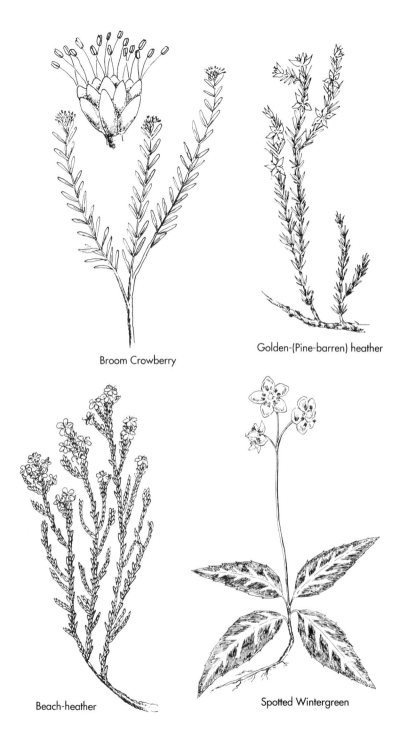

stem, dark green with conspicuous white to greenish-white lining along ribs. Flowers white, slightly pinkish, with green center (ovary), waxy, nodding. Common in dry woods. Flowers late June - mid-July.

Heath Family (Ericaceae)

*Trailing Arbutus *Epigaea repens*   6" - 12" long
Prostrate, trailing. Stems rough-hairy, light brown, creep along ground. Leaves oval, leathery, hairy, evergreen. Flowers tubular, opening into 5 flared lobes, white to pinkish-white, in small clusters at tips of branches, fragrant. Occasional to frequent in dry, sandy woods. Common on pine plains. Flowers early April - early May.

*Teaberry or Checkerberry *Gaultheria procumbens*   2" - 8"
Slender stems arise from creeping, underground stems. Leaves broadly oval, thick, slightly toothed from mid-leaf to tips, evergreen, shiny dark green, clustered at tops of stems. Flowers small, waxy, white, urn-shaped, 5-lobed at rim, nodding, from axils of leaves. Fruit round, bright red berry. Leaves and fruits have wintergreen flavor. Common in dry or moist thickets, woods. Flowers late June - early August. Fruits late September - early October, persisting into winter.

**Bearberry *Arctostaphylos uva-ursi*   6" - 12" long
Prostrate shrub. Stems trailing, with thin reddish bark. Leaves small, shiny, thick, entire, evergreen, spatulate-shaped, round-blunt at tip, tapering toward base. Flowers small, bell- or urn-shaped, white or pinkish-white, pendant in small, terminal clusters. Fruit round, dull-red berry. Frequent to common on exposed dry sands, gravels, especially in pine plains. Flowers late April to mid-May. Fruits August - early September.

Diapensia Family (Diapensiaceae)

**Pyxie *Pyxidanthera barbulata*   6" - 10" long
Prostrate, trailing shrub. Stems creep along ground. Leaves tiny, narrow, linear, needle-like, evergreen, crowded toward ends of branches, forming moss-like mats. Flowers numerous, small, without stalks, flower (corolla) with 5 blunt lobes, white, rarely pale pinkish. Common in dry or damp white sandy, open, woody barrens. Flowers April - early May.

Madder Family (Rubiaceae)

Partridge-berry *Mitchella repens*   6" - 10" long
Stem creeping, prostrate on ground. Leaves opposite, paired, on short stalks, rounded or heart-shaped, entire, evergreen, shiny, green veined with white. Flowers partly joined, in pairs at tips of branches or from outer leaf axils, tubular, 4-lobed, fine-hairy, creamy-white to faint crimson-pink. Fruit

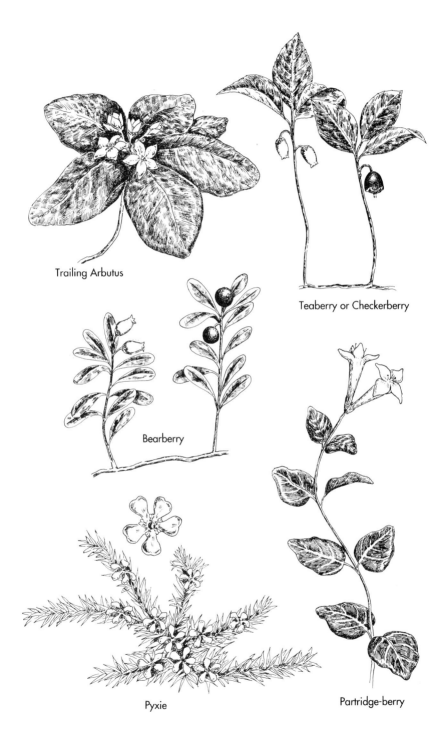

red, berry-like. Occasional to frequent intrusion along wooded stream edges or moist, wooded knolls. Flowers late May - June.

VINES AND VINE-LIKE PLANTS

Vines are shrubs, often woody and perennial, whose stems need support and must either creep along the ground or climb on other vegetation or objects by twining or by means of tendrils. All are true flowering plants (angiosperms) (see page 75). The greenbriers in the lily family, below, are monocotyledons (see page 76). The balance of the vines and vine-like plants described are dicotyledons (see page 76).

Lily Family (Liliaceae)

Greenbrier or Catbrier *Smilax* spp.
Except for the halberd-leaved greenbrier, pine barrens greenbriers are green-stemmed, often evergreen, woody vines with sharp thorns, that climb by tendrils. Leaves alternate, entire, mostly broad, smooth, shiny, usually conspicuously parallel-ribbed. Flowers small, greenish or greenish-yellow, hang from leaf axils. Fruits blue-black (usually) berries.

**Halberd-leaved Greenbrier *Smilax pseudo-china*
Annual, herbaceous vine, without thorns, that climbs by means of tendrils. Leaves green, slightly paler under, triangular-oval, with 3 prominent ribs running into an often bent, twisted tip. Flowers 10 or more in a cluster. Berries blue to black. Occasional in moist, sandy, shady ground. Flowers mid-June to early July. Fruits in early autumn.

*Common Greenbrier or Catbrier *Smilax rotundifolia*
Stout vine with stout thorns. Leaves ovate with rounded bases, green above and below, shining above. 3 middle ribs of leaves prominent underneath. Berries blue or bluish-black. Common in moist to dryish thickets. Blooms May - early June. Fruits by mid-autumn.

**Red-berried Greenbrier *Smilax walteri*
Similar to *S. rotundifolia,* above, except for its coral-red berries and lack of thorns. Local in deep, swampy thickets. Flowers May - early June. Fruits by mid-autumn.

*Glaucous-leaved Greenbrier *Smilax glauca*
Leaves elliptical, less than 2" long, strongly whitish on under sides. Stems dark to purple. Berries blue. Common in dry, sandy soils. Blooms late May - June. Fruits by mid-autumn.

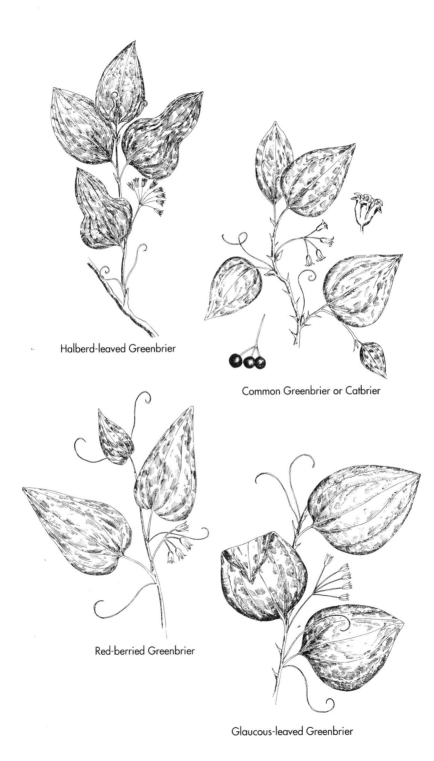

Vines

Laurel-leaved Greenbrier *Smilax laurifolia*
High climbing evergreen vine armed with rigid thorns and having long, thick, glossy evergreen leaves that "hang in festoons from trees and bushes on the edges of deep wooded swamps."[11] Mid-rib of leaf more prominent underneath than the 2-4 lateral ribs. Berries black. Local in wooded swamps. Flowers August - early September. Fruits early autumn.

Rose Family (Rosaceae)

*Dewberry *Rubus flagellaris*
Trailing vine with roundish stems and thorns. Stems root at tips. Leaves compound, 3-5 sharp-pointed leaflets, evergreen, dull. Flowers large, 1", white, 5-petalled, solitary or in small clusters. Fruits compound clusters of black, fleshy seeds, sweet. Occasional in dry fields, borders of thickets. Flowers mid-May to mid-June. Fruits July - August.

*Swamp or Running Blackberry *Rubus hispidus*
Similar to above but stem bristly, not thorny; leaves shiny, not dull; flowers smaller, 1/2"-3/4", fruits reddish to black, sour. Common in moist ditches, thickets, cedar swamps. Blossoms June - early July.

Pea or Pulse Family (Leguminosae = Fabaceae)

*Ground-nut or Wild Bean *Apios americana*
Small herbaceous vine that climbs and twines over low vegetation. Leaves divided into 5-7 broad, pointed leaflets. Flowers velvety, maroon or brownish-purple, fragrant, solitary, paired or in small clusters from leaf axils. Moist thickets, swamps. Flowers July - August.

*Milk-Pea *Galactia regularis*
Small, thin, smooth, trailing, prostrate herbaceous vine with elongate-oval leaflets, in 3's, rounded at each end. Flowers small, reddish-purple on short stalks in small loose clusters, from leaf axils. Dry, open, sandy ground. Flowers July - early August.

Cashew Family (Anacardiaceae)

Poison Ivy *Rhus radicans = Toxicodendron radicans*
Vine-type shrub that spreads over vegetation and climbs tree trunks by running rootlets. Leaves in 3 smooth, light green, ovate leaflets. Flowers whitish-green, in compact clusters from leaf axils. Fruits clusters of small, dull, grayish berries, persist well into winter. Introduced into pine barrens in thickets, fence rows, waste places, disturbed areas. Flowers mid-May to mid-June. Fruits mid-August - September. Entire plant poisonous to touch.

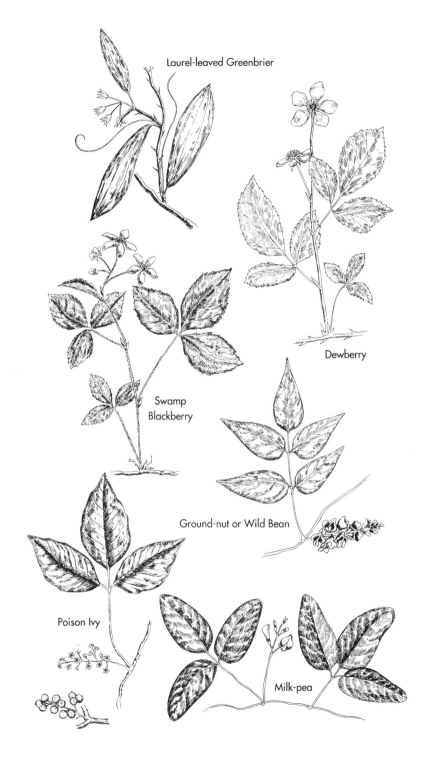

Vines

Vine or Grape Family (Vitaceae)

Virginia Creeper or Woodbine *Parthenocissus quinquefolia*
Shrubby vine that trails along ground or climbs vegetation by means of long tendrils. Leaves divided into 5, rarely 3 or 7, dark green leaflets, arranged radially from top of stem. (Poison ivy climbs by means of rootlets, leaves always divided into 3 leaflets.) Flowers small, clustered, yellowish-green or whitish. Fruits blue berries. Occasional where introduced into pineland woods, thickets. Flowers late June - July. Fruits late September - October.

Summer-grape *Vitis aestivalis*
High climbing, woody vine with easily shredded bark. Twigs hairy, reddish. Climbs by means of tendrils. Leaves large, heart-shaped, shallowly 3-5 lobed, toothed, red-woolly underneath. Flowers small, greenish-white, in small clusters. Fruits black, in clusters. Occasional where introduced into pineland thickets, dry woods. Flowers mid-June to early July. Fruits September - October.

Heath Family (Ericaceae)

**American Cranberry *Vaccinium macrocarpon*
Low, trailing, bog shrub. Stem trails along ground, sends up erect branches. Leaves small, elliptical, blunt-tipped, evergreen, paler underneath than above. Flowers long-stalked, nodding, solitary but 2-6 per upright branch, pinkish-white, with 4 curled-back lobes and conspicuous, pointed cone of apparently fused stamens. Fruits yellow-green, ripening to bright red. Common in wet, sandy, peaty meadows, bogs. This is the commercial cranberry. Flowers mid-June to mid-July. Fruits ripen late September - October.

Morning-glory Family (Convolvulaceae)

**Morning-glory *Breweria pickeringii* var. *caesariense*
= *Stylisma pickeringii* var. *caesariense*
Stems prostrate, creep along ground, freely branching. Leaves very narrowly linear, tapering toward base. Flowers large, showy, flared tubular or funnel-shaped, white, in small clusters of 1-5. Confined to dry sands and dunes. This subspecies known only from New Jersey pine barrens.[12] Rare.[13] Classified as threatened. Flowers late June - August.

Dodder *Cuscuta* spp.
Parasitic, herbaceous vine that lacks leaves and green color. Stems usually yellowish-orange. Vines twine, like long, tangled strings, about other plants and send suckers into them, living on them as parasites. Flowers white, small, in small clusters, August - September. 3 species in pinelands, separated by technical differences. *C. compacta* may be most common.

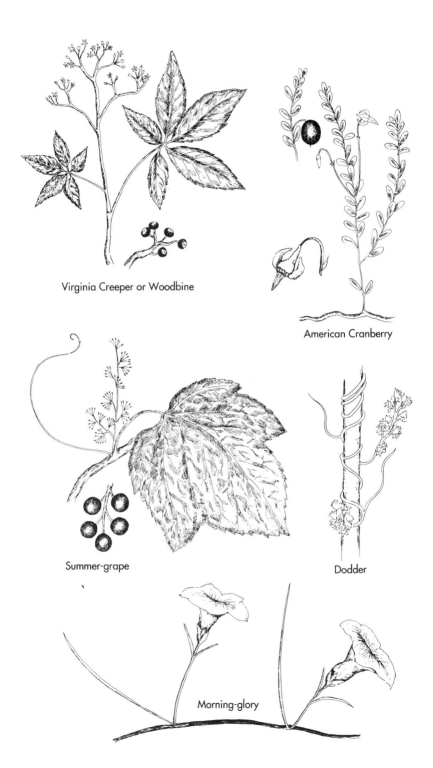

Aquatic Plants

HERBACEOUS PLANTS

Herbaceous plants do not have any persistent, woody tissue. Instead, these plants grow up out of the ground, or out of water, each spring and then die down at the end of the growing season. These may be either perennial, biennial, or annual plants.

In an effort to try to reduce the number of species one would have to sort through if all herbaceous plants were lumped together, these are broken down into smaller groupings based on habitat or structure as follows:
Page

1. Aquatic plants (except aquatic grass-like plants) 144
2. Plants with unusual structures (insectivorous plants and a cactus) 152
3. Grass-like plants (grasses, sedges, and rushes) either land or aquatic based 156
4. Plants with parallel-veined leaves (mainly lilies and orchids) 172
5. Plants with netted-veined or other non-parallel-veined leaves 182
6. Plants with numerous small flowers so arranged that their total appear as single flowers. The composites 206

AQUATIC HERBACEOUS PLANTS

Plants that are entirely or primarily aquatic. These plants grow submerged in water and their roots are embedded in mud at the bottom of water. They bear their leaves under water, or their leaves float on the surface of the water, or some of their leaves may stand up somewhat above water level.

Several of the grass-like plants, mainly sedges and rushes, also are aquatic, semi-aquatic, or nearly so. However, such of these plants that are included in this field guide are described within their family groups and are only listed in this grouping of aquatic herbaceous plants.

The first twelve of the following plants are monocotyledons, the remaining nine are dicotyledons (see page 76).

Bur-reed Family (Sparganiaceae)

**Slender Bur-reed *Sparganium americanum*   up to 3'
Aquatic or marsh plant, usually erect, sometimes floating, with long, linear,

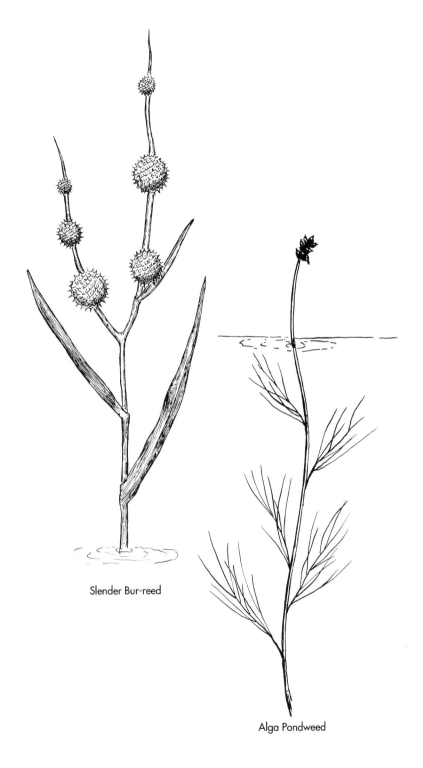

Slender Bur-reed

Alga Pondweed

Aquatic Plants

grass-like leaves which are flat, thin, pliant, barely 1/2" wide. Female flowers form hard, round, bur-like, greenish balls, about 1" thick, at intervals along stem. Male flowers smaller, appear shriveled. Flower stalks not usually branched. Muddy shores, shallow waters in ponds, swamps, streams. Blossoms June - mid-July.

Pondweed Family (Zosteraceae = Potamogetonaceae)

**Alga Pondweed *Potamogeton confervoides*
Plants grow almost totally submersed in water. Roots threadlike, tuberous. Stems jointed, thread-like, flexible, branching. Leaves bunched, flimsy, hair-like. Flowers and seeds tiny, just break surface of water. Flowers late June - September. This species characteristic of and restricted to ponds and streams in pine barrens. (Illus. on pg. 145)

Pondweeds are the largest family of truly aquatic seed plants, with many species. Their dense underwater growths provide cover for aquatic insects, snails, fishes, and other animals. Several kinds of ducks feed almost exclusively on pondweeds.

Water-plaintain Family (Alismataceae)

**Arrowhead *Sagittaria engelmanniana* 10" - 2 1/2'
Erect aquatic or marsh herb with long-stalked, smooth, very long, up to 10", very narrow, arrow-shaped, green leaves with two basal lobes. Flowers 1/2"-1" wide with 3 roundish petals, usually arranged in whorls of 3 on leafless stems. Common along edges of streams, peaty bogs. Flowers mid-June - September.

Grass Family (Gramineae = Poaceae)

Wild Rice *Zizania aquatica*
See page 162 for description.

Sedge Family (Cyperaceae)

**Swaying Rush or Water Club-rush *Scirpus subterminalis*
See page 168 for description.

Arum Family (Araceae)

Arrow-Arum *Peltandra virginica* 1' - 1 1/2'
Erect aquatic herb with long-stalked, smooth, very long, 8"-14", large, broad, arrow-shaped, green leaves with 2 basal lobes. Flowers enclosed in long, 4"-8", tapering, pointed hood or envelope. Fruit a cluster of green berries. Occasional to frequent along muddy edges of streams. Flowers throughout June.

Aquatic Plants

Golden Club *Orontium aquaticum*   1' - 2'
Aquatic or marsh herb. Leaves long-stalked, elliptical, without distinct midvein, lay on surface of water or grow out of water, upper side covered with satiny, grayish film that repels water, underside whitish with many parallel veins. Flowers yellow to orange-yellow on elongate, compact spike or club, atop a thickened, white stalk. Common along sandy, muddy, peaty shores of bogs, ponds. Flowers April - May.

Pipewort Family (Eriocaulaceae)
Bog, marsh, or aquatic herbs with clusters of basal, narrow, grass-like leaves that taper conspicuously from a wide base. Flower stalks tall, slender, leafless, topped with small, solitary, button-shaped heads of minute, grayish-white flowers.

Ten-angled Pipewort *Eriocaulon decangulare*   1' - 3'
As in family description above. Leaves blunt at tips. Flower stalks with 10-12 ribs. Common in shallow bogs, peaty pools. Flowers mid-July - early October. (No illus.)

Flattened Pipewort *Eriocaulon compressum*   6" - 3'
Similar to above but leaves short, not over 5", gradually tapering to sharp point. Short leaves, large grayish flower heads, and early flowering distinguish this species from other two. Frequent in wet bogs, swamps. Flowers mid-May - June. (No illus.)

Seven-angled Pipewort *Eriocaulon septangulare*   1" - 8"
Similar to both above but leaves longer, up to 9", sharp-pointed. Flower stalks with 4-7 ribs. Same habitats as 10-angled but not as common. Flowers early July - early October.

Pickerelweed Family (Pontederiaceae)

*Pickerelweed *Pontederia cordata*   1' - 3'
Marsh or aquatic herbs, often in large patches. Leaves large, heavy, heart- or arrow-shaped, vary from very narrow to wide, dark green, long-stalked, often stand a foot or more above water surface. Flower stalks with one leaf below a 3"-4" spike of violet-blue flowers, rising above level of leaves. Common in shallow water, muddy shores, margins of streams well back up into pinelands. Flowers late June - mid-September.

Rush Family (Juncaceae)

Bayonet Rush *Juncus militaris*
See page 172 for description.

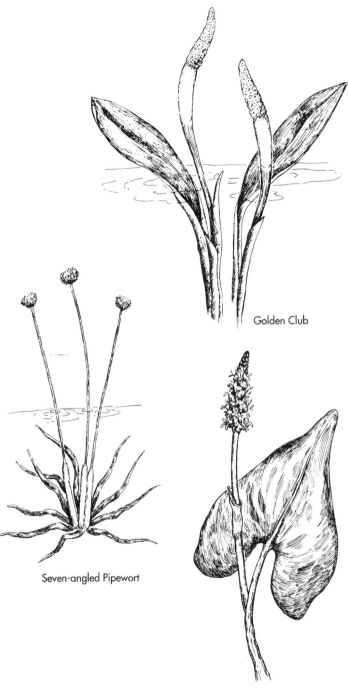

Golden Club

Seven-angled Pipewort

Pickerelweed

Aquatic Plants

Water-lily Family (Nymphaeaceae)

Perennial, aquatic herbs with tuberous roots submerged under water. Long-stalked leaves and flowers float on (usually), or rise, just above water surface.

**Yellow Pond-lily, Spatterdock or Bullhead-lily *Nuphar variegatum* = *N. luteum* ssp. *variegatum*

Leaves dark green, 4"-8", rounded to ovate to lance-shaped, with single, narrow V-notch, float on or stand just above water surface. Flowers nearly globular, 1"-3", consist of large, outer, greenish-yellow sepals tinged with red, with smaller yellowish to reddish petals inside flower cup. Common in ponds, slow streams. Blooms mid-May - early September.

*Fragrant or White Water-lily *Nymphaea odorata*

Leaves firm, 5"-10" across, dark green, pinkish-purple underneath, rounded, with single, narrow V-notch, usually float on water surface. Flowers 3"-5" across, white to pinkish-white, many petalled, fragrant. A common floating aquatic in pinelands ponds, cranberry bog reservoirs, slow streams. Blossoms June - September.

**Water-shield *Brasenia schreberi*

Leaves oval, shield-shaped, entire (no V-notch), 2"-4" across. Leaf stalks attached to middle of undersides of leaves. Flowers small, 3/4", dull red or maroon-purple. Leaves, flower stalks and under sides of leaves slimy, covered with gelatinous film. Occasional in ponds and sluggish streams. Blooms June - early August.

Gentian Family = Bogbean Family (Gentianaceae = Menyanthaceae)

*Floating-heart *Nymphoides cordata*

Perennial, aquatic herb. Leaves broadly ovate or heartshaped, 1"-2" across, long-stalked, lie flat on water surface, water-lily-like, often purplish-red underneath. Flowers small, 5-petalled, white, develop in small, dense clusters from side of leaf stalk at surface of water and rise just out of water. Clusters of short, thick roots dangle in water from stalks, just under leaves. Rare.[14] Classified as threatened (CMP). In ponds, sluggish streams. Flowers July - August.

Bladderwort Family (Lentibulariaceae)

Aquatic or semi-aquatic insectivorous herbs that grow in water or mud. Most have short, smooth, leafless flower stalks that rise from a mass of thread-like, under-water (or submerged in mud) leaves bearing tiny bladders that entrap minute aquatic organisms. Most bear small, yellow, 2-lipped flowers. There are 11 species in the New Jersey pinelands.

**Purple Bladderwort *Utricularia purpurea*   2" - 6"

Plants submersed in water (usually) or mud. Leaves in whorls of 5-7, multi-

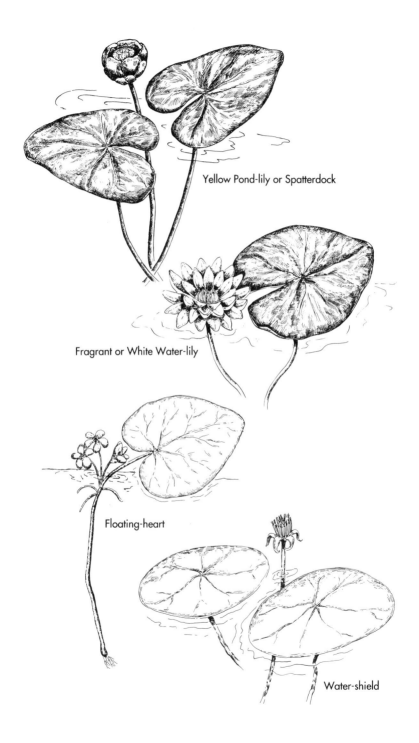

Aquatic Plants

branched, float at or just under water surface. The only bladderwort with a cluster of leaf branches at each joint of stem. Flowers small, 1/2", 2-lipped, single, lavender or pink-purple, rise above water surface from mass of floating leaves and bladders. Our only purple species. Rare[15]. Classified as threatened (CMP). Flowers mid-July - early September.

**Swollen Bladderwort *Utricularia inflata* 4" - 10"
Whorl of several, 4-10, swollen, wedge-shaped arms, like spokes of a rimless wheel, with fine forked ends and inflated stems, floats at or just under water surface, serves as floating support for flower stalk above, as well as for underwater, filament-like, bladder-bearing leaves. Flowers about 2/3", yellow, 3-7 on naked stalks. Occasional in bogs, ponds, often in enriched waters. Flowers mid-May - September.

**Fibrous Bladderwort *Utricularia fibrosa* 4" - 15"
Thread-like stems tangled in mats, floating or creeping on bottom, in shallow water. Flowers large, 3/4", yellow, 2-7 on erect, naked stalk. Common in shallow, sandy ponds. Flowers late May - early September. (No illus.)

**Horned Bladderwort *Utricularia cornuta* 4" - 10"
Non-floating, but in very shallow water or mud, somewhat terrestrial. Leaves tiny, slim, undivided, fringed with speck-sized bladders, half buried in soil, either in or out of water. Flower stalks slender, erect, naked. Flowers 3/4" with 1/2" spur that hangs down from base of flower, or vary to small and permanently bud-like. Frequent in sandy swamps. Flowers late June - August.

**Zig-zag Bladderwort *Utricularia subulata* 4" - 8"
Non-floating, somewhat terrestrial. Growth similar to horned bladderwort, above. Tiny, undivided leaves half buried in soil, either in or out of water. Speck-sized bladders on underground branches. Flowers small, 1/4", yellow, vary from open to permanently bud-like, 3-7 on thin, wiry, zig-zag stalk. Common on wet sand and in shallow, sandy ponds. Flowers late May - early September. (No illus.)

HERBACEOUS PLANTS WITH UNUSUAL STRUCTURES OR HABITS

Plants with unusual leaves: either 1) large, hollow tubes or "pitchers," open at top, usually filled with water, or 2) small reddish leaves covered with tiny hairs, on the tips of which are globules of a sticky liquid, or 3) leaves reduced to small scales on broad, flat, oval, fleshy, thorny joints. All the following are dicotyledons (see page 76).

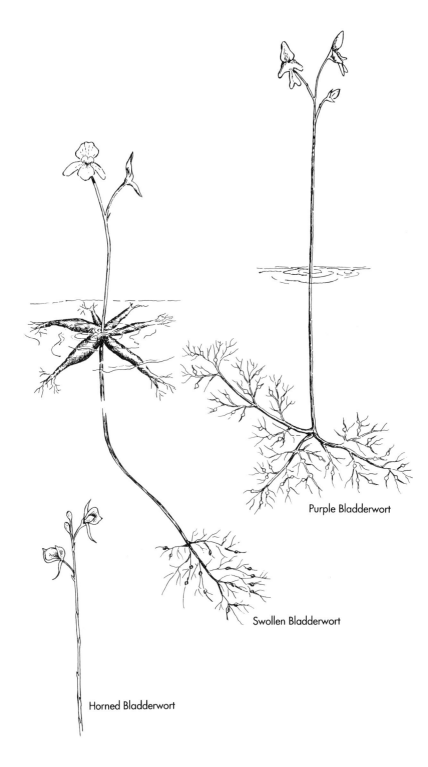

INSECTIVOROUS PLANTS

Herbaceous plants that attract and entrap insects as a supplementary nutrient. Most usually are somewhat suffused with crimson. The pitcher-plant and the three sundews which follow as well as the bladderworts (see under herbaceous plants in previous section) are all insectivorous plants.

Pitcher-plant Family (Sarraceniaceae)

**Pitcher-plant *Sarracenia purpurea*   Flower 1' - 2'
Interesting plant because of its unique, pitcher-shaped, red-purple-veined green leaves, 4"-10" which are open at top and hold water. Glands on leaf attract small insects which fly to and crawl in but can not retreat because of downward and inward pointing hairs on lip of leaf. They fall in, drown, and plant absorbs their nutrients. Flowers solitary, large, 2", dark red, nodding on tops of separate, leafless flower stalks. Common in peaty sphagnum and cedar bogs. Flowers late May - mid-June.

Sundew Family (Droseraceae)
All three of the following sundews catch insects by attracting them to sweet, sticky dewdrops the plants exude on the tips of reddish hairs on their leaves, to which the insects get stuck, die, and the plants then absorb their nutrients. All are locally common on wet sands, in and around edges of sandy and peaty bogs.

*Round-leaved Sundew *Drosera rotundifolia*   4" - 8"
Leaves small, round, on long stalks, lie almost flat on ground in basal rosettes. Flowers white, rarely pinkish, borne on slender stalks. Locally frequent on wet sand, in and around edges of peaty bogs and cedar swamps, often in middle of deep masses of sphagnum moss. Blossoms July - August.

*Spatulate-leaved Sundew *Drosera intermedia*   4" - 8"
Similar to above but leaves oval or spoon-shaped. Flowers white. Common on wet sand around edges of damp bogs. May be most common of the 3 sundews. Blossoms July - August.

**Thread-leaved Sundew *Drosera filiformis*   6" - 16"
Our largest and most conspicuous sundew. Differs from both above in that leaves are long, narrow, upright, and uncurl like a fern "fiddle-head" as they grow. Flowers crimson-pink to dull magenta-purple on separate stalks. Blossoms open in early morning sun, close by noon or shortly after. Common on open wet sand around edges of damp areas. Blooms earlier than other two, from late June - August.

Bladderwort Family (Lentibulariaceae)
See under aquatic herbaceous plants on pages 150-152.

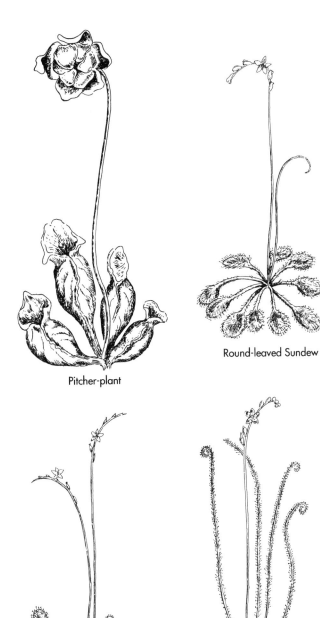

CACTUS

Cactus Family (Cactaceae)

Prickly Pear *Opuntia humifusa* 1' - 2'
Large, fleshy joints of stem erect or nearly prostrate on ground, often spreading, forming large mats. Joints armed with tufts of barbed bristles. Leaves minute, scale-like. Flowers large, 2"-3" across, showy, yellow. Fruits pulpy, red or purplish-red. Dry open sands. Natural only along margins of pine barrens, introduced into other pineland localities. Flowers June - early July.

GRASSES AND GRASS-LIKE (SEDGES, RUSHES) HERBACEOUS PLANTS

To many people, these three families of plants are very confusing and seem to "look alike." Many individuals even fail to recognize these as flowering plants, yet they are, although their flowers are often inconspicuous and their flower parts highly specialized. All members of these families are monocotyledons (see page 76 for description). All have relatively narrow, linear, "grass-like" leaves with parallel veins, and small, inconspicuous flowers.

Identification of these plants from most existing books is based on knowledge of the characteristics of their individual flowers and/or fruits (seeds). However, most of these flowers and fruits are so small that it is difficult to see the necessary characters without a hand lens or a microscope. In addition, a wholly separate terminology has been developed for their flower parts and seeds so that identification requires a fair amount of botanical knowledge. For these reasons, no attempt is made in a field guide of this type to describe and illustrate more than some obvious characteristics such as general shape, color, and texture of a very few of these plants.

For more complete and more accurate identification of these and all the approximately 145 species of grasses, sedges, and rushes in the pine barrens, reference to more technical botanical manuals such as Britton and Brown, 1952, Fernald, 1950, Gleason, 1962, Gleason and Cronquist, 1963, Hitchcock (Chase rev.), 1951, and Pohl, 1954 is recommended.

Grass Family (Gramineae =Poaceae)
Narrow, linear leaves with parellel veins and small, inconspicuous flowers. Stems usually round, mainly hollow except where leaves are attached at joints (nodes). Leaves in two rows on stems. Bases of leaves wrap around stems in sheaths that are split open part way down side of stem opposite

Prickly Pear

Grasses

leaf blade. Flowers (spikelets) arranged on stalks in two rows. Identification depends largely upon structure and arrangement of these spikelets. Fruits in the form of grains. At least 65 species of grasses in pinelands of which Stone, 1911, listed 18 as characteristic. Many pinelands grasses are introduced species that often can be found along roadsides, in lawns, and around old homesites. The following 24 species are characteristic native species.

In the identification of grasses, it is important to know the meaning of the following terms (K.A.):

Awn - a bristle-like extension of the nerves of the lemma, palea, or glumes.

Floret - the flower of a grass, consisting of the male and female flower parts and two (usually) enclosing scales (see lemma and palea, below)

Glumes - a pair of (empty) scales at the base of a grass spikelet.

Lemma - the outer, and usually larger, of the two scales that enclose the flower of a grass.

Palea - the inner, and usually smaller, scale of the floret.

Panicle - a loose, compound arrangement (flower cluster, or inflorescence) in which the florets are borne on stalks that branch off larger stalks.

Spikelet - the "unit of inflorescence" of a grass, formed by a series of grass florets, regularly arranged along a shortened axis. The number of florets in a spikelet can vary from one to many, in different species and genera.

*Six-weeks Fescue-grass *Festuca octoflora* = *Vulpia octoflora* 4" - 15"
Stems solitary, not clump-forming, usually erect. Spikelets flattened, 6-10 flowered, arranged in a slender panicle. Each floret about 1/8" long. Lemmas taper into a straight awn about 1/16"-1/8" long. Annual. Abundant in dry, sandy soil. Flowers mid-May - mid-June. (K.A.)

*Blunt Manna-grass *Glyceria obtusa* 1' - 3 1/2'
Stems usually erect, stiff, smooth, pointing upward. Leaf blades (3-6) flat or folded, wide, smooth below, rough above. Spikelets 3-7 flowered. Flower head (panicle) dense, plump, 3"-8" long, at top of stem. Bogs, streams, wet places, pond margins. Perennial. Blossoms early July - August.

*Wild Oat-grass *Danthonia spicata* 8" - 2'
Stems bunched or single, erect. Leaves very narrow, often rolled into a tube, mostly clustered near base of plant. Spikelets 4-6 flowered, arranged in a short raceme. Glumes 3/8"-1/2" long, longer than lemmas. Lemmas about 3/16" long, ending in two triangular teeth, with an awn about 1/4" long.

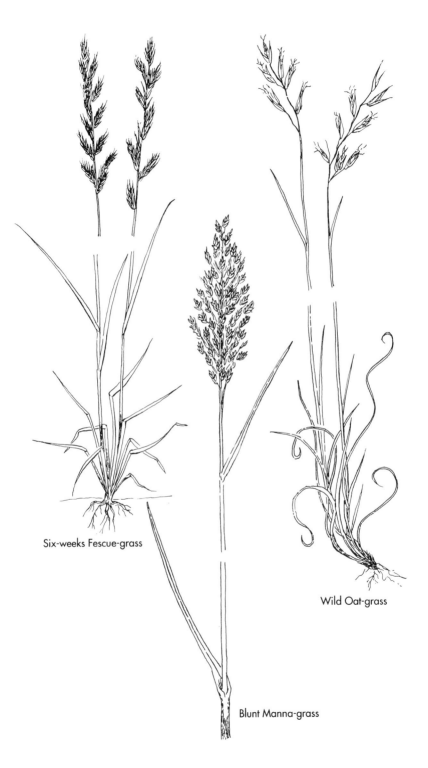

Grasses

Perennial. Common in dry, sandy soil. Flowers late May - late June. (K.A.)

Two other, similar species in pine barrens areas are ***silky wild oat-grass**, *D. sericea,* frequent in dry, open woods, clearings, and sands, and ****smooth wild oat-grass**, *D. epilis,* abundant in, but restricted to wet sphagnum bogs.

*Nuttall's Reedgrass *Calamagrostis cinnoides*   2' - 4'

Stems erect. Leaf blades to 3/8" wide. Spikelets one-flowered, about 1/4" long, arranged in dense panicle up to 8" long. Glumes curved outwardly, slightly longer than lemma. Base of lemma with tufts of hairs, easily seen by pulling lemma free of glumes. Back of lemma bears short awn that originates well below tip of lemma and barely exceeds it. Perennial. Frequent in damp sands. Flowers July - September (K.A.).

**Pine-barrens Reedgrass *Calamovilfa brevipilis*   2' - 4'

Stems slender, erect, rise from short, thick rhizomes. Leaf blades long, linear. Spikelets one-flowered, about 1/5" long, arranged in rather open, purplish panicle up to 10" long. Glumes taper to sharp point, the outer one ovate, the inner one much longer. Lemmas hairy on back. Perennial. A threatened species. Occasional in pine barrens bogs, swamps, and pitch pine lowlands. Flowers early July - late August.

*Rough Hairgrass or Ticklegrass *Agrostis hyemalis*   1' - 3'
= *A. hiemalis*

Stems erect, slender, smooth. Leaf blades erect, threadlike, in clumps at bases of stems. Spikelets small, single-flowered, pointed, at ends of branches. Flowering arrangement (panicle) very open, branched, 6"-2' long, usually purplish. Dry or moist, open, sandy soil. Perennial. Flowers early June - mid-August.

A similar species is **tall bentgrass**, *A. altissima,* frequent in pine barrens swamps.

**Late-flowering Dropseed *Muhlenbergia uniflora*   8" - 16"

Loosely matted and delicate. Stems slender, erect, in dense tufts, often with reclining bases. Leaf blades flat, crowded along lower part of stems. Flower stems rise from axils of old, flattened stems. Spikelets one-, occasionally two-flowered, dark purplish, about 1/2" long, long-stemmed, arranged in very open, diffuse flowering arrangement (panicle). Glumes pointed, 1/2 length of spikelets. Lemmas faintly 3-nerved, acute. Perennial. The common species of drop-seed in damp, sandy, peaty bogs. Flowers mid-August - early September.

**Torrey's Dropseed *Muhlenbergia torreyana*   1' - 2'

Clump-forming, spreading by rhizomes. Stems and sheaths flattened at base. Leaves folded along midrib. Spikelets one-flowered, about 3/32" long,

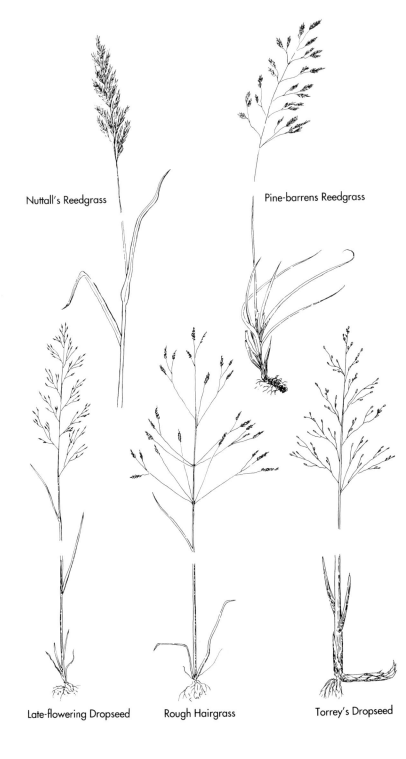

grayish-purple, arranged in very diffuse panicle. Glumes pointed, about equal in length, also equal to lemma. Perennial. Occasional in bogs. Classified as a threatened species. Flowers mid-August - late September. (K.A.)

Black Oat-grass *Stipa avenacea*   1 1/2' - 3'
Stems forming loose clumps. Leaf blades to 12" long but very narrow. Spikelets single-flowered, up to about 3/8" long, arranged in a loose panicle. Lemmas dark brown when mature, fused to the grain, bearing an awn up to 2" long that is twice-bent (twisted) near the middle, unlike any other local grass. Perennial. Occasional in dry, sandy soil. Flowers late May - early June. (K.A.)

*Poverty-grass *Aristida dichotoma*   6" - 1 1/2'
Stems tufted, wiry, branched at base, forked at nodes. Grows in small, shallowly rooted clumps. Terminal panicles elongate, 2"-5", slender, often no more than a loose spike. Other panicles reduced. Spikelets about 1/4", often reddish. Florets have small awns that stick out about 1/4" horizontally on sides. In poor, dry, sandy soil. Annual. Flowers mid-August - mid-October.

Rice Cutgrass *Leersia oryzoides*   2' - 4'
Stems erect or bent at base. Leaf sheaths and leaf edges very rough, with reflexed barbs. Spikelets one-flowered, flattened transversely, lacking glumes, about 3/16" long, growing 3-8 together on branches of a diffuse panicle. Perennial. Common in wet soil along streams and bog edges. Flowers August - September (K.A.)

Wild Rice *Zizania aquatica*   up to 10'
Stems slender to stout, simple or branching, soft. Leaves flat, wide, with long sheaths and blades. Male flowers straw-colored to purplish, droop down. Female spikelets hug stem, fall off easily. Primarily in quiet marshes and brackish river mouths but runs well up into pine barrens along principal water courses. Flowers mid-July - August.

*Slender Paspalum *Paspalum setaceum*   1' - 1 1/2'
Stems slender, erect. Grows in tufts from knotty crown. Leaf sheaths and blades stiffish, hairy. Leaf blades up to 5". Spikelets minute, lined up along one side of very slender stem in somewhat curved, usually solitary, spikelike racemes, which may or may not be branched. Most common of small, fruited paspalums in dry, sandy soil. Flowers early July - mid-October.

*Warty Panic-grass *Panicum verrucosum*   8" - 2'
Stems very slender, not clump-forming. Leaf blades thin, smooth, up to 6" long and 3/8" wide but usually smaller. Spikelets single-flowered, arranged in very open, spreading panicle. Glume and lemma distinctly "warty" (visible with hand lens). Annual. Common in damp ground. Flowers August - September. (K.A.)

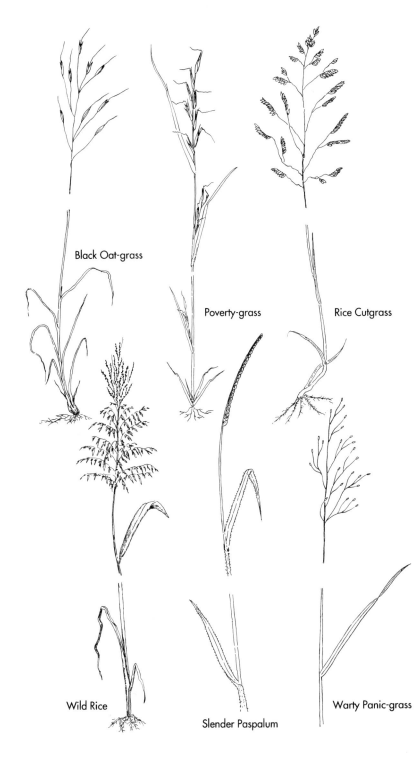

Grasses

*Switchgrass *Panicum virgatum*   3' - 5'
Stems erect, forming large clumps. Leaf blades to 18" long and 1/2" wide. Spikelets one-flowered, as in all *Panicum* spp., arranged in open, pyramidal panicle. Frequent along roadsides and in old fields, sometimes planted. Perennial. Flowers mid-July - September. (K.A.)

*Lindheimer's Panic-grass *Panicum lanuginosum* var.   1' - 3'
lindheimeri = *Dichanthelium acuminatum* var. *lindheimeri*

Small, grows in dense tufts. Slender stem rises from basal clump of short, wide, evergreen leaves. Inflorescence, up to 5", at top of stem, many branched, with small, oval flower cluster at end of each branch. Spikelets single flowered, finely hairy. Plants later develop dense tufts of short, leafy branches along stems, with secondary flower clusters partially concealed among these leaves. Open dry sands. Perennial. Flowers mid-June to mid-August. At least 19 species of panic grasses in pinelands of which 11 are characteristic.

**Pursh's Millet-grass *Amphicarpum purshii*   1' - 2 1/2'
or Peanut-grass
Stems erect, slender, smooth. Sheaths and leaf blades hairy, tufted. Growing underground from crown are slender runners, 1"-2" long, each bearing a single, large spikelet (seed) at its tip. Reproduction is by these underground spikelets ("peanuts") as aerial spikelets in narrow, aerial panicles are usually sterile. Common in damp, sandy areas, especially along dikes around cranberry bogs. Annual. Flowers early August - mid-September.

*Broom Beardgrass *Andropogon scoparius*   1 1/2' - 4 1/2'
or Little Bluestem = *Schizachyrium scoparium*
Grows in clumps. May even form loose sod. Stems stiff, hard. Leaves 6"-1 1/2' long. Branches wiry, scattered along upper half of stem, intermingled with leaves. Spikelets hairy, 5-20 pairs, lined along branches in 1"-2 1/2" long, loose racemes. Glumes each with a spiral awn, bent at point of exsertion. May cover dry soil in many old fields, clearings, open woods with tufted growth of buff or purplish stalks. Perennial. Flowers mid-July to mid-October.

*Broom-sedge *Andropogon virginicus*   2' - 4'
A grass, not a sedge. Grows in clumps. Leaves all along stem. Upper leaves enclose flowers. Spikelets apparently one-flowered, a second flower represented by its pedicel (stem) only. Floret long-hairy. Perennial. Common in old fields, clearings, roadsides. Flowers mid-August - late September. (K.A.)

*Bushy Beardgrass *Andropogon virginicus* var. *abbreviatus*   1 1/2' - 5'
= *A. glomeratus*

Entire inflorescence strongly top-shaped (inversely conical), condensed into dense, broom-like cluster at top of stem, often overtopped by upper leaves.

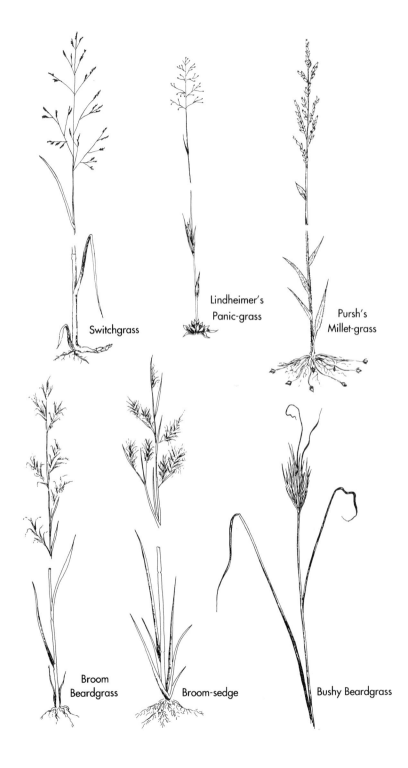

Sedges

Its dry, yellowish head clusters, on slender stems, are characteristic of winter swamps.[16] In peaty, boggy, damp soils, sandy swamps. Perennial. Flowers mid-August - mid-October.

Indian Grass *Sorghastrum nutans* 3' - 8'
Very tall, clump-forming grass. Spikelets apparently one-flowered (second flower represented by its stem only), about 1/4" long, fuzzy, arranged in narrow panicle. Lemma with a single bent awn, to 1/2" long. Frequent in dry, open soil. Perennial. Flowers mid-August - mid-September. (K.A.)

Sedge Family (Cyperaceae)
Grass-like. Stems solid, often more or less triangular, without obvious joints. Leaves, when present, in three vertical rows with leaf sheaths closed around stem, not split open. Flowers (spikelets) clustered spirally on stalks. Fruits small, dry, hard, one-seeded, almost like nutlets. Identification depends chiefly on shape of these fruits, presence or absence of surrounding bristles, and on size and structure of spikelets. There are at least 68 species of sedges in the pinelands, of which 31 are considered characteristic. The following 16 species are a few examples.

Gray's Cyperus (Sedge) *Cyperus grayii* = *C. grayi* 4" - 15"
Stems slender, growing singly or in small clumps, from a basal corm. Leaves stiff, short, folded. Flower heads at ends of branches up to 4" long (often much shorter), above 3-7 long bracts, rounded in general outline, with spikelets radiating in all directions. In dry, sandy soils. Fruits August - September (K.A.).

*Slender Sedge *Cyperus filiculmis* var. *macilentus* 2" - 3'
Grows in small clumps. Stems wiry, triangular, bearing small, rounded flower heads above three long, curly, shriveled bracts. Spikelets arranged in two ranks. In dry, sandy soils. Fruits late June - October.

Three-way Sedge *Dulichium arundinaceum* 10" - 2'
Grows singly. Stems round, hollow, jointed. Leaves numerous, in three ranks (obvious when looking lengthwise along stem). Flower spikes flattened, located in upper axils of leaves. Frequent, growing in shallow water. Fruits July - October (K.A.)

**Triangular-stem Spike-rush *Eleocharis robbinsii* 6" - 2 1/2'
Stems slender, soft, sharply three-angled, often with tufts of additional, thread-like, underwater stems. Spikelets lance-shaped, greenish. Frequent in shallow ponds, slow waters, peaty pools. Fruits mid-July - mid-September. A sedge, not a rush.

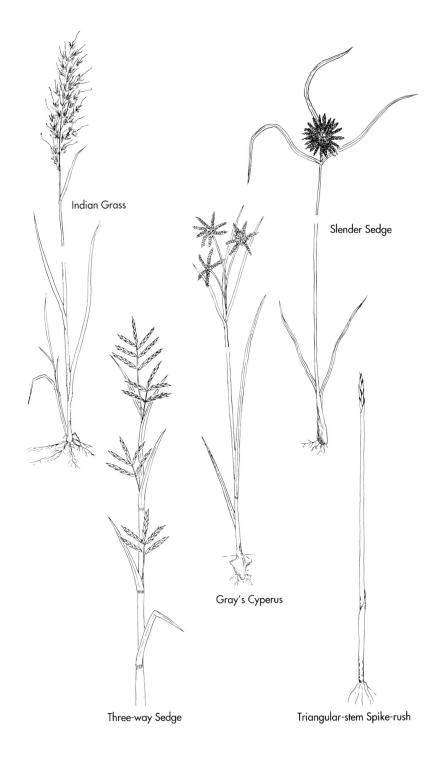

Sedges

*Green Spike-rush *Eleocharis olivacea* 1" - 6"
Stems soft, often spongy, with basal sheaths but no leaf blades, i.e. stems leafless, with single cone-shaped flower cluster at top. Grows in clumps, often forming large mats. Wet sands, peats, damp shores, moist open ground. Fruits late July - October. A sedge, not a rush.

**Swaying Rush or Water Club-rush *Scirpus subterminalis* Up to 2'
Aquatic, with abundant, long, hair-like leaves floating just under surface of water, often growing in great masses, its long stems and leaves swinging in the current. Often associated with *E. robbinsii*, above. A few roundish stems stand with tips out of water. Common in bogs, ponds, slow-moving, shallow streams. Fruits early July - late August. A sedge, not a rush.

*Three-square or Chair-maker's Rush *Scirpus americanus* 1 1/4" - 5"
= *S. pungens*
Leaves elongate, linear. Stems sharply triangular, with concave sides, long tapering to sharp tips above flower cluster of spikelets. Scaly, reddish-brown flower cluster of sharp-tipped spikelets grows out of side of stem near top. Marshes, swamps, stream edges. Fruits late June - early September. A sedge, not a rush.

*Wool-grass *Scirpus cyperinus* 3' - 5'
Grows in clumps, with flowering stalks taller than basal leaves. Stem triangular. Spikelets many-flowered, arranged in a repeatedly branched, somewhat umbrella-shaped panicle, above three unequal, spreading bracts. Leaf blades to 3/8" wide. Frequent in wetlands, including cranberry bog edges, ditches. Fruits July - September. A sedge, not a grass. (K.A.)

*Tawny or Virginia Cotton-grass *Eriophorum virginicum* 1 1/2'-4'
Fluffy, cottony clusters of white or tawny hairs, with long, shriveled bracts, at tops of stems. Stems may or may not have one or more long, thin, linear leaves. Common in bogs, peaty meadows. Flowers mid-August - September. A sedge, not a grass.

Small-headed Beak-rush *Rhynchospora capitellata* 10" - 2 1/2'
Grows in clumps. Stems very slender. Spikelets in two to 12 rounded clusters, one of which is at top of stem. Lateral clusters smaller than terminal cluster. Flowers, scales and fruits dark brown. Fruits have a pointed "beak." Common in bogs and wet sand. Fruits July - September. (K.A.)

The genus *Rhynchospora* is a large one with at least 16 species, of which 11 are characteristic in various wet habitats in the pines. Distinguishing characteristics are thin, weak, vaguely triangular stems, with flower clusters and fruits borne from sides of stems near axils of leaves. Fruits have a tubercle or "beak" on top. These are sedges, not rushes.

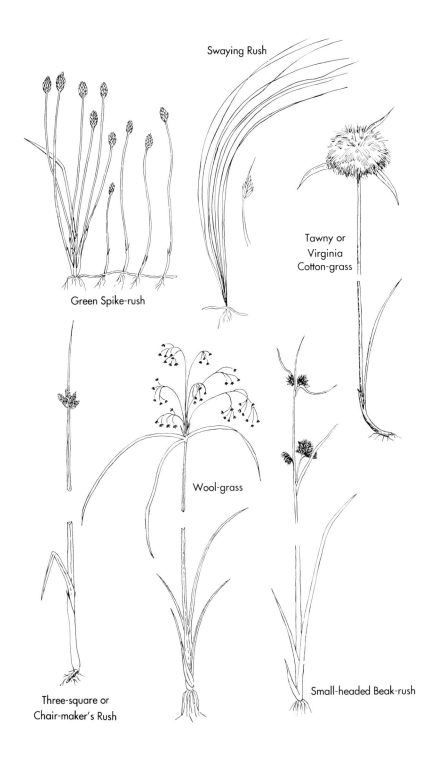

Sedges

**White Beak-rush *Rhynchospora alba* 10" - 20"
Glows in clumps. Stems very slender, usually taller than basal leaves. Spikelets in one to three rounded clusters, with one at top of stem and others, if any, on thin side branches. Flower scales white. Fruits with tubercle or "beak" on top. Common in peat bogs. Fruits August - September. (K.A.)

**Twig-rush *Cladium mariscoides* 1' - 3'
Stems stiff with few channeled leaves and few side branches. Leaves about height and width of stems. Most branches near top where they bear brown, scaly, flower clusters up to 1' high. Common in sandy bogs, marshes, swamps. Fruits mid-July - October. A sedge, not a rush.

*Pennsylvanica Sedge *Carex pensylvanica* under 16"
Grows in tufts to large patches. Leaves thin, persist after flower stalks die, even into next season as dead leaves. Stems triangular, bearing both club-shaped male flowers at tips and female flowers below male ones. Common ground cover in dry, open, oak woodlands. Fruits mid-May - mid-June.

The genus *Carex* is very large with at least 24 species in the pinelands. Distinguishing characteristics are that flowers are unisexual, i.e. male and female flowers borne on separate flower heads, or sometimes together in the same flower cluster.

**Walter's Sedge *Carex walteriana* 16" - 3 1/2'
Grows in large colonies. Stem triangular. One or two male flower spikes on stem well above female spike. Female spike cylindric, stemless, 1"-1 1/2" long. Common at edges of swamps and bogs. Fruits June - July. (K.A.)

*Long Sedge *Carex folliculata* up to 4'
Thin, triangular stem with club-shaped male flower cluster on top. Widely spaced female flower clusters down stem at axils of leaves. Common as clumps in wet, peaty thickets, swampy woods. Fruits early June - mid-July.

**Bull Sedge or Button Sedge *Carex bullata* up to 3'
Stems slender, sharply angled, few to numerous, from creeping rootstalks. Leaves stiff, narrow, elongated. Male spikes usually two, long stalked, well above two to three cylindric female spikes. Beak sharply two-toothed at tip. Common in pine barrens bogs and swamps. Fruits mid-June - September.

Rush Family (Juncaceae)
Grass-like. Stems round, pointed, either solid or, more often hollow, or pithy. Leaves usually wiry, round in cross-section. Flowers very small but somewhat like flowers of a lily with 3 petals and 3 sepals arranged in a circle, clustered in open or dense terminal heads. Fruits small, round, 3-parted seed capsules with 3 persistent sepals that often remain on plant for

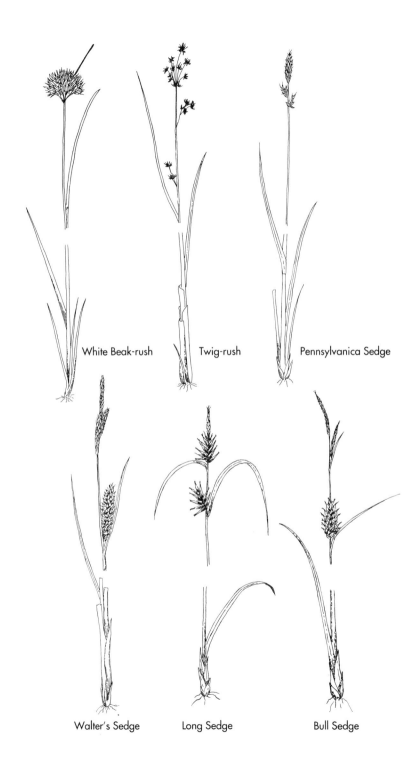

Rushes

most of season. Most species belong to genus *Juncus,* and tend to grow in cool, wet areas. At least 12 species in pinelands of which three are characteristic. The following four species are a few examples.

*Soft or Common Rush *Juncus effusus* 1 1/2' - 6'
Grows in large clumps. Distinguished from other rushes by bushy, greenish to brownish flower cluster that arises from side of stem. Common in peaty swamps, edges of ponds, other wet places. Seed capsules by mid-June - mid-July.

*Canada Rush *Juncus canadensis* 15" - 3'
Stems growing in erect clumps. Leaves tubular and hollow, with transverse partitions. Flowers in dense, rounded clusters. Individual seeds (not capsules) have long "tail-like" appendages at each end - visible with a 10x lens. Common in wet sands, edges of bogs, other wet places. Seed capsules August - September. (K.A.)

**Bayonet Rush *Juncus militaris* 1' - 3'
Stems have a tall mid-stem leaf that rises high above cluster of brownish flowers. Flowers in branches within a cluster. Often there are submersed, thread-like leaves that spread out in the current of streams. Seed capsules late July-August.

*Bog or Brown-fruited Rush *Juncus pelocarpus* 6" - 1'
Stems very slender, erect, with very few slender leaves. Much branched head of greenish flower clusters, each flower with 3 blunt-tipped sepals and 3 similar petals. Underwater plants merely small clumps of tapered leaves. Common in wet sands, shallow shores, damp pools, bogs. Fruits late August - late September.

LAND HERBACEOUS PLANTS WITH PARALLEL-VEINED LEAVES and FLOWER PARTS (in our species) IN 3's or 6's.

These are the remaining herbaceous monocotyledons (see page 76) after having separated out both the aquatic and the grass-like herbs.

Yellow-eyed Grass Family (Xyridaceae)

*Slender or Twisted Yellow-eyed Grass *Xyris torta* 6" - 2 1/2'
Small, rush-like herbs with linear, grass-like basal leaves and erect, leafless flower stalks. Base of plant with bulbous swelling. Leaves and stalks conspicuously spirally twisted. Flowers yellow, small, 1/4" wide, but showy, 3-petaled in conelike scaly head atop stiff stalk. Common in wet, peaty, sandy, swampy soils. Flowers July - August.

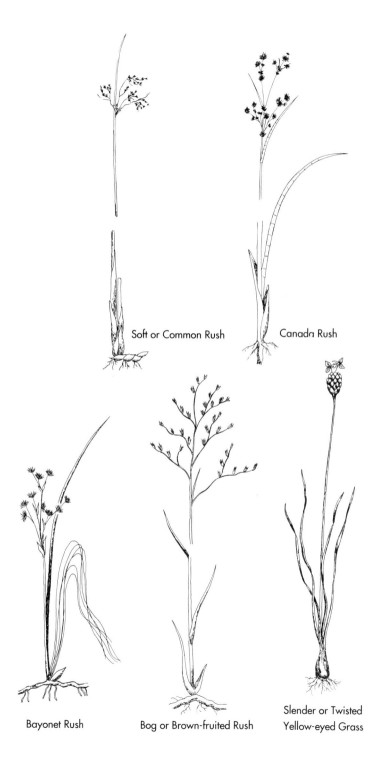

Herbs with parallel-veined leaves

Carolina Yellow-eyed Grass *Xyris caroliniana* 6" - 3'
Similar to above but base lacks bulbous swelling and neither leaves nor stalks are twisted. Flowers mid-July - September. In addition, there are four other yellow-eyed grasses considered characteristic to pine barrens. (No illus.)

Lily Family (Liliaceae)
Herbs (here) with bulbous bases to stems, leaves parallel-veined, flowers showy with 3 petals, 3 sepals, and 6 stamens.

Bog-Asphodel *Narthecium americanum* 10" - 1 1/2'
Leaves small, stiff, narrow, grass-like, mostly growing up like short grass from bases of flower stalks. Flowers small, bright yellow, in short, dense spike at top of erect flower stalk. Local in wet, peaty sands, bogs, chiefly in heart of pine barrens. Rare.[17] Classified as a threatened species (CMP). Blooms mid-June - July.

Turkeybeard *Xerophyllum asphodeloides* 2' - 4'
Leaves long, narrow, wiry, grass-like in thick, basal clump. Few shorter, stiffer leaves ascend flower stalk. Flowers small, white, in dense head 3"-10" high, on top of tall flower stalk. Frequent in pitch pine lowlands. Blooms late May - early July.

*Swamp-pink *Helonias bullata* 1' - 2'
Leaves narrow, oblong, evergreen, in basal rosette. Flowers small, bright pink, lavender-pink, or lilac, with bright blue stamens, in dense spike-like cluster at top of nearly leafless, hollow flower stalk. Classified as a threatened species (CMP) of bogs, swamps. Blooms mid-April - mid-May.

False Asphodel *Tofieldia racemosa* 6" - 1 1/2'
Leaves erect, grass- or iris-like, in 2 vertical rows from near base. Flowers small, star-like, white, "like miniature turkey beard"[18] in short spike atop erect, slightly sticky stalk. Rare.[19] Classified as endangered (CMP) in wet, peaty sands, bogs in heart of pinelands. Blooms late June - mid-July.

Pine-barren Bellwort *Uvularia pudica* var. *nitida* 6" - 15"
= *U. puberula*
Stems angled, usually branched above middle. Leaves alternate, oval or oblong, green on both sides, attached to stem without leaf-stalks. Flowers single, bell-like, yellow, hang from branching stems. Seed capsule sharply 3-angled. Rare[20] and local along wooded edges of bogs, swamps. Flowers late April - May.

*Turk's-cap Lily *Lilium superbum* 3' - 7', 2' - 3' in p.b. swamps
Stout, erect. Leaves smooth, lance-shaped, taper toward bases and tips, in whorls on lower stem, alternate on upper. Nodding flowers arise from up-

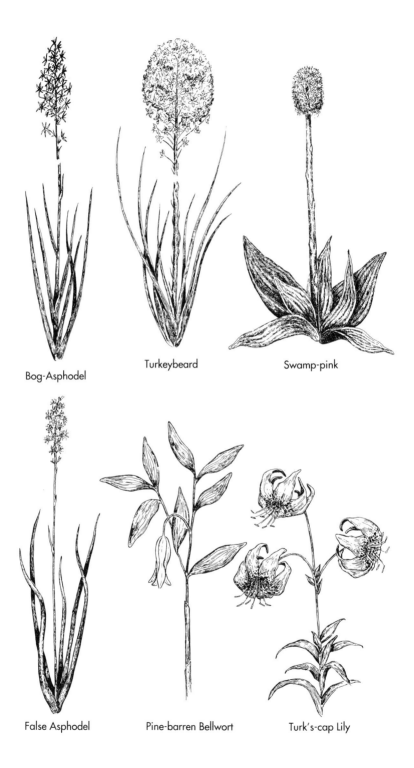

Herbs with parallel-veined leaves

per axils and from summit of stem. Flowers large, showy, orange to orange-red, purple-spotted, remarkable for greatly reflexed petals and sepals, completely exposing stamens. Yellowish-green star in center of flower. Local in peaty meadows, borders of bogs. Blooms July.

*Colicroot *Aletris farinosa*   1' - 2 1/2'
Leaves lance-shaped, in basal rosette. Flowers small, whitish, tubular, granular-surfaced, in cluster on top 4"-12" of stiff, nearly leafless, flower stalk. Occasional in dry or moist, peaty sands. Flowers mid-June - July.

Bloodwort Family (Haemodoraceae)

**Redroot *Lachnanthes tinctoria* = *L. caroliniana*   8" - 2 1/2'
Herbs with fleshy, almost blood-red roots. Stems stout, erect. Leaves narrow, pointed, clustered at base, scattered on stem. Flowers woolly, dull yellowish to rust colored, in dense, branching cluster at top of hairy stalk. Common in swamps, bogs. Pernicious weed in cranberry bogs. Blooms July - August.

Amaryllis Family = Lily Family (Amaryllidaceae = Liliaceae)

*Yellow Stargrass *Hypoxis hirsuta*   3" - 1'
Leaves narrow, grass-like, hairy, often higher than flowering stalk. Flowers small, 6-parted, star-like, greenish-yellow to deep yellow, in small cluster at top of flowering stalk. Occasional in dry, meadow grasses, open sandy woods. Flowers mid-May - August.

Amaryllis Family = Bloodwort Family (Amaryllidaceae = Haemodoraceae)

**Golden-crest *Lophiola americana* = *L. aurea*   1' - 2'
Upper part of stem and flower head whitened with soft, matted wool. Leaves alternate, linear, narrow. Flowers small, golden-yellow, surrounded by dense, downy-white-woolly covering, in branching cluster. Frequent in bogs, swamps, in heart of pine barrens. Blossoms late June - July.

Iris Family (Iridaceae)

*Eastern Blue-eyed Grass *Sisyrinchium atlanticum*   1 1/2' - 2'
Leaves linear, blade- or grass-like, with a waxy bloom, shorter than flower stalks. Flower stalk slender, branches off mid-way up stem at axil of leaf-like bract. Lower stem with slightly flared edges (winged). Flowers 6-parted, light blue to violet, with bright yellow centers. Frequent in open damp meadows, fields. Blooms mid-May - early June.

*Slender Blue Flag *Iris prismatica*   1' - 3'
Leaves linear, very narrow, almost grass-like, usually less than 1/4" wide. Flowers lavender-blue with yellow sepal bases and deep violet veins, on

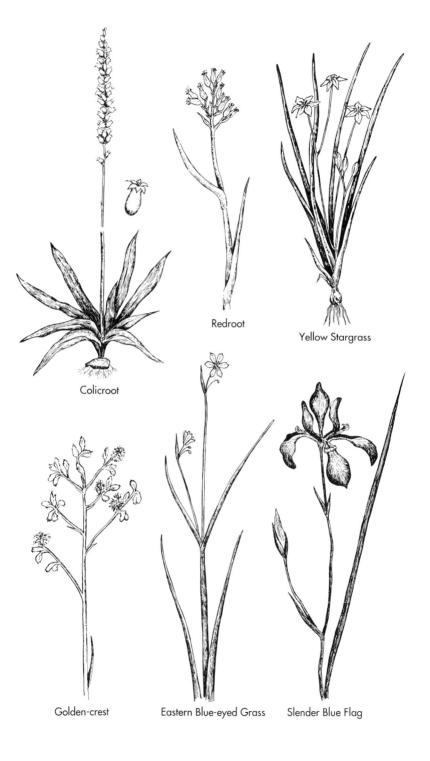

Herbs with parallel-veined leaves

slender stalks. Common in marshes, meadows, bogs, swamps, swales. Blossoms late May - June.

Orchis Family (Orchidaceae)
Perennial herbs with tuberous or bulbous roots and, often, very striking, showy flowers. Leaves entire (untoothed), parallel-veined. Flowers irregular, composed of 6 segments: 3 similar outer ones (sepals), 2 side ones (petals), and a middle one (petal) developed into a lower lip, often fringed or spurred. Well represented in New Jersey pinelands.

*Pink or Stemless Lady's-slipper *Cypripedium acaule*   6" - 1 1/2'
or Moccasin-flower
Two large, 6"-8", opposite, ovate, basal, seemingly stalkless, hairy leaves. Flower large, 1"-2", solitary, split-baglike or moccasin-shaped, pink with reddish veins, on top of leafless flower stalk, with single bract arching forward over flower. Occasionally all white. Frequent in dry or moist, acid, sandy woods, bog edges. Blooms May - early June.

*Green Woodland Orchis *Habenaria clavellata*   6" - 1 1/2'
= *Platanthera clavellata*

Single, lance-shaped leaf 1/4 way up flower stalk, with 2-3 much smaller ones above. Stalk slender, with 5-12 small, almost insignificant, greenish-white, somewhat twisted flowers on short spike. Each flower has characteristic long, slender spur curved upward and around to one side from base. Wet, sandy, woodland bogs, swamps. Blooms late July - mid-August.

**White Fringed Orchis *Habenaria blephariglottis*   1' - 2 1/2'
= *Platanthera blephariglottis*

Stalk leafy with narrow, lance-shaped leaves, 1-3 lower leaves up to 6"-8", upper leaves reduced to bracts. Flowering cluster a large, full, 6"-7" spike of pure- or creamy-white flowers. Flower spur long, up to 1". Lip of flower oblong, finely fringed around edge. Occasional to frequent in wet, peaty meadows, sphagnum bogs. Blooms mid-July - mid-August.

*Yellow Fringed Orchis *Habenaria ciliaris*
= *Platanthera ciliaris*   1' - 2 1/2'

Similar to white fringed orchid, above, but flowers are golden- to orange-yellow. Large spike of flowers. Lip of flower conspicuously fringed. Spur slender, very long, 1" or more. An endangered species of sandy, peaty bogs, swales. Blooms late July - August.

**Crested Yellow Orchis *Habenaria cristata*
= *Platanthera cristata*   1' - 2'

Similar to yellow fringed orchid, above, but flower lip more deeply fringed, spur less than 1/2 as long. An endangered species of damp, open bogs, woods, and cedar swamps. Blooms late July - August.

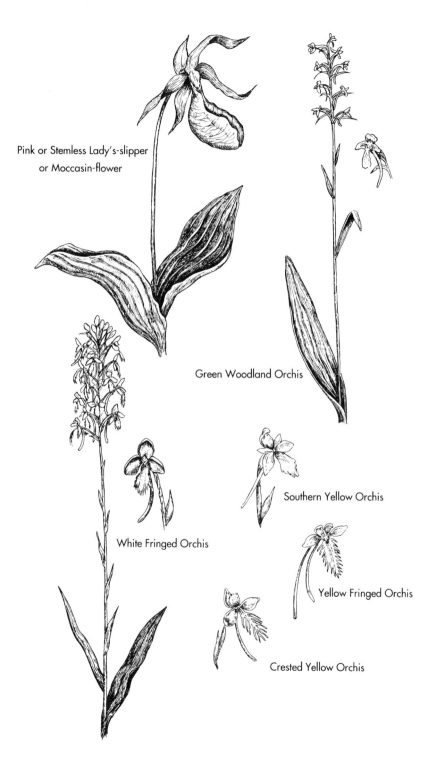

Herbs with parallel-veined leaves

Southern Yellow Orchis *Habenaria integra* 1' - 2'
= *Platanthera integra*
Stalk with several leaves, the lower two larger, lance-shaped. Flowering cluster a round spike of golden, orange-yellow blooms. Lip toothed at edge, or sometimes entire, but not fringed. An endangered species in sphagnum bogs and wet pine barrens. Blooms August - September. (Illus. on pg. 179)

Rose Pogonia or Snakemouth *Pogonia ophioglossoides* 6" - 1 1/2'
Single lance-shaped leaf clasps stem midway up with smaller leaf (bract) just below blossom. Flower large, 3/4", solitary (usually), showy, rose-pink, occasionally white, with distinctive yellow-crested, pink-fringed lip. Common in open sphagnum bogs, wet, peaty meadows. Blossoms June - early July.

*Grass Pink *Calopogon pulchellus* = *C. tuberosus* 6" - 1 1/2'
Usually only 1 narrow, pointed leaf from base of slender stem. Flowers about 1", variable in size, color, in loose cluster of up to 4-6 magenta-pink to pink-purple blooms, each with upright, yellow-crested lip above flower. Frequent to locally common in damp meadows, open sphagnum bogs. Blooms mid-June - July.

*Arethusa *Arethusa bulbosa* 6" - 1'
Single, grass-like leaf develops after flower matures. Flower single, 1"-2", magenta-pink with 3 erect sepals, and hood over purple spotted, yellow-crested lip, on top of nearly leafless stalk. Rare, local in open sphagnum and cedar bogs. Blooms mid-May - mid- June.

*Little Ladies'-tresses *Spiranthes tuberosa (beckii)* 6" - 1'
Leaves basal, short, oval-oblong, usually appear after flowers mature. Flowers small, white in 1-sided, spirally-twisted spike. Rare[21] and classified as threatened (CMP) in dry, sandy fields, edges of woods. Blooms mid-July - early September.

**Jagged Ladies'-tresses *Spiranthes laciniata* 1' - 3'
Leaves narrow, grass- or thread-like, lower ones up to 10"-12", stem leaves smaller, may appear after flowers mature. Flowers white, or slightly yellowish, sometimes veined with green, somewhat downy, scattered in loose spiral, or single row up slender stem. Tip of lip of flowers jagged. This may be a hybrid of two other species, **spring ladies'-tresses**, *S. vernalis,* and **grass-leaved ladies'-tresses,** *S. praecox*. Occasional in damp meadows, swamps. Blossoms August - September.

*Nodding Ladies'-tresses *Spiranthes cernua* 6" - 2'
Basal leaves long, grass-like, usually present. Blossoms in tall spike of small, arched, creamy-white flowers arranged in a double or triple spiral on leafy

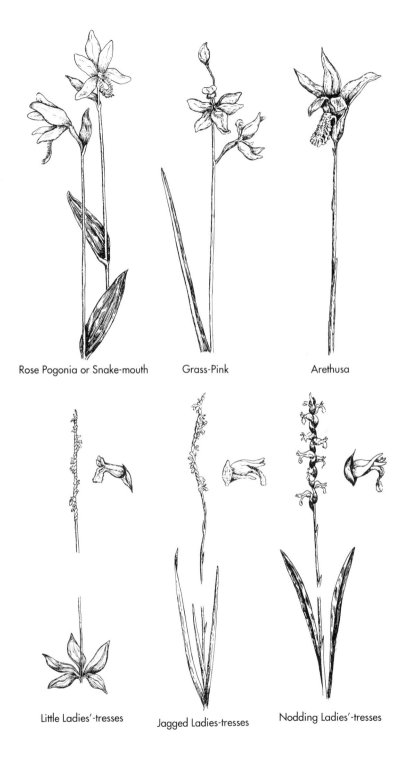

(bracted) stem. Common in moist meadows, marshes, damp ground. Blooms September - mid-October.

LAND HERBACEOUS PLANTS WITH LEAVES MOSTLY NETTED-VEINED (not parallel-veined) and FLOWER PARTS IN 2's, 3's (occasionally), OR 4's, 5's (usually) OR THEIR MULTIPLES.

These are the remaining herbaceous dicotyledons (see page 76) after having separated out the aquatic herbs, unusual structured plants, and the composites (next section).

Sandalwood Family (Santalaceae)

Bastard or Star Toadflax *Comandra umbellata*   6" - 1'
Small herb. Leaves alternate, oblong-elliptical, entire, attached singly. Flowers small, greenish-white, short-tubular, with 5 lobes (sepals), grouped in terminal clusters. A parasite on roots of other plants. Dry, sandy soil. Flowers mid-May - June.

Buckwheat Family (Polygonaceae)

*Jointweed *Polygonella articulata*   6" - 15"
Very slender, branched annual with wiry, jointed stem and tiny, linear leaves. Flowers numerous, tiny, white to pink, in open, branching flower heads, seem to become deeper red after first frosts. Frequent in well drained, sterile, sandy soils, embankments. Flowers September - October.

Pink Family (Caryophyllaceae)

**Pine-barren Sandwort *Arenaria caroliniana*   4" - 1'
                 = *Minuartia caroliniana*
Tiny, linear-awl-shaped, overlapping leaves form dense green mats, like a moss, from a long, stout, vertical tap-root. Slender, branching, slightly sticky flower-stalks bear numerous white, star-like flowers with greenish centers. Frequent on bare patches of dry, white sand. Blossoms June - July.

Pea or Pulse Family (Leguminosae = Fabaceae)

Very large family, well represented in pinelands. Distinguishing characteristics: alternate, usually compound leaves, thus separated into leaflets; irregular, "butterfly-shaped" flowers that are identical side to side (bilaterally symmetrical), and pods that contain seeds that split open in half. Flowers have 5 petals, the 2 lower ones joined into a "keel," 2 side ones appear as "wings," upper one is called a "banner."

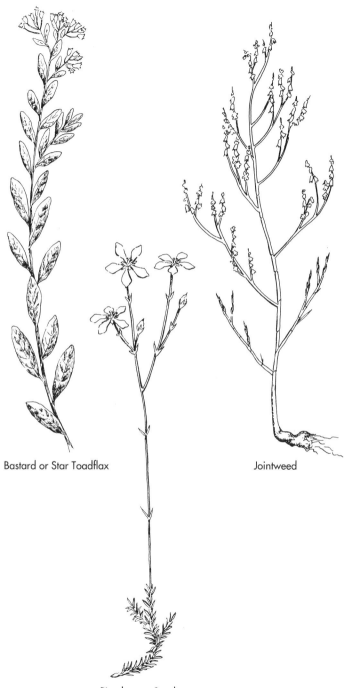

Herbs with netted-veined leaves

*Wild Indigo *Baptisia tinctoria* 2' - 3'

Smooth, bushy, multi-branched herb. Leaves alternate, in leaflets of 3, bluish-gray-green, turn black when dried. Flowers small, pale to bright yellow, in open, loose, groups at tips of branches. Pods small, somewhat roundish, purple. Common in dry, sandy, open clearings, woods. Flowers late June - July.

*Wild Lupine *Lupinus perennis* 1' - 2'

Leaves divided into 7-11 fine, hairy leaflets, radiating, wheel-like, from top of stem. Flowers lavender-blue, clustered on tall, 4"-10", erect, showy spikes. Seed pods oblong, flat, downy. Dry, sandy, sterile soils in thickets, open woods, clearings, roadsides. Flowers May - early June. Seed pods June - early July.

*Goat's-rue *Tephrosia virginiana* 1' - 2'

Entire plant silky-hairy. Leaves divided into 15-27 paired leaflets with 1 leaflet at tip. Flowers relatively large, 3/4", showy, 2-colored, upper petal (banner) yellowish, others purplish-pink, in terminal clusters. Seed pods long, flat, narrow, hairy. Common in dry, sandy woods, clearings. Blooms June - early July. Seed pods August - early September.

Tick-trefoils or Stick-tights *Desmodium* spp. 1' - 3'

Perennial herbs with alternate leaves divided into 3 leaflets. Flowers small, pinkish, purplish, or even white, in loose, terminal clusters. Less noted for their inconspicuous flowers than for their small, hairy, sticky seed pods, usually divided into several joints which break off and adhere to clothing. At least 6 species of these slender, weed-like plants in pinelands, differences determined largely on character of their seed pods.

Two species, **stiff tick-trefoil**, *D. strictum,* and **rigid tick-trefoil**, *D. rigidum (= D. obtusum)* are characteristic in pine barrens areas. *D. strictum* can be recognized by its very slender, linear leaves, small flowers, and few segments to its pods. It is a rare[22] and threatened (CMP) species. *D. rigidum* often has several stems arising from a stout, branched root and its leaves are elliptical to lance-shaped or ovate. Both are found in dry, open woods, roadside thickets. Both flower late July - early September. Seed pods August - October.

Bush-Clovers *Lespedeza* spp.

Leaves alternate, divided into 3 leaflets, somewhat similar to true clovers. Flowers small, white, pink or purplish, in loose, terminal clusters. Seed pods small, oval, 1-2 jointed. Dry, sandy woods, fields. Blossoms August - early September. Seed pods September - early October.

*Wand-like Bush-Clover *Lespedeza intermedia* 1' - 3'

Plant erect. Stem sparsely hairy. 3 leaflets distinctly oval-elliptical. Flowers purplish or pinkish-purple, from axils of leaves, clustered at top of plant.

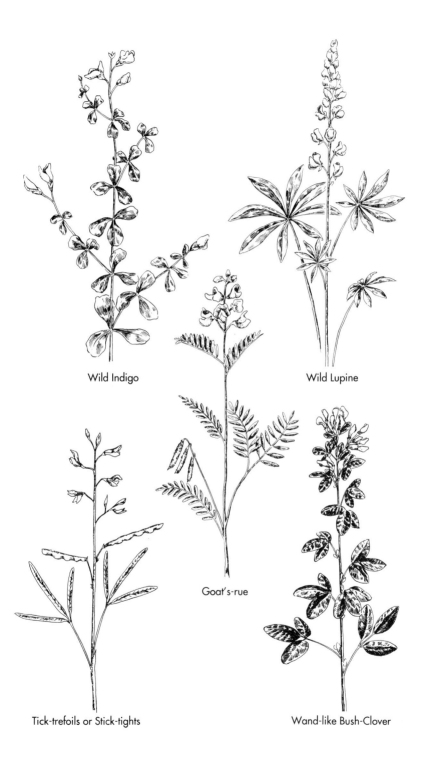

Herbs with netted-veined leaves

*Hairy Bush-Clover *Lespedeza hirta* 2' - 4'
Stem erect, densely hairy. 3 leaflets oval-rounded. Flowers creamy-white with small, purple dot on upper petal (banner), in dense, spike-like terminal clusters.

**Narrow-leaved Bush-clover *Lespedeza angustifolia* 2' - 3 1/2'
Erect herb, often branched, similar to hairy above but 3 leaflets narrow, oblong-linear, much elongated. Flowers creamy-white, purple spotted, in dense clusters from upper leaf axils at top of stem. Flower heads conspicuously stalked. (No illus.)

*Pencil-flower *Stylosanthes biflora* 6" - 1 1/2'
Stems erect, stiff, wiry, branched from base. Leaves in 3 narrow, lance-shaped, bristle-tipped leaflets. Stem, leaves hairy. Flowers orange-yellow to deep orange, single or in small clusters at ends of branches. Occasional and local in sandy ground, edges of dry woods, thickets, clearings. Flowers mid-June - early September.

*Groundnut or Wild Bean *Apios americana*
*Milk-Pea *Galactia regularis*
Both vines in pea family. See page 140 for descriptions.

Flax Family (Linaceae)

Yellow Flax *Linum intercursum* 1' - 2 1/2'
Slender, upright herb with branching stem. Leaves alternate, thin narrow, lance-shaped. Flowers small, barely 1/3", yellow with 5 rounded petals. Occasional in dry, open woods. Flowers late June - August.

*Ridged Yellow Flax *Linum striatum* 1' - 3'
Similar to above but separated by having 3 narrow ridges, or "wings" on stem below each leaf base, stem thus conspicuously angled. Damp sands, peats, low woods. Flowers late June - early August. (No illus.)

Milkwort Family (Polygalaceae)
Large family of small herbs usually with alternate or whorled, narrow, lance-shaped leaves and small, irregular flowers in spikes or, often, clover-like heads. All belong to the genus *Polygala* of which there are 7 pinelands species. Four are described, the other 3 local and/or rare.

*Nuttall's Milkwort *Polygala nuttallii* 6" - 1'
Erect, single or branched stems with small, alternate, lance-shaped leaves. Small, blunt cluster ("clover-head") of tiny, greenish-white to dull purple flowers. Common on damp, open, sandy soils. Flowers July - mid-October.

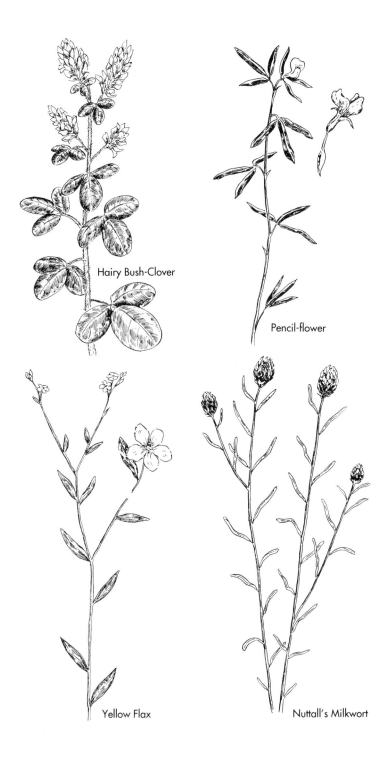

Herbs with netted-veined leaves

Cross-leaved Milkwort *Polygala cruciata* 6" - 1'
Stems smooth, squarish, much branched, with small, narrow, lance-shaped leaves in whorls of 3, or mostly 4, sort of crosslike. Flowers greenish-white to dull magenta to purplish, on very short stalks, in blunt, clover-like heads. Common in damp grounds along edges of low meadows, marshes, bogs, and sandy sites. Flowers late July - early October.

Short-leaved Milkwort *Polygala brevifolia* 3" - 10"
Similar to above but smaller with weaker stem. Leaves shorter, more sparse, in whorls of 3-5. Flowers smaller than above, on longer stalks in clover-like heads, dull magenta-pink to rose-purple. Common in moist, sandy ground in pinelands. Flowers mid-July - early October.

Orange Milkwort *Polygala lutea* 6" - 15"
Stem smooth, simple or sparsely branched, with alternate, narrow, lance-shaped leaves; also a basal rosette of oblong leaves. Flowers bright orange-yellow in dense, showy, clover-like clusters at tops of branches. Common in low, sandy ground in pinelands. Flowers mid-June - mid-October.

Spurge Family (Euphorbiaceae)

Wild Ipecac or Ipecac-Spurge *Euphorbia ipecacuanhae* 6" - 1'
Plant nearly prostrate. Several low, repeatedly branching, spreading stems grow from single rootstalk. Leaves opposite, somewhat fleshy, varied from oval to linear, often arranged in small rosettes. Particularly interesting for color variation in leaves: green on some plants, magenta-crimson or purple on others, sometimes both on same plant. Both stem, leaves with milky juice. Flowers yellow-green, without petals, appear before leaves. Common on dry, open sands. Flowers late April - May.

St. John's-wort Family (Guttiferae = Clusiaceae)
Herbs and shrubs with opposite, paired leaves, alternately arranged on stems. Leaves entire (untoothed) often with tiny black or translucent dots or glands (use hand lens).

St. Peter's-wort *Ascyrum stans* = *Hypericum stans* 1' - 2 1/2'
Low, semi-erect, almost shrubby herb, with opposite, paired, oval-oblong-elliptic leaves that partially clasp stem. Stems with 2 wing-like ridges. Flowers terminal with 4 broad, bright lemon-yellow petals enclosed within 2 pairs of unequal sepals. Common in moist or dry, sandy barrens. Flowers late July - early September.

*St. Andrew's Cross *Ascyrum hypericoides* 4"- 1'
= *Hypericum stragulum*
Low, reclining, sprawling, almost shrubby herb somewhat similar to above but leaves narrower, linear to oblong-lance-shaped, not clasping stem.

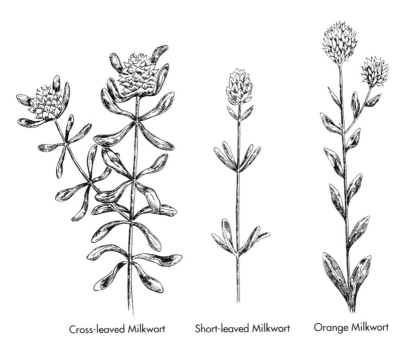

Cross-leaved Milkwort   Short-leaved Milkwort   Orange Milkwort

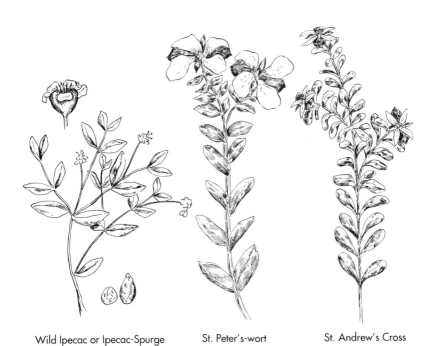

Wild Ipecac or Ipecac-Spurge   St. Peter's-wort   St. Andrew's Cross

Herbs with netted-veined leaves

Flowers in upper leaf axils. Flower petals (4) narrower, forming oblique cross, enclosed within 2 pair of unequal sepals. Common in dry, sandy soils. Flowers July - early September.

**Coppery St. John's-wort* *Hypericum denticulatum* 8" - 2'
Stem smooth, simple, sharply 4-angled. Leaves smooth, narrow to broadly oval, grow upward, nearly erect. Flowers 5-petalled, over 1/3", coppery-yellow, easily identified by color. Occasional to frequent in sandy bogs, swamps. Flowers mid-July - early September.

Canada St. John's-wort *Hypericum canadense* 6" - 2'
Slender, diffusely branched. Branchlets 4-angled. Leaves light dull green, very narrow, elongated, up to 10 times longer than wide. Flowers small, less than 1/4", deep golden yellow, at tips of stems and branches. Seed pods reddish. In dry, more often damp, sandy soils. Most common small St. John's-wort in pines. Blooms July - early September. Seed pods mid-September - October.

Orange-grass or Pineweed *Hypericum gentianoides* 3" - 1'
Erect herb, diffusely branching, apparently leafless. Branches slender, wiry. Leaves opposite, minute, scale-like, pressed close to stem. Flowers very small, orange-yellow, at tips of branches. Common on sandy, sun-baked soils, roadsides, disturbed areas. Flowers mid-July - mid-September.

Marsh St. John's-wort *Hypericum virginicum* 6" - 2 1/2'
= *Triadenum virginicum*
Erect, branching herb. Leaves paired, broad-ovate, blunt, close to stem, 2-3 times longer than wide, dotted with translucent glands (use hand lens). Stem, leaves crimson-purplish late in season. Flowers small, in small, terminal clusters on upper branches, pinkish with 3 orange glands alternating with stamens. Common in bogs, open swamps. Flowers August - early September.

**Shrubby St. John's wort* *Hypericum densiflorum*
A shrub. See page 126 for description.

Rockrose Family (Cistaceae)
Low, often insignificant looking herbs, usually with small, alternate, crowded, scale-like leaves. Flowers 5-petalled.

*Frostweed *Helianthemum canadense* 6" - 1 1/2'
Small herb with slender stem. Leaves alternate, narrow, oblong-lance-shaped, dull green, fine hairy under. Flowers of 2 types: 1) large, 1"-1 1/4" across, solitary (occasionally paired) 5-petalled, golden-yellow, which bloom for only a day; followed by 2) small flowers, without petals, which appear clustered on later developing branches. Common in dry, open

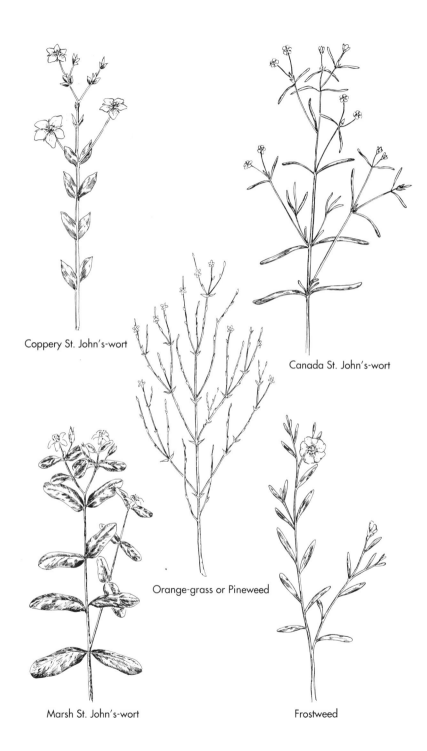

woods, clearings, barrens. Flowers with petals May - June, without petals July - early September.

A similar, less frequent species is **pine-barren frostweed**, *H. propinquum,* with slender, erect stems, either solitary or few together, from slender rhizome. Multiple flowers. Dry sands and barrens.

****Thyme-leaved Pinweed** *Lechea minor* 8" - 20"
Insignificant, finely hairy herb. Stem leaves tiny, linear, numerous. Basal leaves oval-elliptical, appear mid-October, persist over winter. Flowers greenish- or brownish-magenta with 3 tiny, narrow petals. Seed capsule round, like pinhead. Occasional in dry, sandy clearings, open soils. Flowers late August - early October.

****Oblong-fruited Pinweed** *Lechea racemulosa* 4" - 1 1/2"
Similar to above but basal leaves oblong-lance-shaped, appear late September, persist over winter. Seed capsule narrowly elliptical or pear-shaped. Common in dry, sandy ground. Flowers August - mid-September. Three other pinweeds in pinelands, most small, insignificant, difficult to identify without magnification. (No illus.)

Violet Family (Violaceae)
Low herbs. Flowers irregular, with 5 petals, the lowest often widest, veined, extended back into a spur.

Birdfoot Violet *Viola pedata* 4" - 10"
Leaves smooth, deeply palmate, divided into 3-5 or more narrow segments, some of which are subdivided and toothed. Flowers blue-violet to lilac-purple, lower petal partly white, veined with violet. Pine barrens forms vary from typical *V. pedata* in that all five petals are of the same color. Apparently introduced, occasional in dry, open, sandy, pineland fields, roadsides. Flowers late April - mid-May.

***Lance-leaved Violet** *Viola lanceolata* 2" - 6"
Leaves long, narrow, lance-shaped, tapering to stalk. Flowers white, veined with purple. Common in moist, sandy meadows, marshes, roadsides. Flowers late April - early June.

Primrose-leaved Violet *Viola primulifolia* 2" - 10"
Very similar to above but leaves oblong to ovate, abruptly narrowing to stalk. Flowers nearly identical. Occasional in same habitats, at same flowering times.

Loosestrife Family (Lythraceae)

***Swamp Loosestrife** *Decodon verticillatus* 2' - 6'
Slightly shrubby, semi-aquatic herb. Stem smooth, somewhat woody, 4-6

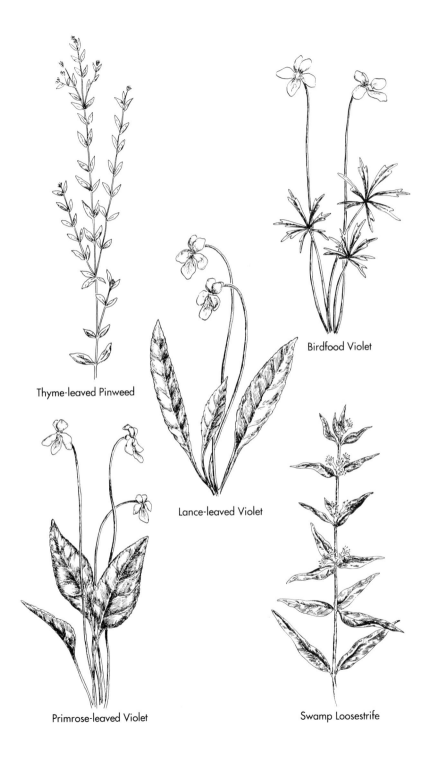

Herbs with netted-veined leaves

angled, often arching, tips reaching ground or water. Part of stem underwater has thick, spongy bark. Leaves lance-shaped, up to 3" long, paired, or mostly in whorls of 3-4. Flowers small, bell-shaped, pale purplish-pink to magenta-lavender, crowded in axils of upper leaves. Common in swamps, edges of enriched shallow ponds. Blooms August - early September.

Melastoma Family (Melastomataceae)

*Meadow-beauty *Rhexia virginica* 1' - 2'
Herb with squarish or angled stem. Leaves opposite, ovate, broad at base, pointed at tip, toothed, with 3 main veins. Flowers showy, 4-petalled, deep rose-magenta to purple, with 8 conspicuous golden stamens. Common in wet, sandy meadows, bogs, swamps. Flowers July - mid-September.

**Maryland Meadow-beauty *Rhexia mariana* 1' - 2'
Very similar to above but plant hairy, stems more round, leaves narrower, lance-shaped, flowers pale pink. Infrequent in same habitats but more common in southern pine barrens. Blooms July - early September.

Evening-Primrose Family (Onagraceae)
Herbs with large, showy (usually) flowers that usually have 4 petals and 4 sepals and, in many species, close after mid-day.

*Seedbox *Ludwigia alternifolia* 2' - 3'
Upright, smooth, with many branches. Leaves alternate, narrow, lance-shaped, pointed. Flowers yellow, on short stalks from axils of upper leaves. Sepals remain flat, not folded back, alternating behind petals. Seed pods squarish. Dried seeds rattle inside. Occasional to common in swamps, ditches. Flowers July - August. Seed pods August - early October.

*Globe-fruited Ludwigia *Ludwigia sphaerocarpa* 2' - 3'
Similar to above but leaves narrower, almost linear. Flowers usually lack petals. Occasional in pineland swamps. Flowers late July - mid-September. (No illus.)

*Fireweed or Great Willow-herb *Epilobium angustifolium* 2' - 6'
Stem tall, smooth, somewhat ruddy. Leaves alternate, lance-shaped, smooth, entire or finely toothed, dark green above, lighter under, with white ribs. Flowers with 4 rounded-petals, rose-pink-magenta in tall, terminal spikes. Seed pods long, slender, curved, tinged red-purplish, open lengthwise to expose silky down. An infrequent introduction into pineland clearings, roadsides, especially after fire burns. Flowers and seed pods June - September.

Common Evening-Primrose *Oenothera biennis* 1' - 4'
Sturdy, upright stem, simple or branched, often tinged with red. Leaves light

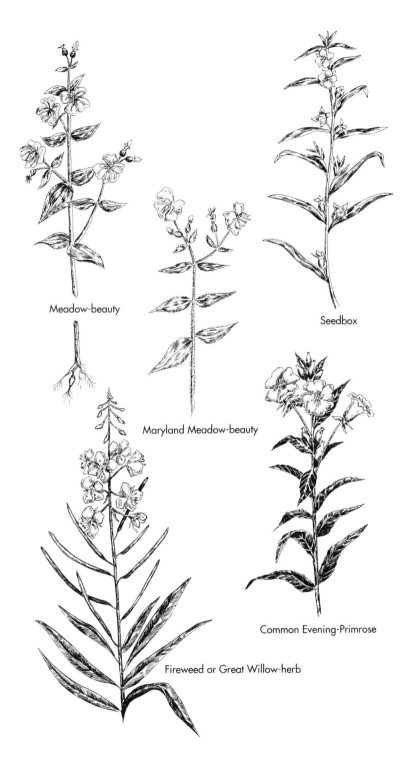

Meadow-beauty

Maryland Meadow-beauty

Seedbox

Common Evening-Primrose

Fireweed or Great Willow-herb

Herbs with netted-veined leaves

green, lance-shaped, wavy-edged, slightly toothed. Flowers large (1" across), showy, pure yellow, with 4 petals, 8 prominent stamens, and X-shaped stigma. Flowers open early evening, close by next mid-day. Biennial. Introduced weed along roadsides. Blooms late June - late September.

Cut-leaved Evening-Primrose *Oenothera laciniata* 1' - 1 1/2'
Upright stem slender, slightly hairy. Leaves alternate, oblong, irregularly and deeply cut and lobed. Flowers small, light yellow, stalked from axils of upper leaves, open toward evening, fade next day. Introduced as weed into sandy, open pinelands. Flowers mid-May - mid-July.

Parsley Family (Umbelliferae = Apiaceae)

**Slender-leaved Cowbane *Oxypolis rigidior* var. *ambigua* 2' - 4'
Tall, slender, smooth herb. Leaves alternate, deep green, long stalked, divided into 3-15 very narrow, lance-shaped, sparsely toothed, variable leaflets. Flowers tiny, dull white, in small clusters. Fruits flat with marginal wings. Apparently restricted to swamps in pinelands. Flowers mid-August - September.

Wintergreen Family = Heath Family (Pyrolaceae = Ericaceae)

Indian-pipe *Monotropa uniflora* 4" - 10"
Stems thick, translucent white, with scaly bracts in place of leaves. Stems often spring up in clusters. Flowers white, nodding, become erect as seed pods mature. Plants turn black when dry or dead. Saprophytic plant without chlorophyll that obtains nutrients from decaying vegetation in soil. Occasional to frequent in pine-oak woodlands, especially in areas with humus. Flowers June - early October.

Pinesap *Monotropa hypopithys* 4" - 1'
Similar to above but stems and flowers buff, yellowish, or reddish. Flowers several and nodding, in curved tips of stems. Seed capsules erect, black. Only occasional above roots of oaks and pines. Flowers early July - September.

Primrose Family (Primulaceae)

*Yellow or Swamp-Loosestrife *Lysimachia terrestris* 8" - 2 1/2'
or Swamp-candles
Stem erect, slender, smooth. Leaves long, narrow, lance-shaped, opposite, in pairs. Flowers small, 5-petalled, yellow with circle of reddish spots on petals, in tall, slender spikes. Small bulblets develop in axils of leaves in late summer. Frequent in open, moist bogs, swamps. Flowers late June - early July.

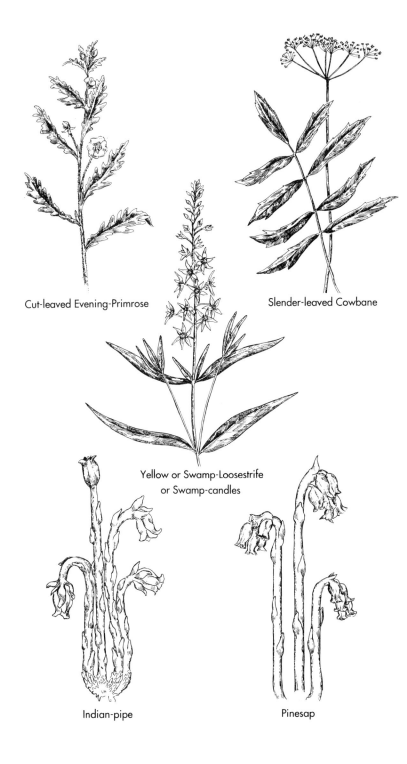

Herbs with netted-veined leaves

*Star-flower *Trientalis borealis*   4" - 1'
Stem thin, nearly bare. Leaves thin, shiny, lance-shaped, taper to both ends, in single whorl of 5-9 at top of stem. Flowers few, small, star-shaped, with 6-7 pointed, white petals, on thread-like stalks from top of leaf-whorl-bearing stem. Occasional, local in low, moist woods, swamps. Flowers May - early June.

Gentian Family (Gentianaceae)
Smooth herbs with leaves usually opposite or basal, stalkless, entire (not toothed). Flowers in 4-12 parts.

**Lance-leaved Sabatia *Sabatia difformis*   1' - 3'
Stem slender, somewhat 4-angled, usually simple, occasionally oppositely branched above. Basal rosette of leaves usually gone by flowering time. Leaves opposite, lance-shaped, with 3-5 ribs. Flowers large, 1" across, 5-lobed, star-like, white, turning yellowish upon fading. Common in pinelands bogs, swamps. Flowers July - August.

**Pine-barren Gentian *Gentiana autumnalis*   6" - 1 1/2'
Stem slender, simple or occasionally branched. Leaves opposite, very narrow, linear, long, up to 2". Flowers solitary or up to 2-3, at tops of stems or branches, large, 5 widely spreading lobes from deep, tubular corolla, bright, light ultramarine blue, brown-speckled inside corolla. An endangered species found in moist, open, sandy barrens, bogs, in heart of pines. Flowers September - early October.

*Upright Bartonia *Bartonia virginica*   4" - 15"
Stem slender, slightly angled, stiff, wiry, simple or sparsely branched, with tiny, opposite, yellow-green scales in place of leaves. Flowers tiny, 4-lobed, greenish-yellow in loose clusters at tops of stems, branches. Petals oblong, rounded. Frequent damp ground, moist, sandy bogs. Flowers mid-July - August.

*Twining Bartonia *Bartonia paniculata*   8" - 15"
Similar to above but variable. Stem less rigid, often flexible, sometimes twining. Petals lance-shaped, tapering to sharp tip. Frequent in low, wet, peaty sands. Flowers late August - September. (No illus.)

Dogbane Family (Apocynaceae)
Herbs with entire, usually opposite leaves, terminal clusters of small, nodding, ball-shaped flowers, flower parts in 5's, slender pods with tufted seeds, and an acid, milky juice throughout plants.

Spreading Dogbane *Apocynum androsaemifolium*   1' - 2 1/2'
Stem sturdy, smooth, branched, tinged with red on sunny side. Leaves ovate-elliptical, paired, without teeth. Flowers small, tubular, terminal, pale pink. Occasional in dry, open fields, roadsides, waste places. Flowers June - August.

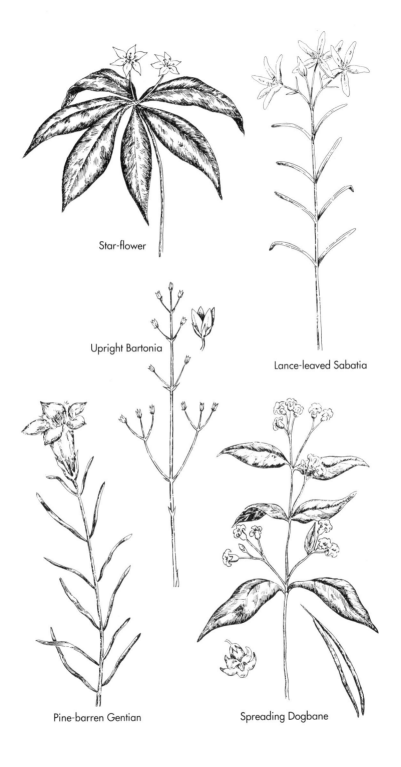

Herbs with netted-veined leaves

Indian-Hemp *Apocynum cannabinum* 1' - 2 1/2'
Similar to above but slightly smaller, less spreading. Leaves narrower, rounded at bases, stalked. Flowers smaller, more greenish-white. Flowers, leaves finely hairy but leaves highly variable in shape and pubescence. Occasional in pines. Same habitats and timing as above.

Milkweed Family (Asclepiadaceae)
Herbs with simple, mostly opposite leaves, flowers in terminal clusters, flower parts in 5's, heavy pods with seeds attached to white tufts of floss, and thick (usually) acid, milky juice.

Butterfly-weed *Asclepias tuberosa* 1' - 2 1/2'
Stem erect, sturdy, hairy, branched near top, almost bushy. Leaves alternate, narrow-oblong, hairy beneath. Flowers bright yellow-orange, in nearly flat topped clusters at tops of branches. Plant juice only slightly milky. Introduced into pine barrens and occasional in dry, open, sandy fields, roadsides, disturbed areas. Flowers late June - early August.

**Red Milkweed *Asclepias rubra* 1' - 4'
Erect, smooth. Leaves broadly lance-shaped, rounded at base, tapering to long tip. Flowers small, in terminal clusters, purplish-red. A rare [23] and endangered (CMP) species in pinelands swamps, bogs. Flowers late June - July.

*Blunt-leaved Milkweed *Asclepias amplexicaulis* 2' - 3'
Erect or reclining. Leaves oblong, rounded at both ends, wavy along sides, clasp stem without leaf-stalks. Flowers greenish, tinged with magenta-purple. Frequent in dry, sandy fields, open woods. Flowers mid-June - mid-July.

Mint Family (Labiatae = Lamiaceae)
Aromatic herbs. Stems usually branched, squarish. Leaves opposite. Flowers small, irregular, often in spikes or in clusters at axils of leaves.

*Bluecurls or Bastard Pennyroyal *Trichostema dichotomum* 6" - 2'
Delicate. Stem erect, stiff, branching, woolly-sticky. Leaves opposite, oblong to lance-shaped, sticky, aromatic. Flowers single or paired at tips of branches with 4 exceptionally long, violet stamens that curl way above blue-violet flower. Common in dry, open, sandy fields, waste areas, and cinders along railroads. Flowers mid-August - mid-September.

**Narrow-leaved Bluecurls *Trichostema setaceum* 6" - 1'
Very similar to above. Leaves very narrow, linear, smooth or nearly so, with only single, central vein. Local, not as common, in same habitats. Same flowering times. (No illus.)

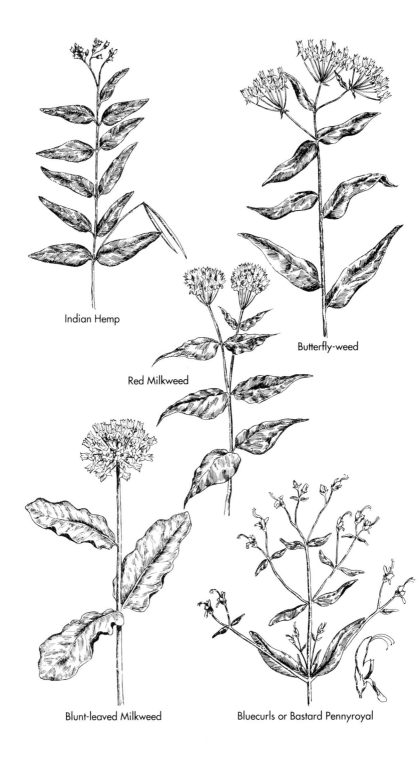

Herbs with netted-veined leaves

Horsemint *Monarda punctata* 2' - 3'
Stem square, branched above. Leaves opposite, lance-shaped, shallow-toothed. Flowers in rosettes in upper leaf axils, just above whorls of whitish to lavender bracts. Flowers yellow with purple spots. Dry, open, sandy fields, roadsides. Flowers late July - early October.

*Mountain-mint *Pycnanthemum verticillatum* 1' - 3'
Stem stout, stiff, square, downy, branched above. Leaves opposite, paired, slender, lance-shaped, tapering at base, entire or with few low teeth, downy on veins beneath, fragrant. Flowers small, white spotted with purple, in numerous flat-topped, branching clusters of rounded heads. Occasional in moist clearings, thickets. Flowers July - early September. A similar species is ***short-toothed mountain-mint**, *P. muticum,*with primary leaves ovate or broadly ovate-lance-shaped, and rounded or almost heart-shaped at base.

*Sessile-leaved Bugleweed *Lycopus amplectens* 1' - 3'
or Water-Horehound
Stem square, smooth. Leaves opposite, paired, oblong-lance-shaped, broad at base, stalkless, with few shallow teeth. Flowers small, in dense clusters encircling stem in upper leaf axils, white. Common in moist, sandy soils, especially around cranberry bogs. Flowers August - early October.

Figwort Family (Scrophulariaceae)
Herbs with alternate, opposite, or whorled leaves. Flowers irregular, with inflated corolla tubes and 2 lips, 1 above, 1 below, or have petals joined at base only. Well represented in pinelands with at least 9 species.

*Blue or Old-field Toadflax *Linaria canadensis* 6" - 2'
Stem very slender, smooth, branched above. Leaves alternate, small, smooth, shiny, stemless, linear, long, narrow. Flowers small, 2-lipped, with long, slender, curved spur at base, blue to pale lavender, lower lip with whitish palate, in loose, open groups. Common in dry, open, sandy soils, waste places. Flowers late April - early July.

*Golden Hedge-hyssop *Gratiola aurea* 6" - 15"
Semi-aquatic herb. Stem creeping with upright branches. Leaves opposite, paired, toothless, ovate, blunt-tipped, without stalks. Flowers long-stalked from leaf axils, appear 4-lobed, bright yellow. Occasional in wet, sandy ground. Flowers late June - September. Might be considered an aquatic plant but usually grows out of water or partly out. Sometimes grows underwater, then only 1"-2" high with sharp-tipped leaves but without flowers.

**Pine-barren Gerardia *Gerardia racemulosa* 1' - 4'
= *Agalinis virgata*
Stem slender, straight, rigid, smooth. Branches spreading. Leaves small, paired, narrowly linear, nearly thread-like, often curling, strongly so when dried. Flowers on short stems from upper leaf axils, large, 1"-1 1/2", broadly

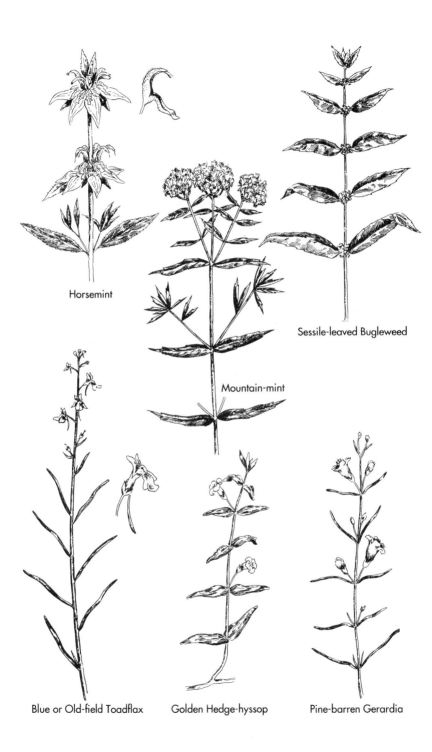

Herbs with netted-veined leaves

funnel-shaped with 5 wide, flaring lobes, deep pink to magenta-purple. Occasional to frequent in damp bogs. Flowers late August - September.

**Bristle-leaved or Thread-leaved Gerardia *Gerardia setacea = Agalinis setacea*   1' - 2 1/2'
Similar to above but flowers smaller, 1/2"-3/4", flower stalks as long or longer than flower length. Frequent in dry sands. Flowers mid-August - mid-September. (No illus.)

Downy False Foxglove *Gerardia virginica= Aureolaria virginica*   2' - 4'
Stem simple, branching. Leaves oblong-lance-shaped, lower ones deeply lobed, with wavy margins on lower half, upper ones shallowly lobed to entire. Stems, leaves downy. Flowers large, 1", showy, bright lemon-yellow, finely hairy, tubular or bell-shaped with 5 broad, spreading, rounded lobes. Partly parasitic on roots of oak trees. Occasional in dry, open woodlands. Blossoms July - early August.

*Fern-leaved False Foxglove *Gerardia pedicularia*   2' - 3 1/2'
= *Aureolaria pedicularia*
Similar to above but usually more branching, even bushy, leaves finely divided, fern-like, into very fine lobes. Both stems and leaves sticky-fine-hairy. Frequent in dry, open woodlands. Flowers mid-August - mid-September.

*Cow-wheat *Melampyrum lineare*   4" - 1 1/2'
Stems slender, smooth, wiry, branching. Leaves opposite, mostly lance-shaped, lower ones without teeth, upper leaves usually with 1 or 2 pointed lobes near base, slightly hairy. Flowers small, tubular, opening into 2 lips, white with straw-colored yellow palate, in upper leaf axils. Common in open, dry, sandy woods where plants are somewhat parasitic on roots of other plants. Flowers late May - August.

**Chaffseed *Schwalbea americana*   1' - 2'
Upright herb, fine-hairy. Leaves alternate, oblong, upper ones gradually reduced to narrow bracts. Flowers 1"-1 1/4" in open, loose, leafy-bracted spikes, dull purplish-yellow. An endangered species in moist, sandy pinelands. Flowers June - early July. (Illus. on pg. 207)

Madder Family (Rubiaceae)

**Pine-barren Bedstraw *Galium pilosum* var. *puncticulosum*   1' - 3'
Stems slender, stiff, square, hairy, branched. Leaves oblong-narrow, hairy, with 1 mid-vein, in whorls of 4. Flowers stalked, greenish-white to purple. Frequent in dry woods. Flowers late June - July. (Illus. on pg. 207)

*Rough Buttonweed *Diodia teres*   6" - 2 1/2'
Stem prostrate, spreading or ascending, finely hairy. Leaves narrow, with-

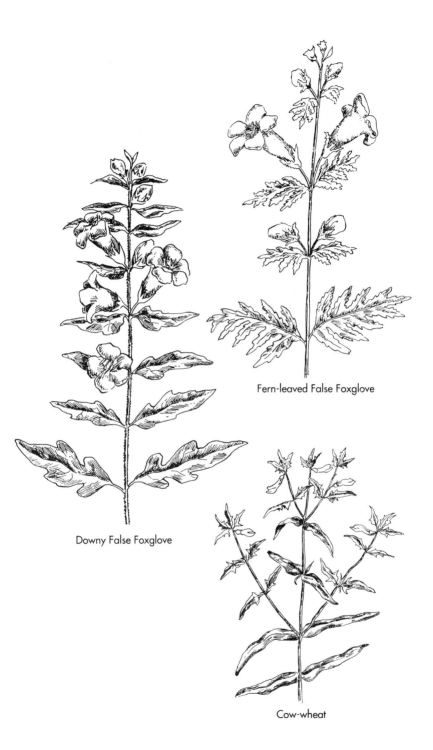

Fern-leaved False Foxglove

Downy False Foxglove

Cow-wheat

out stalks, stiff, in pairs. Flowers small, funnel-shaped, in axils of leaves, white or pale purple. Occasional to common in dry, sandy, open ground and roadsides. Flowers mid-July - September.

Bluebell Family (Campanulaceae)

**Canby's Lobelia *Lobelia canbyi*   1' - 3'
Stem simple or branched, smooth or nearly so. Leaves alternate, entire, thin, linear to lance-shaped. Flowers 2-lipped, deep blue with white throat, bearded lip, in loose, well-spaced, terminal spikes. Rare[24] and classified as endangered (CMP). Restricted to wet, sandy spots in central pine barrens. Flowers late July - September. (No illus.)

**Nuttall's Lobelia *Lobelia nuttallii*   6" - 2 1/2'
Similar to above but not as tall or robust. Stem very slender, sometimes branched. Basal leaves ovate, stem leaves narrow, linear, entire. Flowers small, 2-lipped, smooth, blue with white center. Common in same habitats. Flowers July - early September.

COMPOSITES

Perennial (mostly) and annual herbs remarkable for their usually compound flower heads composed of many closely arranged, smaller flowerets, the whole resembling a single flower. These may be either: 1) All flowers (florets) small, tubular (**discoid**), e.g. boneset, thistle. 2) Both tubular and ray flowers on same head, tubular in center, ray on outside (**radiate**), e.g. aster, daisy, sunflower, goldenrod. 3) All flowers ray- or strap-shaped (**ligulate**), plants with milky or colored juice, e.g. dandelion, hawkweed. Plants in the following section are arranged in this same sequence. All composites are dicotyledons (see page 76). Most pine barren composites flower in the late summer through early fall.

Composite Family (Compositae = Asteraceae)
This is the largest family of flowering plants, of which there are between 48 and 60 in the pine barrens. However, in the barrens, the number of composite species is exceeded by both the number of grasses and the number of sedges.

DISCOID COMPOSITES

Bonesets or Thoroughworts
Tall, erect herbs, branching near tops of stems. Flowers tubular, white, in branched, flat-topped clusters. Differences between species often minute. Seven of eight pine barrens species are described.

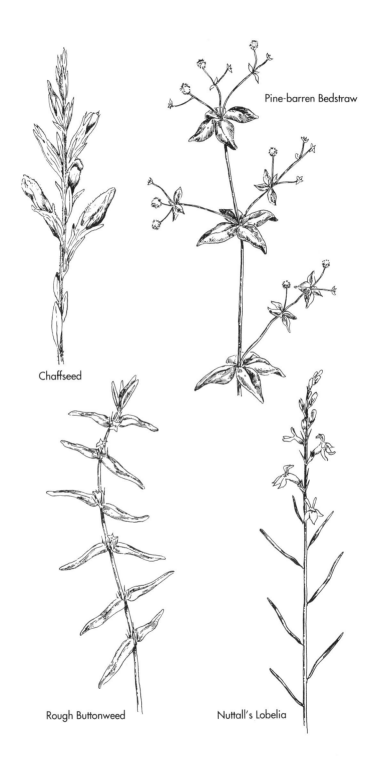

Chaffseed

Pine-barren Bedstraw

Rough Buttonweed

Nuttall's Lobelia

Composites

**White Boneset* *Eupatorium album* 1' - 3'
Stem tall, erect, rough-hairy, branching near top. Leaves opposite, lance-shaped, narrowed to the base, veined, nearly without stalks, coarsely toothed, light green. Flowers tubular, white, on small heads which together form broad, nearly flat, terminal clusters. Common in dry, sandy, open pine barrens. Flowers August - early September.

**White-bracted Boneset* *Eupatorium leucolepis* 1 1/2' - 3'
Similar to white boneset, above. Leaves opposite, paired, narrow, lance-shaped, nearly without stalks, with few small teeth. Occasional smaller leaves in leaf axils. Bracts supporting flower heads white. Flowers tubular, white. Frequent in open damp sands, peats, bogs, swamps. Flowers August - early September.

Hyssop-leaved Boneset *Eupatorium hyssopifolium* 1' - 2'
Similar to white-bracted, above, but leaves very narrow, almost linear, grass-like, entire but sometimes slightly toothed, in whorls of 3's or 4's with clusters of smaller leaves in leaf axils. Flowers tubular, white. Frequent in dry to moist open sandy fields, woods. Flowers August - September.

Rough Boneset *Eupatorium pilosum* 3' - 4'
= *E. rotundifolium* var. *saundersii*
Similar to white boneset, above. Stem hairy, leaves without stalks, larger-lower ones oblong, somewhat blunt, with 3-12 coarse teeth on each margin. Upper leaves nearly entire, linear or lance-like. Flowers tubular, white. Frequent in wet, open sands, bogs. Flowers August - early September.

Hairy Boneset *Eupatorium pubescens* 2' - 4'
= *E. rotundifolium* var. *ovatum*
Similar to rough boneset, above. Stem hairy. Stem, lower leaves oblong-pointed, with 12-25 sharp teeth on each margin. Upper leaves toothed. Flowers tubular, white. Common in moist to dry fields, open woods. Flowers August - early September.

Round-leaved Boneset *Eupatorium rotundifolium* 2' - 4'
Similar to hairy boneset, above, but leaves smaller, almost as broad as long, palmately veined, semi-clasping around stem, teeth fewer, blunt, not sharp. Flowers tubular, white. Common in dry, open, sandy soils, woods. Flowers August - early September.

**Pine-barren Boneset* *Eupatorium resinosum* 2' - 3'
Another white boneset. Stems slender, minutely soft pubescent. Leaves finely pubescent, sometimes resinous underneath, narrowly lance-shaped with toothed edges and bases narrowed and clasping stem. Flowers ten or more in terminal clusters. A rare and threatened species restricted to wet bogs in heart of pine barrens. Flowers mid-August - September. (No illus.)

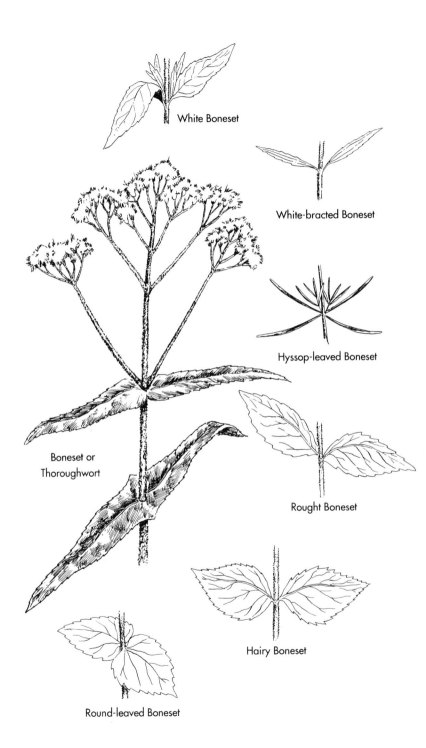

Composites

**Hairy Blazing-star *Liatris graminifolia* 1' - 3'
Stem not usually branched. Leaves narrow, lance-shaped, numerous, singly attached to stem. Flowers small, tubular, 5-lobed, lavender-lilac, in small cylindrical heads. Heads arranged up stem in loose, open, spike- or wand-like clusters. Stems and stalks of flower heads hairy. Frequent in dry to moist sandy pinelands. Flowers August - September.

*Catfoot or White *Gnaphalium obtusifolium* 1' - 2 1/2'
or Sweet Everlasting
Stem cottony-woolly, leafy. Leaves alternate, long, narrow, gray-green above, downy-whitish under. Slightly aromatic. Flowers white, in branching clusters at tops of stems. Flowers not expanding until ready to seed. Common in dry, open, sandy soils. Flowers late August - September.

*Purple Cudweed *Gnaphalium purpureum* up to 1 1/2'
Stem slender. Leaves alternate, widest toward rounded tips, entire, downy underneath. Flowers white, in axils of upper leaves, forming elongate, spike-like clusters. Bracts surrounding flower heads brownish or purplish. Occasional and local in dry, open, sandy tracts, especially about abandoned fields, habitations. Flowers late May - early June.

RADIATE COMPOSITES

*Maryland Golden Aster *Chrysopsis mariana* 1' - 2'
Stem stout, considerably branched. Leaves alternate, lance-shaped, entire, without stalks. Stem, leaves silky-hairy when young, become smooth with age. Flowers large, nearly 1", showy, aster-like, at tips of branches, both central florets and outer ray flowers golden yellow. Common in dry, sandy clearings, open woods. Flowers August - early September.

**Sickle-leaved Golden Aster *Chrysopsis falcata* 6" - 15"
= *Pityopsis falcata*
Similar to above but much smaller. Stems, leaves white-woolly. Leaves stiff, small, narrow, linear, curved or sickle-shaped. Rare[25] and threatened. Occasional and local on bare, open stretches of white sand on eastern edges of pine barrens. Flowers July - early September.

Goldenrods *Solidago* spp.
Large, diverse group of composite family members. Identification difficult, often based on characters too technical for a field book of this type. All have compound flower heads composed of tiny, tubular flowers (florets) on a central disk, surrounded by few to several ray flowers, which in most (except *S. bicolor*) are a shade of yellow. To help field identification of nine of the 12 pine barrens species, these are divided into 3 groups according to their flowering pattern:

Group I, **wand-like**: Flowers on short stalks from axils of upper leaves, with balance of flowers crowded on upper part of stem in long, slender,

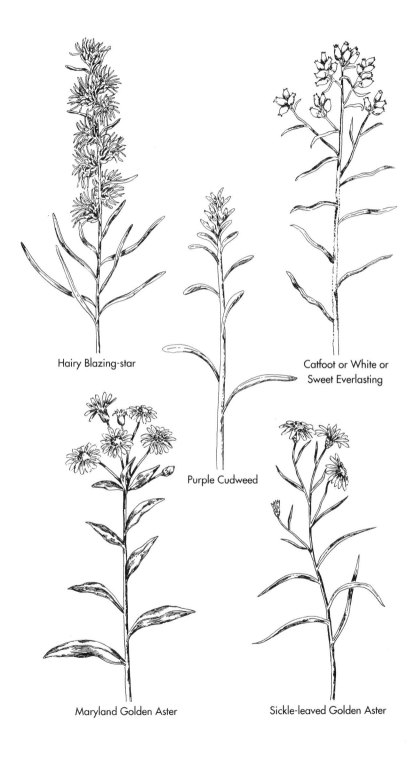

Composites

cylindrical spikes, never in 1-sided clusters. These are *Solidago bicolor, S. puberula, S. erecta,* and *S. stricta*.

Group II, **plume-like**: Flower heads situated along one (upper) side only of long, branching, often curved, spreading, plume-like branches, but in some, heads are at ends of shorter branches, all branches together forming a terminal plume. This group includes all other pinelands species except *S. tenuifolia*.

Group III, **flat-topped**: Flower heads crowded at or near tips of branches, the whole forming a rounded or flat-topped arrangement. Only 1 pinelands species: *S. tenuifolia*

The following 4 species have group I, wand-like arrangement of flowering heads:

*White Goldenrod or Silverrod *Solidago bicolor*   1' - 3'
Stem upright, simple or branched, gray-hairy. Leaves alternate, elliptical, rough-hairy, toothed, feather-veined, ribs hairy underneath. Wand-like flowering pattern. Central, tubular florets creamy-yellow, outer 5-12 ray flowers white. Our only white goldenrod. Frequent in dry, open, sandy woods. Flowers late August - early September.

**Downy Goldenrod *Solidago puberula*   1' - 3'
Stem, leaves fine-hairy, slightly sticky. Stem often purplish. Leaves narrow, lower leaves lance-shaped, stalked, toothed. Wand-like flowering pattern. About 10 bright yellow ray flowers. Bracts surrounding flower heads pointed. Frequent in dry, sterile, generally disturbed, peaty sands. Flowers September - early October.

**Slender Goldenrod *Solidago erecta*   1' - 3'
Similar to above but stem, leaves usually smooth. Upper leaves spreading or ascending. Wand-like flowering pattern. Flowers with 6-9 light yellow rays. Bracts surrounding flower heads blunt. Occasional in dry, open thickets, woods. Flowers August - September.

**Wand-like Goldenrod *Solidago stricta*   2' - 5'
in part = *S. sempervirens* var. *mexicana*

Stem slender, smooth. Lower leaves lance-shaped, entire. Upper leaves small, numerous, bract-like, hug stem. Flower clustering pattern long, slender, wand-like. Flowers with 5-7 yellow rays. Uncommon and local endangered species, somewhat restricted to typical damp, sandy pinelands. Flowers mid-August - September. (See Group I illustration)

The following 4 species have group II, plume-like arrangement of flowering heads:

*Swamp Goldenrod *Solidago uliginosa*   2' - 4'
Stem stout, smooth. Leaves thick, smooth, lance-shaped, shallowly toothed. Basal leaves long, up to 12", with long leaf stalks which sheath stems. Flowering pattern plume-like. Flowers on short stalks, crowded on slender

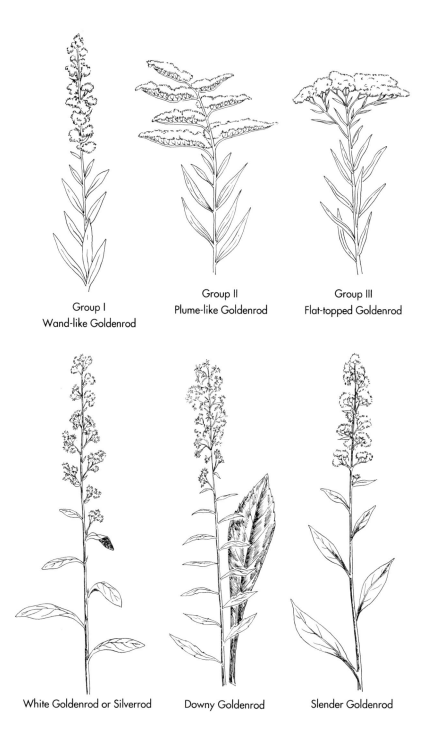

stems, with 5-6 small, light golden yellow, ray flowers. Generally distributed but infrequent in pineland bogs, swamps. Flowers late August - early October.

*Field or Gray Goldenrod *Solidago nemoralis* 6" - 2'
Stem simple, unbranched. Leaves rough, thick, 3-ribbed, dull-toothed. Broad, lance-shaped, larger leaves often widest near tips. Taper at base to long stalk. Small leaflets in axils of larger leaves. Stem leaves much smaller than basal. Both stem, leaves grayish-green, covered with fine hairs. Flowering pattern plume-like. Flowers clustered, forming thickly set, narrow, 1-sided plumes. Flowers with 5-9 golden-yellow rays. Frequent in dry, sterile fields, open woods. Flowers mid-August - September.

*Fragrant Goldenrod *Solidago odora* 2' - 3'
Stem slender, smooth, often reclining. Leaves smooth, narrowly lance-shaped, entire, 3-ribbed, shining, light green, dotted with minute glands (use hand lens). Stems, leaves with anise fragrance when crushed. Plume-like flowering pattern. Flower heads small, in small, 1-sided groups, with 3-5 golden-yellow rays. Frequent, generally distributed in dry, open, sands, woods. First pinelands goldenrod to bloom. Flowers mid-July - August.

**Pine-barren Goldenrod *Solidago fistulosa* 2' - 4'
Stem hairy. Leaves elliptical, feather-veined, entire or sparsely toothed, crowded as they grow up stem, clasp stem, hairy on mid-rib beneath. Flowering pattern plume-like. Flowers with 7-12 yellow rays. Frequent, generally distributed in moist, sandy soils, swamps. Flowers late August - September.

The following species has a group III, flat-topped arrangement of flowering heads:

*Slender-leaved Goldenrod *Solidago tenuifolia* 1' - 2 1/2'
= *Euthamia tenuifolia*

Stem slender, smooth. Leaves very narrow, linear, 3-ribbed, with resin dots (use hand lens). Fragrant. Tufts of small leaves in axils of larger leaves. Flower heads tiny, with 6-12 rays, in numerous groups of 2-3, forming a flat or roundish-topped cluster pattern. A conspicuous flat-topped goldenrod in our pine barrens. Frequent in dry, sandy soils. Flowers late August - early October. A similar species is *S. graminifolia* which has made its way into the pine barrens.

Asters *Aster* spp.
Asters, like goldenrods, make up a large group of composite family members. All have compound heads composed of central, tubular florets and outer ray flowers (radiate). In most cases, outer ray flowers are bluish, lilac, purple, pinkish or white. Many species are exceedingly variable, many hybridize, so identification is difficult without the aid of technical characters not possible in this book. There are at least 11 species of asters in the pine barrens.

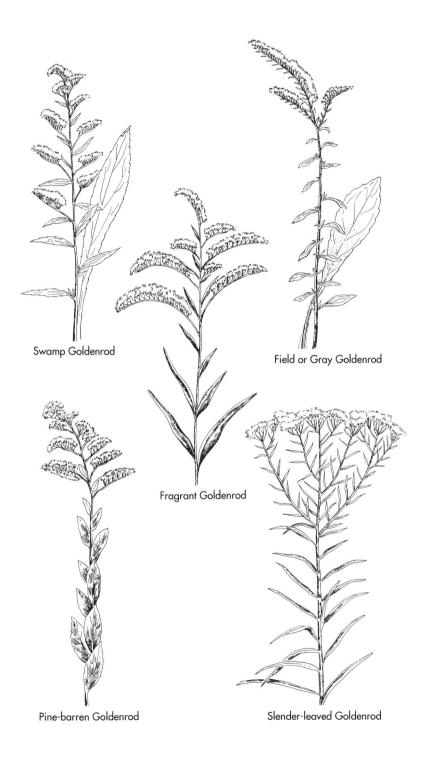

Composites

*Wavy-leaved Aster *Aster undulatus*   1' - 3'
Stems stiff, rough. Lower leaves heart-shaped, on long, broad, winged stalks which clasp stem. Upper leaves not stalked, but also clasp stem. Stems, leaves with fine, whitish down. Ray flowers pale blue-violet. Dry clearings, open woods. Flowers September - mid-October.

*Late Purple Aster *Aster patens*   1' - 3'
Stems slender, rough-hairy, widely branched. Leaves oval-oblong, rough on edges and upper surface, without stalks, heartshaped at base, clasping stem. Flowers with 15-30 light to deep blue-violet to purple rays. Dry clearings, open woods. Flowers mid-August - early October.

*Silvery Aster *Aster concolor*   1' - 2'
Stem slender, smooth. Leaves oblong, entire, silky-hairy on both sides, numerous, without stalks, slightly clasping stem. Flowers in narrow, elongated, wand-like clusters, with few short branches. 8-16 flower rays lilac or blue-violet. Rare[26] and endangered (CMP) species in dry, sandy, barren, open woods. Flowers late August - early October.

**Showy Aster *Aster spectabilis*   1' - 2'
Stems simple, stiff, slightly rough near base. Leaves oblong-lance-shaped, entire, rough. Basal leaves tapering to long stalks. Flowers showy, few in number, with 15-25 long, 3/4" or more, deep blue-violet rays. Common in dry, sandy barrens. Flowers late July - September.

**Bushy Aster *Aster dumosus*   1' - 2 1/2'
Stems smooth or slightly fine–hairy, broadly branched. Leaves small, very narrow, linear, rough. Flowering branches with numerous, very tiny leaves. Flowers small, numerous, on long stalks at ends of branches, with 15-25 blue, pale lilac, or even white rays. Frequent in dry or wet sandy thickets, bogs, woods. Flowers late August - early October.

*New York Aster *Aster novi-belgii*   1' - 3'
Stems smooth or slightly downy. Leaves smooth, long, thin, narrowly lance-shaped to sharp-pointed, small upper ones clasping stem, lower ones slightly toothed. Flowers large, with 15-24 nearly 1/2" long, pale to deep violet, blue-violet, or lilac rays. A highly variable species. Damp meadows, thickets. Flowers September - October.

**Slender Aster *Aster gracilis*   6" - 2'
Small, stiff plant. Several stems may rise from enlarged, rhizome-like base (corm). Leaves elliptical, rough, without teeth, on erect stalks. Flowers few with short, violet or bluish rays. Common in dry, sandy woods. Flowers late July - early September. (No illus.)

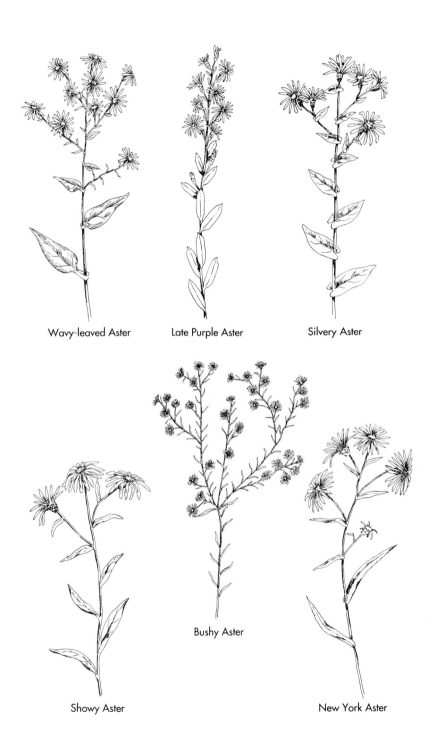

Composites

Bog Aster *Aster nemoralis*   6" - 2'
Stem slender. Leaves very numerous, lance-shaped, narrowly tapering at both ends, with rough margins. Flowers large, solitary or few, with light violet-purple rays. Frequent in peaty bogs, cedar swamps. Flowers mid-August - September.

Stiff-leaved Aster *Aster linariifolius*   6" - 2'
Stems stiff, wiry, downy, branched. Leaves numerous, stiff, narrow, almost linear, entire, 1-ribbed, rough on margins. Flowers solitary or few on tips of branches. Ray flowers light to deep blue-violet or lilac. Dry, sandy soils, open woods. Flowers September - mid-October.

Toothed White-topped Aster *Sericocarpus asteroides*   6" - 2'
= *Aster paternus*

Stems slender, sturdy, branching upward. Leaves alternate, small, finely and sparsely toothed beyond middle, lower leaves in rosettes at base, upper leaves attached to stem without stalks. Flowers with few, 4-6, narrow white or cream-colored rays in circular, flat-topped clusters. Common in dry, sandy woods, clearings. A summer aster, flowering late June - early August.

Narrow-leaved White -topped Aster *Sericocarpus linifolius*   1' - 2 1/2'
= *Aster solidagineus*

Similar to above but taller, leaves narrow, linear, without teeth. More frequent form in dry, sandy woods, clearings. Also a summer aster, flowering mid-June - mid-August.

Woodland Sunflower *Helianthus divaricatus*   2' - 5'
Stems slender, smooth, branching above. Leaves opposite, nearly without stalks, lance-shaped, somewhat heart-shaped at base, tapering to tip, thick, rough above, hairy below. Flowers large, 2" or more, solitary or few, with 8-20 yellow rays. Disk flowers yellow. Occasionally local in dry woods, clearings. Flowers mid July - early September.

Narrow-leaved Sunflower *Helianthus angustifolius*   2' - 5'
Similar to above but leaves thin, narrow, linear, stiff to rigid. Lower leaves opposite. Flowers large, 2" - 3", few, with brownish- to purplish-black central disk florets, and yellow ray flowers. Occasional to frequent in damp thickets, bogs, swamps. Flowers August - September.

Rose-colored Tickseed *Coreopsis rosea*   6" - 2'
Stem leaves paired, very narrow, almost thread-like. Flowers small, under 1", daisy-like, central disk yellow with 4-8 reflexed, rose-pink to white ray flowers. Rare[27] but classified as threatened (CMP) in sandy, pineland swamps. Flowers July - early September. (Illus. on pg. 221)

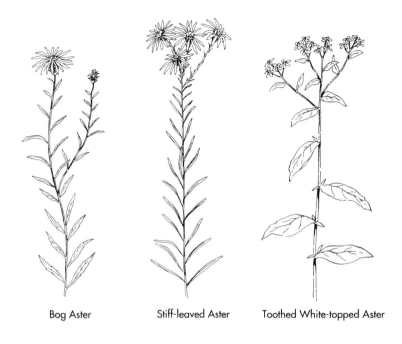

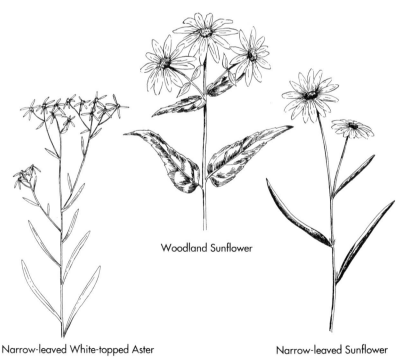

Composites

**Slender-leaved Tickseed-Sunflower *Bidens coronata*   2' - 5'
or Beggar-ticks
Leaves deeply divided into 5-7 narrow lobes or leaflets. Flowers showy, about 2" across, with 6-19 golden yellow rays. Flower heads with 6-8 outer bracts. Seeds wedge-shaped with 2 spines, stick to clothing. Bogs, swamps. Flowers late August - early October. Several other species in pine barrens.

LIGULATE COMPOSITES

*Dwarf Dandelion *Krigia virginica*   2" - 1'
Small, slender. Leaves lobed, toothed, in basal rosettes. Flowers small, all ray-type, dandelion-like, about 1/2", solitary, on slender, usually leafless stalks, golden yellow. Stalks later may become branched, bear few leaves. Frequent in dry, sandy, sterile barrens, disturbed areas. Flowers May - June.

*Rattlesnake-root or *Prenanthes serpentaria*   1' - 4'
Lion's-foot
Stem smooth, leafy. Leaves varied, lower ones thick, broad, bluntly lobed, upper ones deeply cut, uppermost lance-shaped with 2 small, lateral lobes near base. Flowers small, numerous, pendant, bell-shaped, in open, branching, spreading, flat topped, 5-18 headed clusters, dull creamy-white - pinkish, enclosed in hairy-bristly, green-magenta-tinted envelope. Frequent in dry barrens, open woods. Flowers late August - early October.

**Pine-barrens or Slender *Prenanthes autumnalis*   2' - 4'
Rattlesnake-root
Somewhat similar to above but basal leaves with winged stalks, stem leaves without stalks. Flowering heads simple, in slender, unbranched spikes, lower leaves in leaf axils, upper on leafless stem, often all on 1 side, not pendant. An endangered species in open, sandy, central pinelands areas. Flowers September - early October. (No illus.)

*Vein-leaved Hawkweed or *Hieracium venosum*   1' - 2 1/2'
Rattlesnake-weed
Stem smooth, generally leafless, or with few, tiny leaflets, branching into a few to several flowered open cluster. Leaves basal, rosettes, hairy, entire, light green, conspicuously veined with dull magenta-purple on ribs, edges, underneath, edged with fine, long bristles. Flowers dandelion-like, deep yellow to orange-yellow. Dry, open woods, clearings. Flowers late May - early July.

*Hairy Hawkweed *Hieracium gronovii*   1' - 4'
Stem slender, ruddy, rough-hairy. Leaves alternate, entire, finely hairy under, at base of plant and lower half of stem, none above. Flowers dandelion-like, small, in tall, slender, open clusters, yellow, open only in sunshine. Frequent in dry, open fields, thickets, woods. Flowers July - early September.

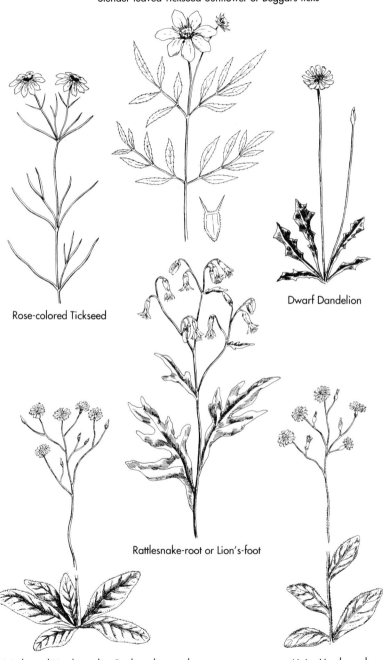

The following are recommended for further identification and reference.

VASCULAR PLANTS

Fairbrothers, D. E. 1979. Endangered, Threatened, and Rare Vascular Plants of the Pine Barrens and their Biogeography. *in* Forman, R.T.T., Jr., Pine Barrens Ecosystem and Landscape. Academic Press, NY

Fernald, M. L. 1950 Gray's Manual of Botany. 8th ed. American Book Co.

Ferren, W. R., Jr., J. W. Braxton, and L. Hand. 1979. Common Vascular Plants of the Pine Barrens *in* Forman, R.T.T., Jr., Pine Barrens Ecosystem and Landscape, Academic Press, NY

Gleason, H.A. and A. Cronquist 1963. Manual of Vascular Plants of north eastern U.S. and adjacent Canada. D. Van Nostrand Co.

Stone, W. 1973. The Plants of Southern New Jersey. Quarterman Publications. Reprint of Part II of Annual Report of the New Jersey State Museum for 1910 entitled *The Plants of Southern New Jersey with especial reference to the Flora of the Pine Barrens and the Geographic Distribution of the Species*. 1911.

TREES AND SHRUBS

Brockman, C.F. 1968. Trees of North America. Golden Press, NY

Harlow, W.M. 1957. Trees of the Eastern and Central United States and Canada. Dover Pub., NY

Keeler, H.L. 1969. Our Northern Shrubs and how to identify them. Dover Publ., NY

Little, E.K. 1979. The Audubon Society Field Guide to North American Trees. Eastern Region. A.A. Knopf, NY

Matthews, F.S. 1915. Field Book of American Trees and Shrubs. G.P. Putnam's Sons, NY

Preston, R.J., Jr. 1961. North American Trees (Exclusive of Mexico and Tropical United States.) M.I.T. Press, Cambridge, MA

Zim, H.S. and A.C. Martin. 1956. Trees. A Guide to Familiar American Trees. Golden Press, NY

GRASSES

Brown, L. 1979. Grasses. An identification Guide. Houghton Mifflin Co., Boston, MA

Chase, A. 1959. First Book of Grasses. The Structure of Grasses explained for Beginners. Smithsonian Inst., 1959

Hitchcock, A.S. Manual of the Grasses of the United States. 2nd Ed. revised by A. Chase. Vols. I & II. Dover Pub., NY

Pohl, R.W. How to Know the Grasses. Wm. C. Brown Co., Dubuque, Iowa

WILD FLOWERS

Mathews, F.S. 1927. Field Book of American Wild Flowers. G. P. Putnam's Sons, NY

Newcomb, L. 1977. Newcomb's Wildflower Guide. Little, Brown and Co., Boston, MA

Niering, W.A. 1979. The Audubon Society Field Guide to North American Wildflowers. Eastern Region. A. A. Knopf, NY

Peterson, R. T. and M. McKenny 1968. A Field Guide to Wildflowers of Northeastern and Northcentral North America. Houghton Mifflin Co., Boston, MA

Rickett, H.W. 1963. The New Field Book of American Wild Flowers. G. P. Putnam's Sons, NY

Wherry, E. T. 1948. Wild Flower Guide - Northeastern and Midland United States. Doubleday and Co., Garden City, NY

Flowering & Fruiting Calendar

APPROXIMATE FLOWERING and FRUITING PERIODS FOR VASCULAR PLANTS

Plant **NON-FLOWERING PLANTS:**	Page	Mar. m.	Apr. e.m.	May e.m.	Jun. e.m.	Jul. e.m.	Aug. e.m.	Sept. e.m.	Oct. e.m.
HORSETAILS									
Horsetail, Common	96		0 0	0					
CLUB-MOSSES									
Carolina	96						0	0 0	0 0
Fox-tail	98							0 0	0 0
Bog	98						0 0	0 0	0
FERNS									
Royal	98			0	0				
Cinnamon	98			0 0					
Curly-grass	100						0 0	0	
Sensitive/Bead	100								0 0
Marsh	100						0	0 0	
Bog/Massachusetts	100						0 0	0 0	
Spleenwort, Ebony	100				0 0	0 0			
Chain-fern, Virginia	102				0	0 0			
Chain-fern, Netted	102						0	0 0	0
Bracken/Brake	102					0 0	0 0	0	
FLOWERING PLANTS									
TREES									
Pine, Short-leaf	106			X X				0 0	0 0
Pine, Scrub/Jersey	108		X	X				0	0 0
Pine, Pitch	108			X X				0	0 0
Cedar, Atlantic White	108		X X					0 0	0 0
Cedar, Red	108	X	X					0	0 0
Birch, Gray	110		X	X				0 0	0 0
Oak, White	110			X X				0	0 0
Oak, Post-	110			X X				0	0 0
Oak, Chestnut-	110			X X				0	0 0
Oak, Black	112			X X				0	0 0
Oak, Scarlet	112			X X				0	0 0
Oak, Southern Red	112			X X				0	0 0
Oak, Black-jack	112			X X				0	0 0
Sweet Bay/Swamp Magnolia	114			X	X X	X	0 0	0 0	0
Sassafras	114		X	X		0	0 0		
Gum, Sweet	114			X	X X			0	0 0

Under months: e - early in month, m - middle of month
X - flowering period, 0 - fruiting period (data mainly from Stone, 1911)

Flowering & Fruiting Calendar

Plant	Page	Mar. m.	Apr. e.m.	May e.m.	Jun. e.m.	Jul. e.m.	Aug. e.m.	Sept. e.m.	Oct. e.m.
TREES									
Cherry, Black	114				X	X	0	0 0	
Holly, American	116				X	X X			0
Maple, Red/Swamp	116	X	X	0 0					
Tupelo/Sour Gum	116				X	X		0 0	0
SHRUBS									
Bayberry/Candleberry	118			X X	X		0	0 0	
Bayberry, Evergreen	118		X	X X	X		0	0 0	
Sweet-fern	118		X	X	0	0			
Alder, Common/Smooth	118	X	X					0 0	
Oak, Dwarf Chestnut-	120			X				0	0 0
Oak, Bear-/Scrub-	120			X				0	0 0
Virginia-willow	120				X X			0 0	0 0
Chokeberry, Red	120		X	X X				0 0	
Chokeberry, Black	122		X	X X		0 0	0		
Shadbush/Serviceberry	122		X X	X	0	0			
Hawthorn, Dwarf	122			X	X			0 0	0 0
Blackberry, Sand	122			X	X X	X 0	0 0		
Beach-plum	122		X	X			0	0 0	0
Sumac, Dwarf/Winged	124					X	X X	0 0	0
Sumac, Poison	124			X	X X		0 0	0 0	0
Winterberry/Black Alder	124				X	X		0	0 0
Winterberry, Smooth	124			X	X			0 0	0 0
Inkberry	124				X	X			0
St. John's-Wort, Shrubby	126					X X	X X	X	0 0
Pepperbush, Sweet	126					X	X X	X	0 0
Azalea, Swamp/Clammy	126				X X	X		0 0	0
Sand-myrtle	126		X	X X	X	0 0	0 0	0	
Laurel, Mountain-	128			X	X X		0	0 0	0
Laurel, Sheep-	128			X	X X		0	0 0	0
Staggerbush	128			X	X X			0 0	0 0
Maleberry	128				X	X		0 0	0 0
Fetter-bush	128			X	X X			0	0 0
Leather-leaf	128		X X	X				0 0	0 0
Huckleberry, Dwarf	130			X	X X	0	0 0	0	
Dangleberry	130			X	X X	0	0 0	0	
Huckleberry, Black	130			X X	X	0 0	0		
Blueberry, Low	130			X X	0	0 0			
Blueberry, Highbush	132			X X		0 0	0		
Blueberry, N. Jersey	132			X X		0 0	0		
Blueberry, Black Highbush	132		X	X	0	0 0			
Buttonbush	132					X X	X		
Witherod	132			X	X		0	0	
Witherod, Naked	134				X	X		0 0	0

Flowering & Fruiting Calendar

Plant	Page	Mar. m.	Apr. e.m.	May e.m.	Jun. e.m.	Jul. e.m.	Aug. e.m.	Sept. e.m.	Oct. e.m.
SUB-SHRUBS									
Crowberry, Broom-	134	X	X		0	0			
Hudsonia, Golden/Pine-barren	134			X	X 0	0			
Hudsonia, Woolly/Beach	134			X	X 0	0			
Wintergreen, Spotted	134				X	X			
Arbutus, Trailing	136		X X	X		0 0			
Teaberry/Checkerberry	136				X	X X	X	0	0 0
Bearberry	136		X	X			0 0	0	
Pyxie	136		X X	X					
Partridge-berry	136			X	X X			0	0 0
VINES									
Greenbrier, Halberd-leaved	138				X	X		0 0	0 0
Greenbrier, Common	138			X X	X			0	0 0
Greenbrier, Red-berried	138			X X	X			0	0 0
Greenbrier, Glaucous-leaved	138			X	X X			0	0 0
Greenbrier, Laurel-leaved	140						X X	X 0	0 0
Dewberry	140			X	X	0 0	0		
Blackberry, Swamp	140				X X	X 0	0 0		
Groundnut	140					X X	X X	0 0	0
Milk-Pea	140					X X	X 0	0 0	0
Ivy, Poison	140			X	X		0	0 0	
Virginia Creeper	142				X	X X		0	0 0
Grape, Summer-	142				X	X		0 0	0
Cranberry, American	142				X	X		0 0	0 0
Morning-glory, Pickering's	142						X X	X X	
Dodder	142						X X	X X	
AQUATIC HERBACEOUS PLANTS									
Bur-reed, Slender	144				X X	X X	0 0	0 0	
Pondweed, Alga	146				X	X X	X X	X	
Arrowhead	146				X	X X	X X	X 0	0
Arrow-arum	146				X X				
Golden Club	148		X X	X X					
Pipewort, 10-angled	148					X	X X	X X	X
Pipewort, Flattened	148			X	X X	X			
Pipewort, 7-angled	148					X X	X X	X X	X
Pickerelweed	148				X	X X	X X	X	
Pond-lily, Yellow	150			X	X X	X X	X X	X	
Water-lily, Fragrant	150				X X	X X	X X	X X	
Water-shield	150				X X	X X	0 0	0	
Floating-heart	150					X X	X X		
Bladderwort, Purple	150					X	X X	X	
Bladderwort, Swollen	152				X X	X X	X X	X	
Bladderwort, Fibrous	152			X	X X	X X	X X	X	
Bladderwort, Horned	152				X	X X	X		
Bladderwort, Zig-zag	152			X	X X	X X	X X	X	

Flowering & Fruiting Calendar

Plant	Page	Mar. m.	Apr. e.m.	May e.m.	Jun. e.m.	Jul. e.m.	Aug. e.m.	Sept. e.m.	Oct. e.m.
INSECTIVOROUS PLANTS and CACTUS									
Pitcher-plant	154			X	X X				
Sundew, Round-leaved	154					X X	X X		
Sundew, Spatulate-leaved	154					X X	X X		
Sundew, Thread-leaved	154				X	X X	X X		
Prickly Pear (Cactus)	156				X X	X			
GRASSES,									
Fescue-grass, 6-weeks	158			X	X X				
Manna-grass, Blunt	158					X X	X X		
Oat-grass, Wild	158			X	X X				
Oat-grass, Wild, Silky	160			X	X X				
Oat-grass, Wild, Smooth	160			X	X X				
Reedgrass, Nuttall's	160					X	X X	X	
Reedgrass, Pine-barren	160					X	X X		
Hairgrass, Rough	160				X X	X X	X		
Bentgrass, Tall	160						X	X X	X
Dropseed, Late-flowering	160						X	X	
Dropseed, Torrey's	160						X	X X	
Oat-grass, Black	162			X	X				
Poverty-grass	162						X	X X	X
Cutgrass, Rice	162						X X	X	
Rice, Wild	162					X	X X		
Paspalum, Slender	162					X X	X X	X X	X
Panic-grass, Warty	162						X X	X X	
Switchgrass	164					X	X X	X	
Panic-grass, Lindheimer's	164				X	X X	X		
Millet-grass, Pursh's	164						X X	X	
Beardgrass, Broom	164					X	X X	X X	X
Broom-sedge	164						X	X X	
Beardgrass, Bushy	164						X	X X	
Indian Grass	166						X	X	
SEDGES									
Cyperus, Gray's	166						0 0	0 0	
Sedge, Slender	166				0	0 0	0 0	0 0	0
Sedge, Three-way	166					0	0 0	0 0	0
Spike-rush, Triangular-stem	166					0	0 0	0	
Spike-rush, Green	168					0	0 0	0 0	0
Rush, Swaying	168					0 0	0 0		
Rush, 3-square	168				0	0 0	0 0	0	
Wool-grass	168					0	0 0	0 0	0
Cotton-grass, Tawny	168						0	0 0	
Beak-rush, Small-headed	168					0	0 0	0	
Beak-rush, White	170						0 0	0	
Twig-rush	170					0	0 0	0 0	0
Sedge, Pennsylvania	170			0	0				

Flowering & Fruiting Calendar

Plant	Page	Mar. m.	Apr. e.m.	May e.m.	Jun. e.m.	Jul. e.m.	Aug. e.m.	Sept. e.m.	Oct. e.m.
Sedge, Walter's	170				0 0	0 0			
Sedge, Long	170				0 0	0 0			
Sedge, Bull	170				0	0 0	0 0	0	

RUSHES

Plant	Page	Mar. m.	Apr. e.m.	May e.m.	Jun. e.m.	Jul. e.m.	Aug. e.m.	Sept. e.m.	Oct. e.m.
Rush, Soft/Common	172				0	0			
Rush, Canada	172						0	0 0	0
Rush, Bayonet	172					0	0		
Rush, Bog/Brown-fruited	172						0	0 0	

OTHER PARALLEL-VEINED HERBACEOUS PLANTS

Plant	Page	Mar. m.	Apr. e.m.	May e.m.	Jun. e.m.	Jul. e.m.	Aug. e.m.	Sept. e.m.	Oct. e.m.
Yellow-eyed Grass, Twisted	172						X X	X X	
Yellow-eyed Grass, Carolina	174					X	X X	X	
Asphodel, Bog-	174				X	X X			
Turkeybeard	174			X	X X	X			
Swamp-pink	174		X X	X					
Asphodel, False	174				X	X			
Bellwort, Pine-barren	174			X X					
Lily, Turk's-cap	174					X X			
Colicroot	176				X	X X			
Redroot	176					X X	X X		
Stargrass, Yellow	176			X	X X				
Golden-crest	176				X	X X			
Blue-eyed Grass, Eastern	176			X	X 0	0			
Blue Flag, Slender	176			X	X X				
Lady's-slipper, Pink	178			X X	X				
Orchis, Green Woodland	178					X	X		
Orchis, White Fringed	178					X	X		
Orchis, Yellow Fringed	178					X	X X		
Orchis, Crested Yellow	178					X	X X		
Orchis, Southern Yellow	180						X X	X	
Pogonia, Rose	180				X X	X			
Grass Pink	180				X X	X X			
Arethusa	180			X	X				
Ladies'-Tresses, Little	180					X	X X	X	
Ladies'-Tresses, Jagged	180						X X	X X	
Ladies'-Tresses, Nodding	180							X X	X

NETTED-VEINED HERBACEOUS PLANTS

Plant	Page	Mar. m.	Apr. e.m.	May e.m.	Jun. e.m.	Jul. e.m.	Aug. e.m.	Sept. e.m.	Oct. e.m.
Toadflax, Bastard	182			X	X X				
Jointweed	182							X X	X X
Sandwort, Pine-barren	182				X X	X X			
Indigo, Wild	184				X	X X	0 0	0	
Lupine, Wild	184			X X	X 0	0			
Goat's-rue	184				X X	X	0 0	0	
Tick-trefoil(s)	184					X	X X	X 0	0
Bush-Clover, Wand-like	184						X	X 0	0

Flowering & Fruiting Calendar

Plant	Page	Mar. m.	Apr. e.m.	May e.m.	Jun. e.m.	Jul. e.m.	Aug. e.m.	Sept. e.m.	Oct. e.m.
Bush-Clover, Hairy	186						X	X 0	0
Bush-Clover, Narrow-leaved	186						X X	X 0	
Pencil-flower	186				X	X X	X X	X 0	0
Flax, Yellow	186				X	X X	X X	0 0	
Flax, Ridged Yellow	186				X	X X	X 0	0	
Milkwort, Nuttall's	186					X X	X X	X X	X
Milkwort, Cross-leaved	188					X	X X	X X	X
Milkwort, Short-leaved	188					X	X X	X X	X
Milkwort, Orange	188				X	X X	X X	X X	X
Spurge, Ipecac	188	X	X X	0 0					
St. Peter's-wort	188					X	X X	X	0
St. Andrew's Cross	188					X X	X X	X 0	0 0
St. John's-wort, Coppery	190					X	X X	X 0	0 0
St. John's-wort, Canada	190					X X	X X	X 0	0 0
Orange-grass/Pineweed	190					X	X X	X 0	0 0
St. John's-wort, Marsh	190						X X	X 0	0
Frostweed	190			X X	X X	X X	0 0	0 0	0
Frostweed, Pine-barren	192			X	X X	X X			
Pineweed, Thyme-leaved	192						X	X X	X
Pineweed, Oblong-fruited	192						X X	X	
Violet, Birdfoot	192	X	X						
Violet, Lance-leaved	192	X	X X	X				0 0	0
Violet, Primrose-leaved	192	X	X X	X			0	0 0	0
Loosestrife, Swamp	192						X X	X 0	0
Meadow-beauty	194					X X	X X	X 0	0
Meadow-beauty, Maryland	194					X X	X X	X 0	0
Seedbox	194					X X	X X	0 0	
Ludwigia, Globe-fruited	194					X	X X	X 0	0
Fireweed	194				X X	X X	X X	0 0	0
Evening-Primrose, Common	194				X	X X	X X	X 0	0
Evening-Primrose, Cut-leaved	196			X	X X	X 0	0		
Cowbane, Slender-leaved	196						X	X X	0 0
Indian-pipe	196				X X	X X	X X	X X	X
Pinesap	196					X X	X X	X X	X
Loosestrife, Yellow	196				X	X			
Star-flower	198			X X	X				
Sabatia, Lance-leaved	198					X X	X X		
Gentian, Pine-barren	198							X X	X
Bartonia, Upright	198					X	X X		
Bartonia, Twining	198						X	X X	
Dogbane, Spreading	198				X	X X	X		
Hemp, Indian	200				X X	X X	X X		
Butterfly-weed	200				X	X X	X		
Milkweed, Red	200				X	X X	X		
Milkweed, Blunt-leaved	200				X	X X			
Bluecurls	200						X	X	
Bluecurls, Narrow-leaved	200						X	X	
Horsemint	202					X	X X	X X	X

Flowering & Fruiting Calendar

Plant	Page	Mar. m.	Apr. e.m.	May e.m.	Jun. e.m.	Jul. e.m.	Aug. e.m.	Sept. e.m.	Oct. e.m.
Mountain-mint	202					X X	X X	X	
Mt.-mint, Short-toothed	202					X X	X X	X	
Bugleweed, Sessile-leaved	202						X X	X X	X
Toadflax, Blue/Field	202		X	X X	X X	X			
Hedge-hyssop, Golden	202				X	X X	X X	X X	
Gerardia, Pine-barren	202						X	X X	
Gerardia, Bristle-leaved	204						X	X	
False Foxglove, Downy	204					X X	X		
False Foxglove, Fern-leaved	204						X	X	
Cow-wheat	204			X	X X	X X	X X		
Chaffseed	204				X X	X			
Bedstraw, Pine-barren	204				X	X X	0	0 0	0
Buttonweed, Rough	204					X	X X	X 0	0 0
Lobelia, Canby's	206					X	X X	X X	
Lobelia, Nuttall's	206					X X	X X	X	
COMPOSITES									
Boneset, White	208						X X	X	
Boneset, White-bracted	208						X X	X	
Boneset, Hyssop-leaved	208						X X	X	
Boneset, Rough	208						X X	X	
Boneset, Hairy	208						X X	X	
Boneset, Round-leaved	208						X X	X	
Boneset, Pine-barren	208						X	X X	
Blazing-star, Hairy	210						X X	X X	
Everlasting, White	210						X	X X	
Cudweed, Purple	210			X	X X	X			
Golden Aster, Maryland	210						X X	X	
Golden Aster, Sickle-leaved	210					X X	X X	X	
Goldenrod, White	212						X	X X	
Goldenrod, Downy	212							X X	X
Goldenrod, Slender	212						X X	X X	
Goldenrod, Wand-like	212						X	X X	
Goldenrod, Swamp	212						X	X X	X
Goldenrod, Field	214						X	X X	
Goldenrod, Fragrant	214					X	X X		
Goldenrod, Pine-barren	214						X	X X	
Goldenrod, Slender-leaved	214						X	X X	X
Aster, Wavy-leaved	216							X X	X
Aster, Late Purple	216						X	X X	X
Aster, Silvery	216						X	X X	X
Aster, Showy	216					X	X X	X X	
Aster, Bushy	216						X	X X	X
Aster, New York	216							X X	X X
Aster, Slender	216					X	X X	X	
Aster, Bog	218						X	X X	
Aster, Stiff-leaved	218							X X	X
White-topped Aster, Toothed	218				X	X X	X X		

Flowering & Fruiting Calendar

Plant	Page	Mar. m.	Apr. e.m.	May e.m.	Jun. e.m.	Jul. e.m.	Aug. e.m.	Sept. e.m.	Oct. e.m.
White-topped Aster, Narrow-leaved	218				X	X X	X		
Sunflower, Woodland	218					X	X X	X	
Sunflower, Narrow-leaved	218						X X	X X	
Tickseed, Rose-colored	218					X X	X X	X	
Tickseed-Sunflower, Slender-leaved	220						X	X X	X
Dandelion, Dwarf	220			X X	X X				
Rattlesnake-root	220						X	X X	X
Rattlesnake-root, Pine-barren	220							X X	X
Hawkweed, Vein-leaved	220			X	X X	X			
Hawkweed, Hairy	220					X X	X X	X	

The Animal Kingdom

Briefly, animals are divided into two major groups:
1) Invertebrates - animals that have an external or outside skeleton (exo-skeleton) instead of having an internal skeleton with a backbone.
2) Vertebrates - animals that have an internal skeleton, the main feature of which is an internal backbone.

In the pine barrens, most of the readily observable forms of invertebrates belong to the group (phylum) known as arthropods (see page 292), while all five classes of the vertebrate phylum: fishes, amphibians, reptiles, birds, and mammals, are present (see pages 232 to 292).

FAUNA OF THE NEW JERSEY PINE BARRENS

Although not as significant, ecologically, as the flora, there is much of interest in the fauna of the New Jersey pine barrens. Zoologists who have studied the animals of the pine barrens have noted that for some groups, particularly some small butterflies and skippers (see pages 362 and 366), a couple of moths (the Herodias underwing and the buck moth), and certain amphibians and reptiles (see below), the region is "rich" or at least "unique" in species, whereas for others, notably birds, earthworms, and calcareous animals like snails, it is monotonously "poor." [28]

Again, one of the more important factors is the zoogeographical distribution of many species, with numerous forms having southern affinities while a lesser number have northern relationships. Referring to the insect fauna, Prof. John B. Smith, formerly a professor of entomology at Rutgers University (1899-1912) and state entomologist of New Jersey (1894-1912) stated in 1909 that the "insect species on the whole resemble those of more southern states, and Georgian or even Floridian forms are not uncommonly met with, and yet the only trace of real Boreal (northern) species have been found in the cold, damp swamps of Ocean County." [29]

The fish fauna is sparse and most species are relatively small, "aquarium-size" forms, with few "sportsman" species. Few are of any significant ecological importance. This paucity of aquatic vertebrates

Mammals

seems related to the high acidity of pine barrens "cedar waters."

Both the amphibian and reptilian faunas, however, are more interesting and significant. Chiefly to be noted among the amphibians are the **pine barrens tree frog**, an abundant but endangered, almost endemic species, and the **carpenter frog**, another form distinctive to the New Jersey pine barrens. Reptiles are a substantial and diverse fauna, with the **northern pine snake** as perhaps the most significant species.

Although both the avian and mammalian faunas are reasonably diverse and interesting, there are few known endemic or significant populations of any species in either group within the pine barrens. Of the mammals, the **white-tail deer** is clearly the most common and conspicuous form to be seen within the pines.

MAMMALS

Mammals are warm blooded, vertebrate animals, with skin more or less covered with hair, that bear live young and nurse their young with milk secreted by mammary glands.

The pine barrens has a limited mammalian fauna. McCormick (1970) listed 34 species. Wolgast (1979) describes 34 species. Of these, 27 of the most common, or conspicuous species are described here. In these descriptions, the term "body" includes head and body.

Opposum *Didelphis virginiana*   Body 15 - 20" Tail 9 - 13"
Grayish with white face and long pointed nose. Paper thin black ears often tipped with white. Naked, scaly, rat-like, prehensile tail. Five toes on each foot, inside toe on hind foot opposable and without a claw. A marsupial: female carries young in fur lined pouch under body. Lives in any sheltered site. Active mostly at night. Feeds on nearly anything. Occasional to common in pine barrens.

Masked Shrew *Sorex cinereus*   Body 2 - 2 1/2" Tail 1 1/4 - 2"
Mouse-like with soft fur, small bead-like eyes, and long pointed nose. Ears concealed or nearly so. Five toes on each foot. Grayish brown, pale underneath. Two-colored tail. Lives in tunnels beneath surface of ground. Feeds on insects. Occasional throughout pine barrens, more common in open, lowland sites.

Short-tailed Shrew *Blarina brevicauda*   Body 3 - 4" Tail 3/4 - 1 1/4"
Similar to above but larger, with shorter tail. External ears not apparent. Small eyes barely discernible. Lead colored. Lives in tunnels, feeds on insects, as above. Occasional throughout pine barrens except in dwarf plains areas.

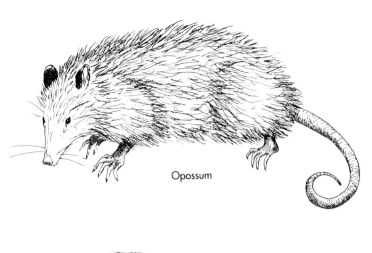
Opossum

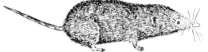
Masked Shrew

Short-tailed Shrew

Eastern Mole

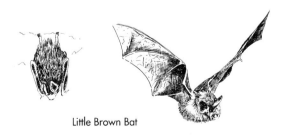
Little Brown Bat

Mammals

Eastern Mole *Scalopus aquaticus* Body 4 1/2 - 6 1/2" Tail 1 - 1 1/2"
Mouse-like with soft, thick fur. Pointed nose naked on tip with upward-opening nostrils. Pinhead size eyes. No discernible external ears. Naked tail. Front feet broad with outward-facing palms. Burrows under surface of ground. Presence apparent from low ridges pushed up from below. Feeds on insects and other small arthropods. Common throughout upland pinelands except in dwarf plains areas. (Illus. on pg. 233)

Little Brown Bat *Myotis lucifugus* Body 1 1/2 - 2" Forearm 1 1/2"
Small flying mammal with forelimb formed into a wing and a membrane of skin extending between forearms, sides of body, and hind legs; also between hind legs and tail. Light to dark brown, underside lighter. Hairs have long, glossy tips. In hollow trees and buildings by day. Flies at night. Feeds on insects. Occasional in pine barrens. (Illus. on pg. 233)

Eastern Cottontail *Sylvilagus floridanus* Body 14 - 17" Ear 2 1/2 -3"
Brown to gray with conspicuous cottony white tail. Most often seen in open feeding on vegetation during morning and evening hours but needs nearby cover of brush and woods. Found in all pine barrens habitats.

Eastern Chipmunk *Tamias striatus* Body 5 - 6" Tail 3 - 4 1/2"
Grayish on head and down middle of back. Two dark stripes, separated by white, along sides of head and body, ending at reddish rump. Runs with tail straight up in air. Dens in burrows in ground, often near human shelters. Feeds on seeds, nuts and other vegetative matter. Common only in disturbed areas around fringes of pine barrens, not in central pinelands.

Woodchuck *Marmota monax* Body 16 - 20" Tail 4- 7"
Reddish brown, underneath paler. Feet darker. Heavy bodied. Short legged. Dens in burrows in ground. Feeds on grasses and other vegetation. Probably rare in pines. Occasional only in disturbed areas and around fringes of pine barrens.

Gray Squirrel *Sciurus carolinensis* Body 8 - 10" Tail 7 3/4 - 10"
Grayish, lightly tinged with reddish brown. Bushy tail bordered with white-tipped hairs. Arboreal. Jumps from tree to tree. Also feeds on ground, stumps, or logs. Occasional to common in oak climax forests in pinelands.

Red Squirrel *Tamiasciurus hudsonicus* Body 7 - 8" Tail 4 - 6"
Similar to above but smaller. Rusty red, underneath white, with black line along sides in summer. Bushy tail reddish. Noisy chatterer. Arboreal. Has habit of sitting on tree limb and "scolding." Feeds on acorns of oaks and seeds of pines. Piles of pine cone cobs and bracts on stumps or ground indicate favorite feeding spots. Common throughout pinelands.

Mammals

Southern Flying Squirrel *Glaucomys volans*
Body 5 - 6"
Tail 3 1/2 - 4 1/2"
Thick, glossy fur. Olive brown, white underneath. Eyes large for size of animal. Tail flattened. A folded layer of loose skin joins front and hind legs allowing animal to glide from one tree down to another. Nocturnal. Sleeps in hollow trees and other shelters during day. Seldom seen. Feeds on both small vegetative and small animal matter. Occasional to common in pine barrens.

Beaver *Castor canadensis*
Body 25 - 30" Tail 9 - 10"
Dense rich brown fur. Naked, scaly, paddle-like tail and webbed feet. Found along streams and small bodies of water. Fells trees and builds dams and houses of sticks and mud and creates impoundments of water. Feeds on inner bark of trees. Common in all suitable pine barrens habitats.

White-footed Mouse *Peromyscus leucopus*
Body 3 1/2 - 4 1/4"
Tail 2 1/4 - 4"
Reddish brown. Underneath white with white feet. Tail brown above, white below. Prefers woods and brushy areas. Nests in sheltered spots. Feeds on berries, seeds, and small arthropods. Common in all pine barrens habitats except cedar swamps.

Red-backed Vole or Mouse *Clethrinonomys gapperi*   Body 3 1/2 - 4 1/2"
Tail 1 - 2"
Mouse-like. Long, soft fur. Reddish on back, grayish on sides and underneath. Tail two colored. A forest ground dweller. Nests under roots or wood. Feeds on berries and seeds. Common in damp areas from cedar swamps to open sphagnum bogs in pinelands.

Meadow Vole or *Microtus pennsylvanicus*
Meadow Mouse
Body 3 1/2 - 5"
Tail 1 1/4 - 2 1/2"
Mouse-like. Brown above. Silvery gray underneath. Tail two colored. Nests above or below ground. Feeds on seeds, vegetation, and insects. Occasional to common in pinelands lowland areas. On occasions may become a pest in cranberry bogs.

Pine Vole or Pine Mouse *Pitymys pinetorum*
Body 2 3/4 - 4"
Tail 2/3 - 1"
Mouse-like. Thick, soft fur, chestnut to auburn brown. Ears small. Tail short. Mounds over its tunnels are conspicuous in upland sites. Feeds on seeds and underground vegetation. Common in open oak and pine woods but absent from dwarf plains.

Muskrat *Ondrata zibethicus*
Body 10 - 14" Tail 8 - 11"
Thick, dark brown fur above. Silvery underneath. Long naked tail flattened

from side to side. Inhabits marshes. Often swims along surface of water. Lives in burrows in embankments and/or builds rounded houses of marsh grasses. Omnivorous. Probably rare in pinelands bogs and marshes. Occasional only around fringes of pinelands.

Southern Bog Lemming *Synaptomys cooperi* Body 3 1/2 - 4 1/2"
Tail 1/2 - 7/8"
Mouse-like. Brownish gray, underneath grayish. Ears short, nearly concealed. Shallow groove near outer edge of each upper gnawing tooth. Feeds on berries and vegetation. Occasional in open lowland bogs with sphagnum, sedges, and heaths.

Meadow Jumping Mouse *Zapus hudsonius* Body 3 - 3 1/4"
Tail 4 - 5 3/4"
Olive yellow to brown, with paler sides. Underneath white to yellowish. Tail long, sparsely haired. Hind feet large. Jumps through grasses like a frog. Feeds on berries, seeds, and insects. Probably throughout pine barrens, more numerous in open bogs and near streams.

Red Fox *Vulpes fulva* Body 22 - 25" Tail 14 16"
Reddish yellow. White underneath. Long, bushy, white tipped tail. Dens in burrows in ground. Omnivorous. Not as common in pines as gray fox.

Gray Fox *Urocyon cinereoargenteus* Body 21 - 29" Tail 11 - 16"
Salt and pepper gray, shading to buffy underneath. Gray face. White throat. Bushy tail with median black stripe, tipped with black. Dens in burrows in ground. Omnivorous. Common throughout pine barrens.

Raccoon *Procyon lotor* Body 18 - 28" Tail 8 - 12"
Grizzled gray with black mask over eyes. Alternating rings of yellowish white and black on tail. Generally lives near water. Dens in hollow trees or logs. Nocturnal. Omnivorous. Common throughout pine barrens.

Long-tailed Weasel *Mustela frenata* Body 8 - 10 1/2" Tail 3 - 6"
Distinguished by its long, slender body, its long, black-tipped tail, and short legs. Upper parts brown; under parts creamy white. May den in old stumps, hollow logs, or abandoned animal burrows. A predator that feeds on insects, small amphibians, reptiles, birds, and mammals. Occasional in pinelands. (Illus. on pg. 241)

Mink *Mustela vison* Body 12 - 17" Tail 5 - 9"
Rich dark brown with white chin patch. Really an aquatic weasel. Always found near water. Feeds on fish and other small animals. Occasional along pine barrens streams. (Illus. on pg. 241)

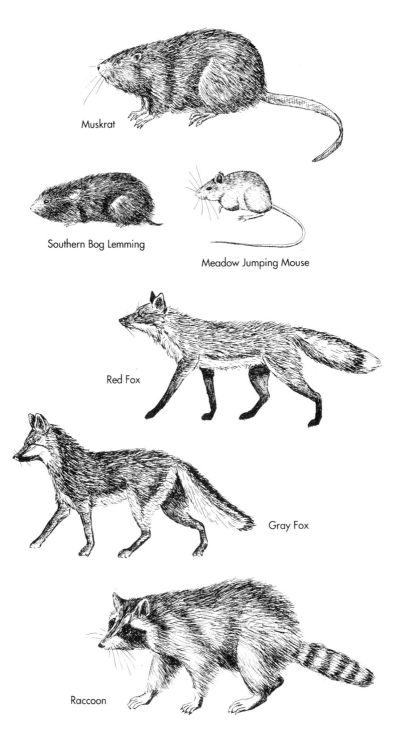

Mammals

Striped Skunk *Mephitis mephitis* Body 13 - 18" Tail 7 - 10"
Black. White stripe on forehead. White patch on back of neck. Two (usually) white stripes down back. Presence more often known by odor than by sight. Dens in burrows or under buildings. Omnivorous. Not common in pines. Occasional only in disturbed areas and around fringes of barrens.

River Otter *Lutra canadensis* Body 26 - 30" Tail 12 - 17"
Rich brown above. Silvery underneath. Broad nose. Small ears. Tail thick at base. Feet webbed. Found only in or near streams and lakes. Swims in loping manner, with frequent dives. Occasional in suitable pines habitats.

White-tail Deer *Odocoileus virginianus* Height 3 - 3 1/2'
Reddish in summer, grayish in winter, under surface of tail white. Raises and wags tail back and forth like a white flag when running away. Males have antlers which are grown and shed annually. Common throughout pine barrens.

To complete the mammalian fauna of the pine barrens according to McCormick, 1970, and Wolgast, 1979, the following eleven less common and/or less conspicuous species are merely listed without descriptions.

Least Shrew *Cryptotis parva*
Star-nosed Mole *Condybura cristata*
Big Brown Bat *Eptesicus fuscus*
Eastern Pipestrelle *Pipestrellus subflavus*
Keen's Bat *Myotis keenii*
Red Bat *Lasiurus borealis*
Hoary Bat *Lasiurus cinereus*
Silver-haired Bat *Lasionycteris noctivagens*
Rice Rat *Oryzomys palustris*
Norway Rat *Rattus nervegicus*
House Mouse *Mus musculus*

The following are recommended for further identification and reference.

Burt, W. H., and R. P. Grossenheider. 1952. A Field Guide to the Mammals, giving field marks of all Species found north of the Mexican Boundary. Houghton Mifflin Co., Boston, MA.

Connor, P. F. 1953. Notes on the mammals of a New Jersey pine barrens area. Jour. Mammal. 34: 227-235.

Palmer, E. L. 1957. Fieldbook of Mammals. E. P. Dutton & Co.

Van Gelder, R. G. 1984. The Mammals of the State of New Jersey: A Preliminary Annotated List. New Jersey Audubon Soc., Bernardsville, N.J.

Wolgast, L. J. 1979. Mammals of the New Jersey Pine Barrens *in* Forman, R. T. T., Jr., Pine Barrens Ecosystem and Landscape. Academic Press, N.Y.

Mink

Long-tailed Weasel

Striped Skunk

River Otter

White-tail Deer
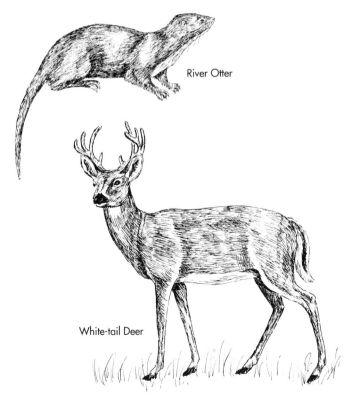

BIRDS

Birds are warm blooded vertebrate animals that have feathers and their forward or upper limbs adapted to enable them (at least all in our area) to fly. Feathers provide body insulation, and wing and tail feathers are vital in flight. Birds feed on both vegetative and animal food.

The avifauna of the pine barrens is not rich. McCormick, 1970, stated that "the Pine Barrens habitat apparently lacks the variety required to support a diverse population of birds." Leck, 1979, states that "the avifauna of the New Jersey Pine Barrens is remarkably simple." McCormick listed 144 birds as occurring in the pine barrens while Leck mentions 73 different species in his chapter on the birds of the pine barrens. Of these, the following 90 different species of the most common and conspicuous birds are briefly described here.

Great Blue Heron *Ardea herodias* 48"
Head whitish, with two black plumes in breeding plumage. Neck brownish-gray. Shoulder patches black. Rest of plumage grayish-blue. Bill yellowish. Our largest heron. Stands motionless or walks slowly in shallow waters and marshes fishing for food. Mainly summer resident but some may remain year-round. A threatened species (CMP).

Green-backed Heron *Butorides striatus* 18"
Top of head greenish-black. Neck chestnut-red with white in front, but appears almost black at a distance. Upperparts greenish. Rest of plumage bluish-green. Legs yellow or orange (breeding). May raise shaggy crest when alarmed. Stands motionless along shore or on low branch over water poised for a quick strike for food. Summer resident.

Tundra Swan *Cygnus columbianus* 52"
All white with black bill. On flooded cranberry bogs in winter feeding on roots of redroot, *Lachnanthes tinctoria*. Winter visitor.

Canada Goose *Branta canadensis* 24"- 44"
Black head and neck with white "chin-strap." Rest of plumage brownish-gray. Flocks fly in noisy, "honking" V-formations. On lakes, ponds and bogs. Formerly both a spring-fall migrant and winter visitor but increasingly becoming year-round resident.

Mallard *Anas platyrhynchos* 23"
Male with glossy green head, white neck band, and chestnut breast. Female mottled brown. In any small body of water. Feeds on vegetation. Mainly summer but often year-round resident.

Great Blue Heron

Green-backed Heron

Canada Goose

Tundra Swan

Mallard

American Black Duck

Wood Duck

Birds

American Black Duck *Anas rubripes* 23"
Dark blackish-brown, with paler brown head and upper neck. In any body of water. Feeds on vegetation. Mainly winter but may be year-round resident. (Illus. on pg. 243)

Wood Duck *Aix sponsa* 18"
Male boldly patterned in maroon, green, purple, and white. Only surface feeding duck with a crest. Female less colorful but with dew-drop shaped white eye patch. Woodland ponds, swamps, streams. Mainly summer but a few year-round. (Illus. on pg. 243)

Turkey Vulture *Cathartes aura* 29" Wingspread 6'
All black except red head in adults. Flies (soars) with two-toned, blackish wings held at an upward angle (dihedral). Feeds on carrion and refuse. Year-round resident.

Sharp-shinned Hawk *Accipter striatus* 10" - 14"
Slim-bodied with slim tail and short, rounded wings. Smallest of accipiter hawks. Adult has dark back and rusty barred breast. Spring-fall migrant, uncommon winter resident, and possible rare breeder in pine barrens.[30] Preys on small birds. Occasional at feeders.

Red-tailed Hawk *Buteo jamaicensis* 19" - 25"
Wings broad, fairly rounded. Adult birds rufous on upper side of tail. Open country and nearby woodlands. Preys on rodents. Year-round resident.

Broad-winged Hawk *Buteo platypterus* 15"- 17"
Broad wings. Broad tail. Tail has two or three broad, white bands alternating with equally broad black bands. Prefers tall oak woodlands. Feeds on small animal life. Summer resident.

Bald Eagle *Haliacetus leucocephalus* 30" - 40"
Adult has white head, white tail, and large, yellow bill. Occasional along lower Mullica-Batsto and Wading-Oswego river ecosystems. One nest site (1987) in Bear Swamp in Cumberland County. An endangered (CMP) year-round resident.

American Kestrel *Falco sparverius* 9" - 11"
Two vertical black stripes against white background on sides of face. Upper back and tail rufous. Wings of male blue-gray; of female rusty. Smallest and most common falcon. Often watches for rodents and insects from poles and wires. Also "hovers" for prey. Year-round resident.

Birds

Ruffed Grouse *Bonasa umbellus*   16" - 17"
Grayish- or reddish-brown. Chicken-like, ground dwelling, woodland bird. Tail fan-shaped with broad, black band near tip. Male with slight crest. Male tries to attract female by beating wings to make a hollow drumming noise. Feeds on berries, vegetation, and insects. Year-round resident.

Northern Bobwhite *Colinus virginianus*   9" - 10"
Reddish-brown. Chicken-like, ground dwelling bird. Conspicuous white throat and white eyebrow (male), buffy in female. Short tail. Call a clear whistled "bob-white." Brushy areas, hedgerows, and edges of woods. Feeds on seeds, vegetation, and insects. Year-round resident.

Spotted Sandpiper *Actitis macularia*   7 1/2"
Olive brown above, white underneath, with round breast spots in summer. Clear underneath in winter with white wedge line near shoulder. Teeters up and down. Along shores of ponds, small lakes, and streams. Summer resident.

Mourning Dove *Zanaida macroura*   12"
Buffy-gray with bluish cast, pinkish wash underneath. Tail long, tapering, pointed, with white edges on outer feathers. Call a low, moanful "coo-ah, coo." Open woods, farms, backyards, roadsides. Year-round resident.

Yellow-billed Cuckoo *Coccyzus americanus*   11" - 13"
Brownish above, white below. Lower mandible of curved bill yellow. Large, paired, white spots under long tail. Rufous on wings in flight. Open woodlands. Summer resident.

Eastern Screech-owl *Otus asio*   8" - 10"
Two color phases: red or rufous, and gray or brownish-gray. Breast streaked. Eyes yellow. Our only small owl with ear tufts. Call a low, mournful, tremulous whinny, on same pitch or descending. Open woodlands. Feeds on rodents, insects and other small animals. Year-round resident.

Great Horned Owl *Bubo virginianus*   18" - 25"
Dark brown, heavily barred and streaked with black. White throat. Our only large owl with ear tufts. Call 3 to 5 deep, resonant hoots. Woodlands. Feed on small mammals and birds. Year-round resident.

Whip-poor-will *Caprimulgus vociferus*   9" - 10"
Mottled brown. Male has white band below black throat, and shows large, white tail patches in flight. Long, rounded tail. More often heard than seen. Call at dusk and during night a loud "whip-poor-will" with accent on last syllable. Open woodlands and field edges. Summer resident.
(Illus. on pg. 249)

Birds

Common Nighthawk *Chordeiles minor* 9" - 9 1/2"
Mottled gray- to dark brown. Undersides barred. Male has broad, white throat band. Wings long, pointed, with conspicuous white band across wing seen in flight. Tail slightly forked. Erratic flight at dusk and night, occasionally by day. Call a nasal "peent." Summer resident.

Chimney Swift *Chaetura pelagica* 5" - 5 1/2"
Swallow-like, with long, curved, pointed wings. Flies like a small "flying cigar." Most often seen flying over or near habitations. Roosts and nests in chimneys, barns, and hollow trees. Summer resident.

Ruby-throated Hummingbird *Archilochus colubris* 3" - 3 3/4"
Male with ruby throat (blackish in poor light) and iridescent green back. Underneath whitish. Tail forked. Female has white throat and broad tail with white spots. In our area, can be confused only with some hummingbird moths (Sphingidae). Along edges of swamps feeding on nectar of azaleas[31] and sweet pepperbush. Spring-fall migrant and summer resident.

Belted Kingfisher *Ceryle alcyon* 13"
Upper parts, including head, crest, and breast band, grayish-blue. Underneath and neck collar white. Female with chestnut band on lower chest and flanks. Perches on dead branches and other lookout spots and dives down to catch fish. Call a loud, dry rattle, usually in flight. Seen along shore lines of any decent sized body of water. Feeds on fish and other small animal life. Mainly a summer resident. A few may remain year-round.

Northern Flicker *Colaptes auratus* 11" - 13"
Brown, barred back. Black crescent on breast. Small red crescent on back of head. In undulating flight has conspicuous white rump patch and golden yellow under wings and tail. Call a loud, repeated "wik, wik," and "wicker, wick-er." Open woods. Feeds on fruits, berries and insects. Mostly summer resident, a few may overwinter.

Red-bellied Woodpecker *Melanerpes carolinus* 9" - 9 1/2"
Black and white barred ("zebra") back. White upper tail coverts (rump patch). Crown and back of head (nape) red in males. Only nape red in females. Faint reddish tinge on belly. Occasional but becoming more common in local woodlands.

Red-headed Woodpecker *Melanerpes erythrocephalus* 8 1/2" - 9 1/2"
Entire head red, both male and female. Back solid black with large white patches on wings, conspicuous both at rest and in flight. Woodlands. Few pairs have been nesting in recent years in Lebanon State Forest. Occasional to rare elsewhere. Summer resident.

Birds

Hairy Woodpecker *Picoides villosus* 9" - 10"
Black and white above and on wings. Middle of back white. Male has red patch on back of head. Large, heavy bill. Call a short "peenk." Open woods. Feeds on wood borers and other insects. Year-round resident.

Downy Woodpecker *Picoides pubescens* 6 1/2" - 7"
Almost identical to above except for smaller size and small bill. Call a light "pic" and a rapid, high whinny descending in pitch. Year-round resident.

Eastern Kingbird *Tyrannus tyrannus* 8" - 9"
Dark slate-gray back, black head. Conspicuous white band across tip of tail. A large flycatcher that perches and flies out to catch flying insects. Somewhat pugnacious, occasionally attacking crows and hawks. Call a series of harsh twitterings. Edges of woods, often near water. Summer resident.

Great Crested Flycatcher *Miarchus crinitus* 8" - 9"
Dark olive-brown above with brown crest. Pale gray throat and breast, yellow belly. Rufous on tail and wings. Large bill. A large, woodland flycatcher. Flies out to feed on flying insects. Call a loud, raucous "wheap, wheap." Summer resident.

Eastern Phoebe *Sayernis phoebe* 6 1/2" - 7"
Olive-brown-gray above. Head, wings, and tail darker gray. Underparts generally whitish. Tail with slight notch. Song a raspy "phoe-be." Woodlands, particularly near streams. Often nests under bridges or eaves of buildings. A flycatcher, so feeds on flying insects. Summer resident.

Eastern Wood-Pewee *Contopus virens* 6" - 6 1/2"
Dark olive-gray above with two narrow, whitish wing bars. Throat dull white. Breast and under sides grayish. Calls its own name, a plaintive "pee-a-wee." A woodland flycatcher. Feeds on flying insects. Summer resident.

Tree Swallow *Tachycineta bicolor* 5" - 6"
Steely blue-green above. Clear white below. Female somewhat duller. Habitat open country and woods near water, especially where there may be dead trees for nest sites. Often skims over water surface catching flying insects. Spring-fall migrant and summer resident. Often gathers in great numbers during fall migration.

Barn Swallow *Hirundo rustica* 6" - 7 1/2"
Blue-black above. Cinnamon throat and underparts. Long, deeply forked tail with white spots. Voice a soft twittering. Semi-open fields and marshes. Often builds its nest inside barns and sheds. Summer resident.

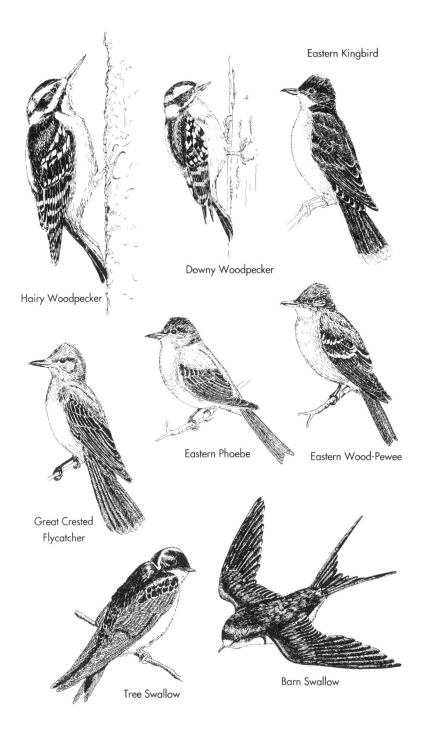

Birds

Purple Martin *Progne subis* 7" - 8"
Male glossy purplish-blue-black, may look black at distance. Female lighter underneath. Forked tail. Our largest swallow and only one all dark underneath (male). Voice a succession of soft gutterals and gurgling. Semi-open areas, often near water. In flight, feeds on flying insects. Summer resident.

Blue Jay *Cyanocitta cristata* 10" - 12"
Bright blue, with distinct crest. White patches on wings and outer tail feathers. Blue-black necklace on gray throat. Gray underneath. Call a noisy shriek, "jay-jay," or scream. Woods and field edges. Omnivorous. Year-round resident.

American Crow *Corvus brachyrhynches* 17" - 20"
All black, including large, heavy, black bill and black legs. Rounded, fan-shaped tail. Call a loud "caw." Woods, fields. Omniverous. Year-round resident.

Fish Crow *Corvus ossifragus* 16" - 20"
Almost identical to American crow but slightly smaller. Identify mainly by voice. Fish crow call a short, nasal "car," "ca," or "ca-ha," not the full "caw" of the American. Inland throughout pinelands as well as along river valleys and near coast. Year-round resident.

Carolina Chickadee *Parus carolinensis* 4 1/4" - 4 3/4"
Black cap and bib. White cheeks. Light buffy sides. Rest of plumage gray above, whitish below. Calls "chick-a-dee-dee", "fee-bee, fee-bay." Open deciduous and pine woods and edges. Feeds on insects, insect eggs, larvae, seeds and berries. Year-round resident.

Tufted Titmouse *Parus bicolor* 5 1/2" - 6 1/2"
Above mostly mouse-gray with a gray, tufted crest. Underneath whitish with rusty flanks. Call a loud, clear, whistled "peter, peter." Often in deciduous woods with flocks of chickadees. Same general habitats and food habits as chickadees. Year-round resident.

White-breasted Nuthatch *Sitta carolinensis* 4 1/4" - 4 3/4"
Blue-gray above with black cap and white face and breast. Chestnut under tail. Female paler. Call a nasal "yank, yank." Song a rapid series of nasal notes all on one pitch. Open deciduous and pine woods and edges. Moves down and around trees head first. Often travels with mixed flocks of chickadees and titmice. Year-round resident.

Red-breasted Nuthatch *Sitta canadensis* 4"- 4 1/2"
Similar to above but smaller, with white eyebrow stripe and black line

Birds

through eye. Underneath rusty. Call higher pitched than white-breasted. A winter resident.

Brown Creeper *Certhia americana* 4 3/4" - 5 1/4"
Streaked brown above with white eyestripe. White underneath. Slender, decurved bill. Creeps up tree trunks using stiff tail as prop. Feeds on insects, insect eggs and larvae. Winter resident and rare local summer breeder.

House Wren *Troglodytes aedon* 4 1/2" - 5"
Small, grayish to reddish brown, with finely barred wings and tail. Energetic, perky bird with cocked tail of most wrens. Exuberant, bubbling, gurgling song. Open woods, brushlands, and back yards. Summer resident.

Carolina Wren *Thryothorus ludovicianus* 4 3/4" - 5 3/4"
Reddish brown above. Prominent white eye stripe. Buffy underneath. Bill rather long and curved. Cocked tail. Call a clear, loud, 2 or 3 syllabled "wheedle, wheedle." Tangles, brushy undergrowth, and brush piles. Year-round resident.

Northern Mockingbird *Mimus polyglottos* 9" - 11"
Light gray above, darker on wings and tail. Lighter gray to whitish below. White wing patches, conspicuous in flight. White outer feathers on long tail. Song varied and repetitious. Thickets and woodland edges. Feeds on insects, wild fruits, and berries. Year-round resident.

Gray Catbird *Dumetella carolinensis* 7 3/4" - 9"
Plain, dark, slate-gray with black cap, blackish tail, and chestnut under base of tail. Call a distinctive, cat-like mewing. May be most common bird in dense, moist thickets, hedges, and underbrush near edges of ponds, swamps. Food as for mockingbird. Summer resident.

Brown Thrasher *Toxostoma rufum* 11" - 12"
Reddish-brown above, lighter but heavily streaked below. Long tail. Curved bill. Yellow eye. Song varied, each phrase repeated once. Thickets, shrubs, brush, and woodland edges. Food as for mockingbird. Summer resident.

American Robin *Turdus migratorius* 9" - 11"
Dark slate-gray back. Brick-red breast. Dark head. Yellow bill. Female paler. Song a series of short, clear phrases like "cheer-up, cheerie." Open woodlands, fields, parks. Feeds on worms, insects, grubs, berries, and seeds. Summer resident, although a few may remain overwinter.

Wood Thrush *Hylocichla mustelina* 7" - 8"
Dark olive- to reddish-brown above, bright reddish-brown on head. Promi-

nent white eye ring. Underneath creamy-whitish with bold, round, dark spots on throat, breast, and sides. Song a series of liquid, bell-like phrases usually ending in a trill. Low, mixed woodlands and edges of cedar and other swamps. Food as for robin. Summer resident.

Eastern Bluebird *Sialia sialis* 6" - 7"
Deep, bright blue above with rusty red breast and white belly. Female softer colored. Song a soft, mellow warble. Open woodlands and field edges, nesting in tree holes, fence posts and man made boxes. Food mainly insects. Spring-fall migrant and summer resident.

Golden-crowned Kinglet *Regulus satrapa* 3 1/2" - 4"
Olive greenish-gray above with white stripe over eye and two distinct white wing bars. Black bordered bright orange (male) or yellow (female) crown. Underneath whitish. Call a high, thin "tsee, tsee." Coniferous woodlands. Often moves in mixed flocks with chickadees and nuthatches. Feeds on insects and their eggs. A winter visitor.

Ruby-crowned Kinglet *Regulus calendula* 4" - 4 1/4"
Olive greenish gray above with conspicuous broken white eye-ring and two distinct white wing bars. Male with small, bright red crown patch, usually concealed and difficult to see. Underneath whitish. Coniferous and other woodlands. A spring-fall migrant and winter visitor.

Cedar Waxwing *Bombycilla cedrorum* 7" - 7 1/4"
Sleek, crested, brownish, with yellow band at tip of tail. Gregarious, often seen in flocks in open woodlands. Feeds on berries of cedar, mulberry, and others. Occasional year-round resident.

European Starling *Sturnus vulgaris* 7" - 8"
A short tailed, black bird. In spring, iridescent black with a yellow bill. In fall and winter, heavily speckled with dark bill. Call a variety of squeaks, chirps, whistles, and mimics. Field edges and around habitations. Omnivorous, but feeds especially on insects, grubs and worms. Year-round resident.

White-eyed Vireo *Vireo griseus* 5" - 6"
Grayish-greenish-olive above with two white wing bars and a white breast. Pale yellowish sides. Distinct yellow "spectacles" around whitish eyes. Song a loud, clear, repeated five to seven note phrase ending with a sharp "chick." Dense moist thickets and woods edges. Feeds on insects and berries. Summer resident.

Red-eyed Vireo *Vireo olivaceus* 5 1/2" - 6"
Olive back with darker wings and tail. Crown bluish gray. Distinct, black-

Birds

bordered white stripe over eye. Underneath white. Song a continuing repetition of short phrases, somewhat robin-like, sometimes sung as often as 40 times per minute. Open woodlands. Feeds on insects and berries. Summer resident.

Black-and-white Warbler *Mniotilta varia* 4 1/2" - 5 1/2"
Patterned in lengthwise, black stripes on white background, with bold black and white stripes on head. Female, and male in winter, paler and more white underneath. Creeps up, down, and around tree trunks and branches like a small nuthatch. Song a thin, wiry "zee, zee, zee." Open deciduous woodlands. Main food, as with most warblers, insects. Spring-fall migrant and summer resident.

Prothonotary Warbler *Pronotaria citrea* 5 1/2"
Head and underparts of male golden yellow. Wings blue-gray, no bars. Female duller. Nest in tree cavities and underbrush in wooded swamps as along the Batsto River. Summer resident.

Northern Parula *Parula americana* 3 3/4" - 4 1/2"
Gray-blue above, but upper back a suffused yellowish-green. Two bold, white wing bars. Broken white eye ring. Yellow throat and breast. Male with reddish and dark bluish breast band. Song a rising trill, ending in an abrupt "zip." Damp woodlands and edges of cedar and other swamps. Uses old man's beard lichen, *Usnea barbata* in the pine barrens[32] as nesting material. Spring-fall migrant and summer resident.

Yellow Warbler *Dendroica petechia* 4" - 5"
All yellow, darker above, lighter below. Yellow tail corners. Male with rusty reddish streaks on breast and sides. Song a clear, variable, "sweet, sweet." Open, damp woodlands, shrubbery, brushy stream edges. Summer resident.

Yellow-rumped Warbler *Dendroica coronata* 4 3/4" - 5 3/4"
Blue-gray (male) or brownish (female) above. Conspicuous bright, yellow rump patch. Bright yellow on crown and sides of breast. Inverted U-shaped black band on white breast. Call a loud "check." Song a somewhat junco-like trill, rising or falling at end. Mixed woods and thickets. Spring-fall migrant and winter resident.

Black-throated Green Warbler *Dendroica virens* 4 1/2" - 5"
Yellow face with solid black throat (male) and olive-green crown. Female with much less black on throat. Apt to stay high in trees in coniferous or mixed forests. Occasional summer resident.

Blackpoll Warbler *Dendroica striata* 4 1/2" - 5 1/2"
Male in spring black striped olive-gray above, with solid cap, white cheeks,

and black stripes on under sides. Females and fall males pale greenish-yellow-gray above. White wing bars in all plumages. Song a one-pitched "zee, zee, zee," becoming louder in middle and fading away at end. In coniferous and other woods. A spring-fall migrant, more likely to be observed during fall migration.

Pine Warbler *Dendroica pinus* 4 3/4" - 5 1/2"
Olive-green above, wings and tail darker gray. Unstreaked back. White wing bars and white tail corners. Throat and breast light yellow, darker streaks on sides. Belly white. Song a somewhat chipping sparrow-like trill but slower and softer. In pine woodlands. Spring-fall migrant and summer resident.

Prairie Warbler *Dendroica discolor* 4 1/2" - 5"
Olive-green above, with faint chestnut streaking on upper back. Two black face stripes, one through eye, other below eye. Bright yellow below with bold, black striping on sides. Wags tail. Song a thin "zee, zee, zee" rising in a chromatic scale. Open woodlands, low pines and woods margins. Summer resident.

Ovenbird *Seiurus aurocapillus* 5" - 6"
Olive-brown above. Black bordered orange streak on crown, seen only at close range. White eye ring. Underneath white with dark brown streaks. Bird walks along ground in underbrush. Seldom seen. More often heard. Song a loud, repeated "teach-er, teach-er, teacher-er," increasing in volume. In brushy, deciduous woodlands. Summer resident.

Common Yellowthroat *Geothlypis trichas* 4 1/2" - 5 1/2"
Upperparts dark olive to olive-brown. Bright yellow throat. White belly. Male has black face mask bordered above with gray. Song a rolling "whitch-i-ty, whitch-i-ty." In wet thickets along edges of swamps and marshes. Summer resident.

American Redstart *Setophaga ruticilla* 4 1/2" - 5 1/4"
Black (male) or olive-brown (female) with bright orange (male) or yellow (female) patches on sides, wings, and tail. Often spreads tail, fan-like. Song a thin series of single or double notes. Open, second growth woodlands. Spring-fall migrant and summer resident.

House Sparrow *Passer domesticus* 5 1/4" - 6 1/4"
Streaked grayish- to reddish-brown above. Male with gray cap, chestnut behind head, white cheeks, and black bib. Female with buffy eye line and unstreaked, pale, grayish breast. Present everywhere in populated areas, especially where there are horses. Song a noisy "chirp, chirp." Feeds on seeds and grain. Year-round resident.

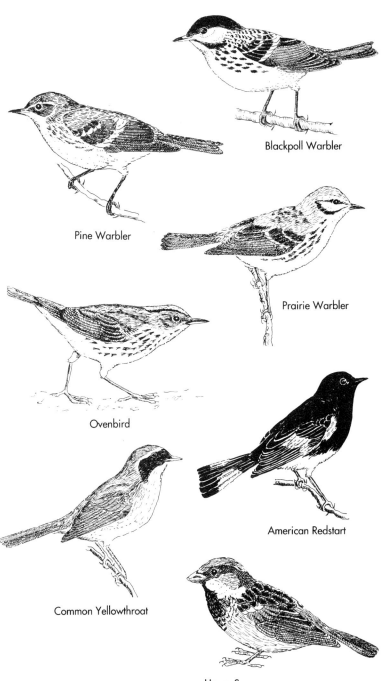

Birds

Red-winged Blackbird *Agelaius phoeniceus*   7 1/2" - 9 1/2"
Male glossy black, with yellow bordered red shoulder patches. Female like a large, dark sparrow, heavily streaked below, with a light stripe over eye. Song a liquid "konk-a-ree." Gregarious, with numbers inhabiting marshes and nearby areas. Feeds on insects, seeds, and grain. Mainly a spring-fall migrant and summer resident but a few may remain overwinter.

Northern Oriole *Icterus galbula*   7" - 8 1/2"
Bright orange and black, with solid black head (male). Female in tones of yellow without the black head. Open deciduous woodlands as around Pakim Pond in Lebanon State Forest. Song a series of loud, clear, musical calls. Summer resident.

Common Grackle *Quiscalus quiscula*   10" - 13"
Male black with iridescent hues of purple or bronze on head, neck, and upper back. Female duller and smaller. Both with yellow eye and keel-shaped tail. Voice a medley of harsh squeakings. Open woods, fields, parks. Feeds on insects, small animal life, and grain. Year-round resident.

Brown-headed Cowbird *Molothrus ater*   6 1/2" - 7 1/2"
Male black with some greenish reflections. Head and neck brown. Female all grayish, darker above. Call a characteristic squeak, like a rusty gate hinge. Flocks with starlings and blackbirds in fields and open woods. Feeds on insects, spiders, seeds, grain, and berries. Parasitic: lays its eggs in nests of other birds. Spring-fall migrant and summer resident but some may overwinter.

Scarlet Tanager *Piranga olivacea*   7"
Sharp contrast of brilliant, flame red head and body with jet black wings and tail in male. Female dull greenish above, yellowish below with brownish wings. Open deciduous, especially oak, woods and forests. Summer resident.

Northern Cardinal *Cardinalis cardinalis*   7 1/2" - 9"
Male all red with a red crest, a heavy reddish bill, and a black face and throat. Female buff-brown with traces of soft red on crest, wings, and tail. Song a loud, clear, whistled "cheer, cheer." Thickets and edges of woods. Feeds on seeds, berries and buds. Year-round resident.

Rose-breasted Grosbeak *Pheucticus ludovivianus*   7" - 8 1/2"
Black and white with large triangle of rose-red on breast (male) and a heavy, pale yellow bill. Female streaked like a large sparrow with broad, white eyebrow stripe and heavy bill. Spring-fall migrant in open, deciduous woodlands as at Pakim Pond, Lebanon State Forest.

Birds

Evening Grosbeak   *Coccothraustes vespertinus*   7 1/4" - 8"
Male golden-yellow and bronze with yellow forehead and eyebrow, and with black wings and tail. Female grayish-tan in place of yellow-bronze. Large, white wing patches conspicuous in flight. Heavy bill. Short, notched tail. Call a loud "chirp." Open, mixed woods. Winter visitor at feeders.

House Finch   *Carpodacus mexicanus*   5" - 6"
Male with bright, brick-red head, throat, breast, and conspicuous rump patch, and with dark streaks on sides. Female streaked, sparrow-like, with pale eyestripe. Heavy bill. Song a continuous, melodious warble. Relatively recent (early 1940s) introduction from west and southwest. Has "taken-over" at house feeders. Mainly a winter resident now but may be becoming year-round.

Pine Siskin   *Carduelis pinus*   4 1/4" - 5"
Heavily and darkly streaked, with a flash of yellow on wings and at base of tail. Two thin wing bars. Tail notched. Thin bill. Call a buzzy, rising "shreee." Gregarious, often in flocks with goldfinches. Seed eater. Winter visitor at feeders.

American Goldfinch   *Carduelis tristis*   4 1/4" - 5"
Summer male bright golden yellow with black forehead, wings, and tail. Female: yellow replaced by brownish olive-yellow, no black forehead. Winter male like female. Song somewhat canary-like medley of trills. Flies in undulating, or "roller-coaster," flight. Feeds on seeds, particularly those of thistles and dandelions. Weedy fields, second growth woodlands, roadsides, and feeders. The state bird of New Jersey. Year-round resident.

Rufous-sided Towhee   *Pipilo erythrophthalmus*   7" - 8 1/2"
Male with black upperparts, black hood, rufous sides, and white underparts. Female: black replaced by a softer brown. White wing patches and white tips of outer tail feathers conspicuous in flight. Long, rounded tail. Call a loud "che-wink." Song a "drink-your-teee" phrase. Scratches among dead leaves on ground in open mixed woods, undergrowth and brushy edges. Perhaps the most common and most characteristic bird of our New Jersey upland pine barrens. A summer resident but a few may remain over winter.

Dark-eyed Junco   *Junco hyemalis*   5 1/4" - 6 1/2"
Slate-gray all over except for sharply set-off white belly and white, outer tail feathers. Pinkish bill. Female somewhat lighter than male. Song a light, musical trill. A ground and low brush feeder on seeds. In undergrowth and brushy edges of woods. Winter resident.

Chipping Sparrow   *Spizella passerina*   4 3/4" - 5 1/2"
Streaked brown above. Bright red crown. Prominent, white, eyebrow stripe with thin, black line through eye. Two thin, white wing bars. Breast

Birds

unstreaked grayish white. Song a simple, one pitched trill. Open, mixed and coniferous woods, woodland edges, and roadsides. Summer resident.

Field Sparrow *Spizella pusilla* 5" - 5 3/4"
Streaked brown above. Brownish-red crown. Narrow, white eye ring. Pink bill. Clear, unstreaked, buffy breast. Song a series of clear notes gradually speeding up into a trill. Open, brushy woodlands and fields. Summer resident.

White-throated Sparrow *Zonotrichia albicollis* 5 3/4" - 7"
Streaked rusty-brown. Black and white striped crown. Small yellow spot between eye and base of bill. Conspicuous white throat. Breast gray. Belly white. Song a series of clear, plaintive whistles, usually one or two clear notes followed by three drawn-out, quivering notes on a different pitch. Forages like a chicken on ground in woodland undergrowth and thickets. Winter resident.

Swamp Sparrow *Melospiza georgiana* 5" - 5 3/4"
Stout, rusty, with grayish breast, white throat, and reddish cap. Song a loose trill somewhat similar to chipping sparrow. In grasses and tussocks of marshes and swamps. Summer resident.

Song Sparrow *Melospiza melodia* 5" - 6 1/2"
Streaked brownish above with faint, grayish eyebrow stripe. Underneath whitish with bold streaking on sides, streaks usually converging into central breast spot. Tail long, rounded. Pumps tail in flight. Song a series of three or four short, clear notes, followed by a buzz and a trill. Brushy areas and streamside thickets. Year-round resident.

The following listings are offered as an aid to those who would like to know which birds are most likely to be seen in different seasons of the year. This arrangement of the above 90 species is only a generalization and must be interpreted very flexibly, for occasionals may occur at almost any season of the year.

Each year, between mid-December and early January, a group of "birders," many of whom are members of the Burlington County Natural Sciences Club, conduct the "Pinelands" Bird Count for the National Audubon Society. The count area is a seven and one half mile radius circle centered in Shamong Twp. and includes all or major parts of Medford, Shamong, Tabernacle, and Washington Twps. in south central Burlington County, and portions of Hammonton Twp. in northwestern Atlantic Co., and Waterford Twp., Camden Co. Since the "Pinelands" bird counts began in 1968,

the number of different species seen has ranged from 52 (1968) to 85 (1987) and the number of individual birds counted from 2,275 (1969) to 12,269 (1981). In the following listings, birds seen rather regularly, in numbers of six or more, on the annual "Pinelands" Christmas Bird Counts, are indicated by an asterisk (*).

Year-round Residents

*Great Blue Heron (?) (S)
*Mute Swan[1]
*Canada Goose (?) (W)
*Mallard (?) (S)
*American Black Duck (?) (W)
Wood Duck (?) (W)
*Turkey Vulture
*Sharp-shinned Hawk (?) (M-T)
*Red-tailed Hawk (?) (W)
Bald Eagle
*Northern Harrier[1]
*American Kestrel
*Ruffed Grouse
*Northern Bobwhite
*Killdeer[1] (?) (S)
*Herring Gull[1]
*Ring-billed Gull[1] (?) (W)
*Rock Dove[1]
*Mourning Dove
*Eastern Screech-owl
*Great Horned Owl
*Belted Kingfisher (?) (S)
*Northern Flicker (?) (S)
*Red-bellied Woodpecker (?) (S)
*Hairy Woodpecker
*Downy Woodpecker

*Blue Jay
*American Crow
*Fish Crow
*Carolina Chickadee
*Tufted Titmouse
*White-breasted Nuthatch
*Carolina Wren
*Northern Mockingbird
*American Robin (?) (S)
*Eastern Bluebird
*Cedar Waxwing
*European Starling
*House Sparrow
*Red-winged Blackbird (?) (S)
*Common Grackle
*Brown-headed Cowbird (?) (S)
*Northern Cardinal
*House Finch (?) (W)
*American Goldfinch
*Rufous-sided Towhee (?) (S)
*Chipping Sparrow (?) (S)
*Field Sparrow (?) (S)
*Fox Sparrow (?) (M-T)
*Swamp Sparrow (?) (S)
*Song Sparrow (?) (S)

(?) Questionable as a year-round species.
(S) Primarily a summer species
(M-T) Primarily a spring-fall migrant
(W) Primarily a winter species

Birds

Summer Residents

(small numbers of some species may remain all year)

*Great Blue Heron
Green-backed Heron
*Mallard
Wood Duck
Broad-winged Hawk
*Killdeer[1]
Woodcock
Spotted Sandpiper
Yellow-billed Cuckoo
Black-billed Cuckoo
Whip-poor-will
Common Nighthawk
Chimney Swift
Ruby-throated Hummingbird
*Belted Kingfisher
*Northern Flicker
*Red-bellied Woodpecker
Eastern Kingbird
Great Crested Flycatcher
Eastern Phoebe
Eastern Wood-pewee
Tree Swallow
Barn Swallow
Rough-winged Swallow
Purple Martin
House Wren
Gray Catbird
Brown Thrasher
*American Robin
Wood Thrush
*Eastern Bluebird
White-eyed Vireo
Red-eyed Vireo
Black-and-white Warbler
Prothonotary Warbler
Northern Parula
Yellow Warbler
Black-throated Green Warbler
Pine Warbler
Prairie Warbler
Ovenbird
Common Yellowthroat
American Redstart
*Red-winged Blackbird
Northern Oriole
*Brown-headed Cowbird
Scarlet Tanager
Rose-breasted Grosbeak
Rufous-sided Towhee
*Chipping Sparrow
*Field Sparrow
*Swamp Sparrow
*Song Sparrow

Winter Residents

(small numbers of some species may remain all year)

*Tundra Swan
*Canada Goose
*American Black Duck
*Red-tailed Hawk
*Ring-billed Gull[1]
*Red-breasted Nuthatch
*Brown Creeper
*Golden-crowned Kinglet
*Evening Grosbeak
*Purple Finch[1]
*House Finch
*Pine Siskin
*Dark-eyed Junco
*Tree Sparrow[1]
*White-throated Sparrow

Spring-Fall Migrants

(small numbers of some species may be present during other parts of year)

*Sharp-shinned Hawk
*Hermit Thrush[1]
*Ruby-crowned Kinglet
*Rusty Blackbird[1]

Yellow-rumped Warbler
Blackpoll Warbler
*Fox Sparrow[1]

1 Not described or illustrated in text.

To complete the avifauna of the pine barrens, according to McCormick, 1970, and Leck, 1979, the following 61 less common and/or less characteristic pine barrens species are merely listed without descriptions.

Pied-billed Grebe *Podilymbus podiceps*
Great Egret *Casmerodius albus*
Black-crowned Night Heron *Nycticorax nycticorax*
Hooded Merganser *Lophodytes cucullatus*
Red-shouldered Hawk *Buteo lineatus*
Golden Eagle *Aquila chrysaetos*
Northern Harrier *Circus cyaneus*
Osprey *Pandion haliaetus*
Merlin *Falco columbarius*
Ring-necked Pheasant *Phasianus colchicus*
Wild Turkey *Meleagris gallopavo*
King Rail *Rallus elegans*
Killdeer *Charadrius vociferus*
American Woodcock *Philohela minor*
Common Snipe *Capella gallinago*
Solitary Sandpiper *Tringa solitaria*
Rock Dove *Columba livia*
Black-billed Cuckoo *Coccyzus erythropthalmus*
Barn Owl *Tyto alba*
Short-eared Owl *Asio flammeus*
Saw-whet Owl *Aegolius acadicus*
Chuck-will's-widow *Caprimulgus carolinensis*
Yellow-bellied Sapsucker *Sphyrapicus varius*
Acadian Flycatcher *Empidonax virescens*
Willow Flycatcher *Empidonax traillii*
Least Flycatcher *Empidonax minimus*

Birds

Horned Lark *Eremophila alpestris*
Rough-winged Swallow *Stelgidopteryx ruficollis*
Cliff Swallow *Petrochelidon pyrrhonota*
Black-capped Chickadee *Parus atricapillus*
Marsh Wren *Cistothorus palustris*
Hermit Thrush *Catharus guttatus*
Swainson's Thrush *Catharus ustulatus*
Gray-cheeked Thrush *Catharus minimus*
Veery *Catharus fuscescens*
Blue-gray Gnatcatcher *Polioptila caerulea*
Yellow-throated Vireo *Vireo flavifrons*
Golden-winged Warbler *Vermivora chrysoptera*
Blue-winged Warbler *Vermivora pinus*
Tennessee Warbler *Vermivora peregrina*
Magnolia Warbler *Dendroica magnolia*
Cape May Warbler *Dendroica tigrina*
Black-throated Blue Warbler *Dendroica caerulescens*
Palm Warbler *Dendroica palmarum*
Northern Waterthrush *Seiurus noveboracensis*
Yellow-breasted Chat *Icteria virens*
Hooded Warbler *Wilsonia citrina*
Wilson's Warbler *Wilsonia pusilla*
Eastern Meadowlark *Sturnella magna*
Orchard Oriole *Icterus spurius*
Indigo Bunting *Passerina cyanea*
Purple Finch *Carpodacus purpureus*
Common Redpoll *Carduelis flammea*
Red Crossbill *Loxia curvirostra*
White-winged Crossbill *Loxia leucoptera*
Henslow's Sparrow *Ammodramus henslowii*
Sharp-tailed Sparrow *Ammospiza caudacuta*
Seaside Sparrow *Ammospiza maritima*
Vesper Sparrow *Pooecetes gramineus*
American Tree Sparrow *Spizella arborea*
Fox Sparrow *Passerella iliaca*

The following are recommended for further identification and reference.
Boyle, W. J., Jr. 1986. A guide to Bird Finding in New Jersey. Rutgers Univ. Press, New Brunswick, NJ
Bull, J. and J. Farrand, Jr. 1977. The Audubon Society Field Guide to North American Birds Eastern Region. A.A. Knopf, N.Y.
Leck, C. F., 1979 Birds of the Pine Barrens *in* Forman, R. T. T., Jr., Pine Barrens

Ecosystem and Landscape. Academic Press, NY.

Peterson, R. T. 1960. A Field Guide to the Birds. Houghton Mifflin, Co., Boston, MA

Robbins, C. S., B. Bruun, and H. S. Zim. 1983. A Guide to Field Identification. Birds of North America. Golden Press, NY

Scott, S. L., ed. 1983. Field Guide to the Birds of North America. Nat'l. Geog. Soc., Washington, D.C.

Stone, W. 1937 Bird Studies at Old Cape May. An Ornithology of Coastal New Jersey. 1965 Reprint. Dover Publ., NY

REPTILES

Reptiles are cold-blooded (their body temperatures close to that of their surroundings), vertebrate animals that, except for snakes and a few lizards, have four legs with each foot having from three to five clawed toes. Their skin is dry, usually covered with horny scales or, sometimes with bony plates. None have gills at any time during their lives so that all must obtain their oxygen directly from the atmosphere. Even those that live in aquatic habitats must come to the surface periodically to obtain air. Most lay eggs with soft, leathery skins, but some lizards and turtles lay eggs with hard shells and a number of snakes and lizards give birth to live young.

The pine barrens supports a varied and interesting array of reptilian fauna. McCormick (1970) listed 30 species. Conant (1979) also lists 30 as occurring in the pine barrens. Of these, 19 of the most common, or conspicuous, or characteristic species are described here. In the following descriptions of turtles, the term "carapace" refers to the upper shell and "plastron" to the under shell. In the descriptions of snakes and the fence lizard, a keeled scale is one that has a raised, central ridge, or "midrib."

Common Snapping Turtle *Chelydra serpentina* 8" - 18"
Carapace oval with several sharp notches (serrated) on rear margin. Tail long, saw-toothed, especially nearer base. Color varies from dark brown to black, often heavily encrusted with moss and mud. Large head. Plastron small, dull white or yellow. Has a fighting disposition. Will bite or "snap" to defend itself. Can strike with rapidity and precision. Is generally omnivorous, feeding on fish, frogs, young waterfowl, and small animals, but also up to 50% of its food may be aquatic vegetation. Snapping turtles are believed to reach ages up to 60-75 years. These are the turtles that are commercially trapped and sold in markets for snapper stew or soup. Habitat is any permanent body of fresh water. Common in pine barrens.

Musk Turtle or Stinkpot *Sternotherus odoratus* 3" - 4 1/2"
Carapace oblong, arched, and narrower in front. Dull brown to black but

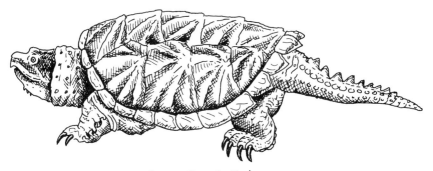

Common Snapping Turtle

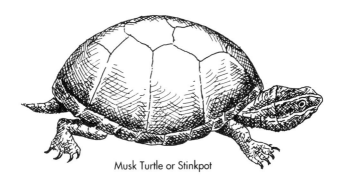

Musk Turtle or Stinkpot

Eastern Mud Turtle

often so coated with moss and mud that color is not seen. Two prominent yellow stripes on each side of head, though these may be obscured. Plastron very small with a tiny hinge that permits motion of the forward third. Head large for size of turtle. Small fleshy projections under chin and throat. Feet strongly webbed. Called "stinkpot" for musky secretion it emits from glandular openings on body. Is a bottom feeder on small living organisms and a scavenger on dead animal matter. Abundant in ponds, slow-moving streams and cranberry bogs throughout pine barrens.

Eastern Mud Turtle *Kinosternon subrubrum* 3" - 4"

Carapace oblong, smooth, olive brown to black, but often heavily coated with moss or mud. Head and neck speckled with greenish yellow but lacks the stinkpot's yellow stripes on side of head. Otherwise, looks somewhat like a stinkpot above, but has a large plastron with two hinges, one in front of, the other behind a central bridge that connects the plastron to the carapace. Plastron yellowish to dark brown, spotted with black. Is a bottom feeder. Inhabits quiet ponds, bogs, and small lakes. Numerous in pine barrens. (Illus. on pg. 273)

Spotted Turtle *Clemmys guttata* 3 1/2" - 4 1/2"

Carapace moderately arched, smooth, edges evenly rounded. Dull black with evenly spaced yellow to orange yellow dots. Orange to yellow dots on neck. Plastron yellowish brown with black shadings. Feeds on both plant and animal matter including insect larvae, fish spawn, and carrion. Often seen sunning itself on a log. More frequently seen in spring than in other seasons. Prefers bogs, swamps, marshes, and small ponds. Abundant in pine barrens.

Eastern Box Turtle *Terrapene carolina* 4 1/2" - 6"

Carapace highly arched and globular, front and rear edges curl upwards. Plastron with a hinge across middle, dividing it into two movable lobes so it can be drawn up tightly against carapace in front and rear for protection. Carapace dark brown with numerous irregular, variable, yellow markings. Plastron usually dark yellow. Usually, eyes of males are red; those of females yellow. Basically a terrestrial turtle. Omnivorous, feeding on both animal food as worms, grubs, and insects, and on a wide variety of foliage and fruits. Common in pine barrens.

Eastern Painted Turtle *Chrysemys picta* 4 1/2" - 6"

Carapace oval, smooth, moderately arched, olive brown to greenish black. Plates on shell in rows across back, each plate margined with pale yellow. Under edges of carapace bordered with bright red bars and crescents, blotched with black. Plastron uniform yellow. Fleshy parts black with bright greenish yellow stripes. Feeds on both plant and animal foods including worms, insect larvae, snails, salamanders, fish spawn, and small fish. Often seen "sunning" on logs and mud banks. Found in lakes, streams, shallow ponds, marshes, and ditches. Abundant in pine barrens.

Spotted Turtle

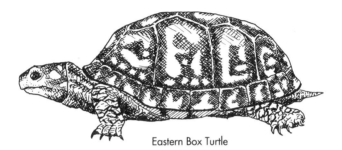
Eastern Box Turtle

Eastern Painted Turtle

Red-bellied Turtle

Reptiles

Red-bellied Turtle *Chrysemys rubriventris*   10" - 13"
Carapace oval, elongate, broadest at rear end, moderately arched. Dusky brown, marked with blotches and spots of deep red. Many pine barrens specimens nearly black. Plastron dusky yellow at center, shading to reddish along edges. Feeds mostly on aquatic vegetation (75%) and on some animal food (25%) including worms, crustaceans, insect larvae, and small fish. At home in ponds, rivers, and relatively larger bodies of fresh water. Common in pine barrens. (Illus. on pg. 275)

Northern Fence Lizard *Sceloporus undulatus hyacinthinus*   4" - 7"
Grayish, tan, or brownish, usually with a series of narrow, wavy, darker cross bands on back. Moderately large, keeled, and pointed scales give it a spiny appearance. Tail long, tapering, easily broken off. Legs sturdy. Toes long with stout claws. Males with prominent blue to black patch on each side of belly and with some blue on throat. Feeds largely on flies and other small insects. Often seen running on sand or on stumps, logs, or fences. Abundant throughout pine barrens.

Northern Water Snake *Natrix sipedon*   24" - 42"
Body stout, heavy in proportion to length. Head large, flattened, somewhat triangular. Pale brown with tinge of red, crossed by irregular, dark, wavy bands. Bands broadest on back, break up into blotches near tail. Markings most clear on young specimens and those that have recently shed their skin. Underneath brightly splashed with crimson and black. Scales strongly keeled. Has a quick temper and will readily bite but is not poisonous. Feeds largely on frogs and small fish. Bears live young. Only large water snake in pinelands. Often seen sunning itself along edges of nearly any pond, swamp, or stream. Abundant in pine barrens.

Eastern Garter Snake *Thamnophis sirtalis*   18" - 26"
Variable in color and markings. Usually greenish olive or dark brown to black. Normally three yellowish stripes, one on mid-line from back of head to tail, and one on each side of body. Usually a double row of black spots between stripes. Underneath yellow to greenish gray with two rows of indistinct, partially hidden black spots. Scales keeled. Feeds on toads, frogs, salamanders, snails, worms, grubs, and other insect larvae. Bears live young. Found in meadows, marshes, upland fields, and woodlands. Occasional in pine barrens.

Eastern Hognose Snake *Heterodon platyrhinos*   18" - 30"
Thick, heavy body. Large, triangular head with distinctly upturned snout. Color variable in shades of yellow to dark brown or black. Dorsal surface blotched with black along edges. Underneath greenish yellow or orange. Scales prominently keeled. Has habit of playing dead when attacked by a predator or when disturbed. It may first flatten or spread out its head and neck, hiss, and inflate body with air. If this does not scare off offender, it

Northern Fence Lizard

Northern Water Snake

Eastern Garter Snake

Eastern Hognose Snake
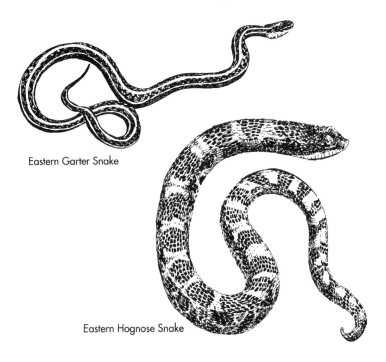

may turn over on its back, open its mouth, and play dead, but if it is turned right side up, it may promptly roll over again. Feeds chiefly on frogs and toads. At home in woods, sandy plains, open fields, and edges of swamps. Locally common in pine barrens.

Northern Black Racer *Coluber constrictor* 36" - 60"
Uniformly slate black above and gray below. Little white patches on chin and throat. Overall dull, satiny finish. Scales smooth. Scientific name a misnomer because this is not a constricting snake. Feeds on insects, lizards, frogs, small snakes, rodents, and birds and their eggs. A good climber but spends most of its time on ground. Prefers open dry country and open fields bordered with brush. Locally common in pine barrens.

Rough Green Snake *Opheodrys aestivus* 22" - 32"
Slender. Uniformly leaf green on back. Underneath yellow. Scales distinctly keeled. Feeds almost exclusively on spiders and insects like crickets, grasshoppers, grubs, and caterpillars. Good climber and commonly moves about on foliage of bushes and low trees. Common in pine barrens.

Corn Snake *Elaphe guttata* 30" - 48"
Attractive. Reddish or yellowish brown with a series of bright red, black bordered blotches. Underneath white with a checkerboard pattern of black. Scales feebly keeled on back, smooth on sides. Is a constrictor. Feeds largely on mice and rats, also on birds and small rabbits. A good climber but most likely seen on ground. Spends considerable time under leaf litter or underground in rodent burrows and other subterranean passages. In pine barrens, known only from Burlington, Cumberland, and Ocean Counties. Listed as an endangered species.

Black Rat Snake *Elaphe obsoleta* 42" - 72"
Robust with large, broad head. Color rich, glossy black. Some scales along sides edged with white giving speckled appearance. Chin and throat white. Underneath grayish. Scales feebly keeled. This is a constricting snake. Feeds on frogs, lizards, small snakes, birds and eggs, rodents, and small rabbits. Good climber of trees. Prefers open country. Locally common in pine barrens.

Northern Pine Snake *Pituophis melanoleucus* 48" - 66"
Ground color a dull grayish white with a series of deep, black blotches on back, distinct at tail end, often running together nearer head. Underneath grayish white, possibly with some black markings. Head rather small. Snout decidedly pointed. Scales strongly keeled. Tip of tail ends in a hard spine. Is a constrictor snake. Feeds on rodents and birds and their eggs. One of

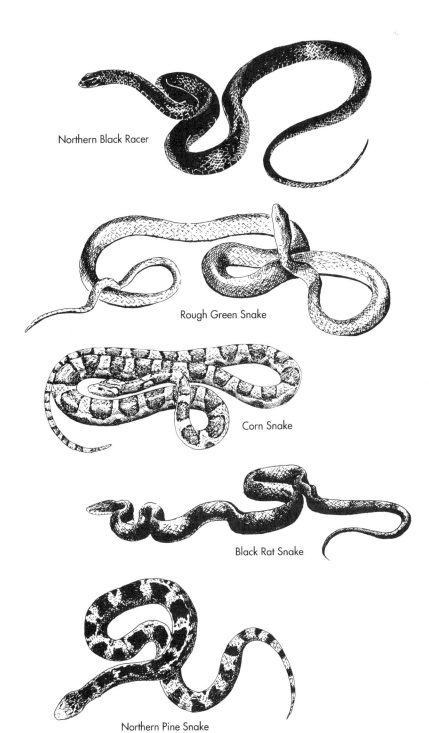

Reptiles

few snakes that excavates its own nesting burrow in open, sandy fields. Prefers flat, dry, sandy pine country. Locally common in pine barrens but classified as a threatened species in New Jersey.

Eastern Kingsnake *Lampropeltis getulus* 36" - 48"
Shiny black with variable narrow yellow or white bands that divide on sides and connect with one another to form a chain-like pattern. Underneath black with blotches of yellow and white. Scales smooth. A strong constrictor. Feeds largely on other snakes but also on lizards, birds, and rodents. A snake of timbered pine belt, preferring stream banks and borders of swamps. Locally common in pine barrens.

Eastern Milk Snake *Lampropeltis triangulum triangulum* x *elapsoides* (intergrade) 24" - 36"
Dull gray to tan, with pattern of large red or brown, irregular, saddle-like blotches. From above, appears to be ringed with gray. Blotches are bordered with black, with smaller black rimmed blotches on sides, alternating with larger ones on back. Underneath grayish white, heavily spotted with black in a checkerboard pattern. Scales smooth. A strong constrictor. Feeds on field and meadow mice, other rodents, lizards, and small snakes. Most favored habitat is in neighborhood of old barns and outbuildings, especially under rotting planks and boards. Numerous and locally common in pine barrens.

Timber Rattlesnake *Crotalus horridus* 36" - 54"
Stocky, with broad head that is distinct from neck. Top of head covered with scales. Black or dark brown crossbands on a ground color of yellow, gray, brown, or even black. Toward head, crossbands break up to form a row of dark spots down back, plus a row along each side. In black specimens, crossbands are barely discernible. Tail usually black, terminating in a horny appendage of loosely fitting rings: its "rattle." Feeds exclusively on warm blooded prey such as shrews, moles, other rodents, rabbits, and birds. The only poisonous snake in the pine barrens. Prefers second growth timber lands where rodents are common. Hibernates in cedar swamps along edges of streams. Becoming scarce in the pine barrens; now classified as an endangered species in New Jersey.

To complete the reptilian fauna of the pine barrens according to McCormick, 1970, and Conant, 1979, the following eleven less common or less conspicuous species are merely listed without descriptions.

Wood Turtle *Clemmys insculpta*
Bog Turtle *Clemmys muhlenbergi*
Ground Skink *Scincella lateralis*
Five-lined Skink *Eumeces fasciatus*
Northern Brown Snake *Storeria dekayi*

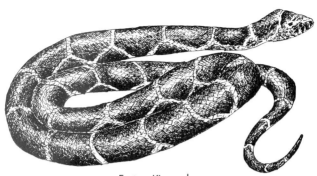

Eastern Kingsnake

Eastern Milk Snake

Timber Rattlesnake

Northern Red-bellied Snake *Storeria occipitomaculata*
Eastern Ribbon Snake *Thamnophis sauritus*
Eastern Earth Snake *Virginia valeriae*
Ringneck Snake *Diadophis punctatus*
Eastern Worm Snake *Carphophis amoenus*
Northern Scarlet Snake *Cemophora coccinea copei*

The following are recommended for further identification and reference.

Anderson, K. 1984. Turtles. The Reptiles of New Jersey. New Jersey Outdoors. Vol. II, No. 2. M/A '84. pp. 17-20.

Babcock, H.L. 1971 re-publication of 1919 edition. Turtles of the Northeastern United States. Dover Publications.

Behler, J.L. and F.W. King. 1979. The Audubon Society Field Guide to North American Reptiles and Amphibians. A. A. Knopf, N.Y.

Conant, R. 1975. A Field Guide to Reptiles and Amphibians of the United States and Canada East of the 100th Meridian. Houghton Mifflin Co., Boston, MA

Conant, R. 1979. A Zoogeographical Review of the Amphibians and Reptiles of Southern New Jersey, with Emphasis on the Pine Barrens *in* Forman, R.T.T., Jr., ed., Pine Barrens Ecosystem and Landscape. Academic Press.

Smith, H.M. and E.D. Brodie, Jr., 1982. Reptiles of North America. A Guide to Field Identification. Golden Press, NY.

AMPHIBIANS

The word amphibian comes from a Greek word meaning "two lives." Thus, amphibians are cold-blooded (their body temperatures are close to that of their surroundings) vertebrate animals that pass their immature lives in fresh water and breathe by means of gills, then change to air-breathing adults and (usually) live upon land, although some may remain in water all their lives. Amphibians possess (usually) four limbs and have a moist skin which may be soft, smooth, rough, or granular. They can easily be separated from reptiles by the fact they do not have scales on their bodies, nor do they have claws on their toes.

Typically, an amphibian is hatched from an egg laid in water, spends its aquatic life like a fish, at which time it has a long, eel-like tail, and obtains oxygen from the water by means of gills. Example: a tadpole. As growth progresses, first the hind legs then the forelegs appear, the tail (of frogs and toads) is absorbed into the body, and lungs develop to replace the gills. The amphibian now is an air breather and can live on land, but at breeding time most must return to water to lay their eggs. However, there are some notable exceptions to this rule, as the red-backed salamander. All adult am-

Amphibians

phibians are completely carnivorous though some tadpoles are herbivorous.

The pine barrens of southern New Jersey supports a relatively large and varied amphibian fauna. McCormick (1970) listed 23 species. Conant (1979) lists 24 as occurring in the pine barrens. Of these, 13 of the most common, conspicuous, or characteristic species are described here.

Marbled Salamander *Ambystoma opacum* 3 1/2" - 4 1/2"
Stocky, rounded head, short, thick body with short, heavy tail and sturdy legs. Upper surface deep bluish black, marbled with irregular broad, wavy bands and blotches of gray or silvery white. Underneath uniform gray or bluish gray. This is one of the few species that breeds in the fall (eggs laid in September - November). Larvae develop in late May or early June of following year. Found in moist, sandy areas. Locally common along edges of pine barrens. (Illus. on pg. 285)

Red-spotted Newt or Red Eft *Notophthalmus viridescens* 2 3/4" - 4"
Has unusual life cycle: begins life as an aquatic larva, changes to become a terrestrial immature ("red eft"), then finally changes to become an aquatic adult. Adults vary from olive green to dark greenish brown with scattered round, black dots and, on each side, a line of red spots trimmed with black. Immature land stage ("red eft") is orange red to bright red, with two rows of black bordered red spots along sides. Underneath paler reddish. Immature terrestrial efts are found under leaf litter, logs, and debris in surrounding woods. Aquatic adults live in ponds, ditches, and marshes. Only a few locality records in pine barrens. (Illus. on pg. 285)

Red-backed Salamander *Plethodon cinereus* 2 1/4"- 3 1/2"
Reddish or yellowish red stripe down back extending out onto top of tail. Several color variations, including an all black variety. Underneath mottled gray and white. Young go through larval stage within the egg. Has no aquatic larval stage. Terrestrial, mostly in wooded areas, usually hidden under objects such as logs, bark or trash. Sometimes found far from water. Abundant in pine barrens. (Illus. on pg. 285)

Four-toed Salamander *Hemidactylium scutatum* 2" - 3 1/2"
Reddish orange, blotched with brown, with small dots of red and black on sides of back. Underneath enamel white, with bold, black spots. Marked constriction at base of tail. Hind feet have only four toes on each (other salamanders five). Females stay with eggs until they hatch and drop into water. Found under damp sphagnum, covered bark, stumps and rotting logs. Numerous records in pine barrens. (Illus. on pg. 285)

Amphibians

Northern Red Salamander *Pseudotriton ruber* 4 1/4" - 6"
Red or reddish orange, profusely dotted with irregular, rounded black spots. Found under moss or other objects in or near clear, cool streams, or even trickles. Abundant in pine barrens.

Eastern Spadefoot Toad *Scaphiopus holbrooki* 1 3/4" - 2 1/4"
Body short, broad, with short, thick limbs. Ashy brown, sometimes yellowish or greenish with variable markings, often with a pale, wavy line extending back from each eye. On inside of each hind foot is a conspicuous black, horny process used in burrowing. Digs backward into hole as it is dug. Disappears rapidly out of sight, often digging deeply into earth. Breeds only during heavy rainfalls. Locally common in pine barrens.

Fowler's Toad *Bufo woodhousei fowleri* 2" - 3"
Greenish gray. Several, usually three or more, warts on each of the largest dark spots. Distinct light line down middle of back. Chest and underneath generally unspotted. Breeds in rain-filled pools. Abundant.

Pine Barrens Treefrog *Hyla andersoni* 1" - 1 1/2"
Small, delicate treefrog with short body. Relatively short legs with well developed pads on fingers and toes. Bright emerald green with broad band of lavender bordered by white. Insides of hind legs tinged with orange. Restricted in pine barrens to white cedar and sphagnum bogs and swamps. Call of males in breeding season from mid-May to mid-June is distinctive low, nasal "quonk." May be locally common to abundant in protected pinelands habitats but is classified as an endangered species, due to threat of habitat destruction. (Illus. on pg. 289)

Northern Spring Peeper *Hyla crucifer* 3/4" - 1 1/4"
Tiny, delicate treefrog with well developed pads on fingers and toes. Light ash gray to olive brown. Dark triangular spot on head and two dusky lines on back that touch to form a rough cross or "X." Underneath yellowish or pinkish white. Voice, a high piping call heard at dusk, is one of first amphibian indicators of spring. Found in small, temporary or semi-permanent ponds or swamps in low, brushy, second growth woodlands. Abundant in pine barrens. (Illus. on pg. 289)

New Jersey Chorus Frog *Pseudacris triseriata kalmi* 3/4" - 1 1/2"
Tiny, delicate treefrog with very small toe discs. Pale gray or dull greenish olive to light brown above. Underneath white with few dark spots. Three well defined dark stripes down back and conspicuous silvery white line along jaw above upper lip. Best identified during mating season (mid-March - April) by its vibrant call, a repeated rolling "creek" or "prreek," rising in pitch toward the end. Sound may be somewhat imitated by running a finger along teeth of a pocket comb. Silent during balance of year. Very secretive and difficult to locate. Found in small, shallow bodies of water during

Red-backed Salamander

Four-toed Salamander

Marbled Salamander

Red Eft

Northern Red Salamander

Eastern Spadefoot Toad

Fowler's Toad

Amphibians

Green Frog *Rana clamitans melanota* 2 1/4" - 3 1/2"
Stout body with muscular legs. A ridge of skin on each side of back. Bright green on head and shoulders, shading to dusky olive brown on back, maybe more brown than green. Usually with numerous dark greenish brown blotches on back and sides. Underneath light, almost white, with some dark spots. Throat of adult male bright yellow. Abundant in shallow lakes, ponds, streams and other fresh waters throughout pine barrens.

Southern Leopard Frog *Rana utricularia* 2" - 3 1/2"
Slender body. Long legs. Pointed snout. Conspicuous ridge or fold of skin on each side of back. Brassy bronze to bronze green, with irregular, rounded, dark olive spots margined with yellowish white. Some spots may run together. Thighs and legs green. Underneath white. In all types of marshes, meadows, and fresh water habitats. Abundant in pine barrens.

Carpenter Frog *Rana virgatipes* 1 1/2" - 2 1/2"
Slender, with narrow head. Upper surface yellowish brown with a distinct yellowish stripe from just behind each eye to the groin, and a second yellow stripe on each side, running beneath eye and along upper lip to snout for a total of four yellow stripes. Underneath whitish, tinged with yellow, boldly marbled with black toward sides and hind legs. Call sounds like two carpenters hitting nails a fraction of a second apart: "pu-tunk."[33] When a colony of these frogs is calling, it sounds like a gang of carpenters hammering in the distance. Inhabits sphagnum and cranberry bogs and wooded swamps and ponds. Locally common in pine barrens habitats.

To complete the amphibian fauna of the pine barrens according to McCormick, 1970, and Conant, 1979, the following eleven less common species are merely listed without descriptions.

Spotted Salamander *Ambystoma maculatum*
Eastern Tiger Salamander *Ambystoma tigrinum*
Northern Dusky Salamander *Desmognathus fuscus*
Slimy Salamander *Plethodon glutinosus*
Eastern Mud Salamander *Pseudotriton montanus*
Northern Two-lined Salamander *Eurycea bislineata*
Northern Cricket Frog *Acris crepitans*
Gray Treefrog *Hyla versicolor*
Bullfrog *Rana catesbeiana*
Pickerel Frog *Rana palustris*
Wood Frog *Rana sylvatica*

Pine Barrens Treefrog

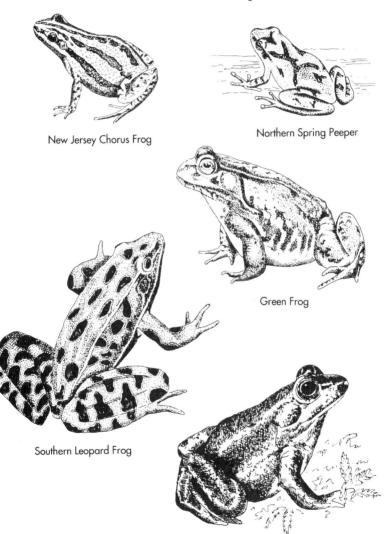

New Jersey Chorus Frog

Northern Spring Peeper

Green Frog

Southern Leopard Frog

Carpenter Frog

The following are recommended for further identification and reference.

Behler, J. L. and F. W. King. 1979. The Audubon Society Field Guide to North American Reptiles and Amphibians. A.A. Knopf, N.Y.

Conant, R. 1958. A Field Guide to Reptiles and Amphibians of the United States and Canada East of the 100th Meridian. Houghton Mifflin Co., Boston, MA

Conant, R. 1979. A Zoogeographical Review of the Amphibians and Reptiles of Southern New Jersey, with Emphasis on the Pine Barrens *in* Forman,R. T. T., Jr., ed., Pine Barrens Ecosystem and Landscape. Academic Press, N.Y.

Dickerson, M.C. 1969 republication of 1906 edition. The Frog Book. North American Toads and Frogs. Dover Publications.

Smith, H.M. 1978. Amphibians of North America. A Guide to Field Identification. Golden Press, N.Y.

FISH

Fish are backboned (vertebrate) animals that have fins in place of limbs, that live in water, and breathe by means of gills.

Largely because of the relatively high acidity and dissolved iron content, and the low concentrations of nutrients in pine barrens waters, the fish fauna is quite limited. McCormick (1970) listed 24 species but Hastings (1979) reduced the number of species indigenous to typical pine barrens acid waters to only 16. The fish of pine barrens streams and ponds are relatively dull and quiet mannered and many are darker in color and more strongly patterned than corresponding specimens from other areas.[34]

Since, in general, fish are not among the more easily or commonly seen fauna on most pine barrens field trips, no attempt is being made here to describe these species and only a few are illustrated. The following listing, together with brief notations, is drawn from Hastings' (1979) text.

Redfin Pickerel *Esox americanus*
Smaller and less common than next species. (No illus.)

Chain Pickerel *Esox niger*
Only large native game fish in pine barrens.

Eastern Mudminnow *Umbra pygmaea*
Abundant in pine barrens. Almost always associated with vegetation. Can survive in low oxygen waters.

Ironcolor Shiner *Notropis chalybaeus*
Typical open water minnow. Feeds on small crustaceans and insects.

Chain Pickerel

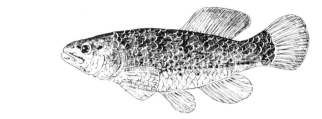

Eastern Mudminnow

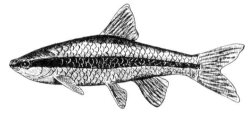

Ironcolor Shiner

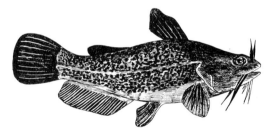

Yellow Bullhead

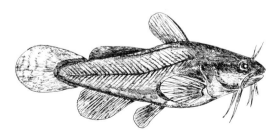

Tadpole Madtom

Fish

Creek Chubsucker *Erimyzon oblongus*
An open water, free swimming fish. A major source of food for the chain pickerel in pine barrens. (No illus.)

Yellow Bullhead *Ictalurus natalis*
Characteristic of sluggish streams and standing water with dense vegetation. More tolerant of acid waters than next. (Illus. on pg. 289)

Brown Bullhead *Ictalurus nebulosus*
More tolerant of polluted waters but less common in pine barrens than above. (No illus.)

Tadpole Madtom *Noturus gyrinus*
Characteristic of dense vegetation. Relatively abundant in pine barrens where aquatic plants are present in slow to moderate currents. (Illus. on pg. 289)

American Eel *Anguilla rostrata*
Been taken in substantial numbers in some pine barrens locations. (No illus.)

Pirate Perch *Aphredoderus sayanus*
Characteristic of sluggish streams and standing water with dense vegetation. Numerous in pine barrens.

Mud Sunfish *Acantharchus pomotis*
Common in pine barrens streams. Similar in habits to next three species but larger, with larger mouth, and more voracious predator.

Blackbanded Sunfish *Enneacanthus chaetodon*
The most characteristic pine barrens fish. Has striking black and white bands. Common in most pine barrens lakes and streams, especially in dense vegetation.

Banded or Sphagnum Sunfish *Enneacanthus obesus*
Nearly restricted to the pine barrens. Distribution as above.

Bluespotted Sunfish *Enneacanthus gloriosus*
In sluggish water with dense vegetation. Less numerous in pine barrens than other sunfish. (No illus.)

Tessellated Darter *Etheostoma olmstedi*
Less common in pine barrens than next. (No illus.)

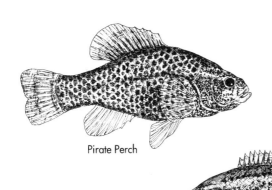
Pirate Perch

Mud Sunfish

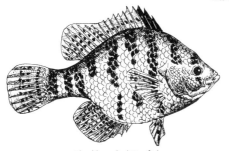
Blackbanded Sunfish

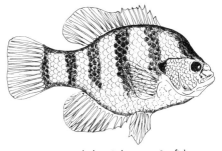
Banded or Sphagnum Sunfish

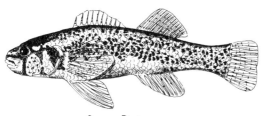
Swamp Darter

Arthropods

Swamp Darter *Etheostoma fusiforme*
Occurs over mud or organic matter in slow, weedy, brown stained back waters. (Illus on pg. 291)

The following are recommended for further identification and reference.
Hastings, R. W. 1979. Fish of the Pine Barens *in* Forman, R.T.T., Jr., Pine Barrens Ecosystem and Landscape. Academic Press, N.Y.
Thompson, P. 1985. Guide to Fresh Water Fishes. Houghton, Mifflin Co., Boston.
Zim, H.S. 1956. Fishes, a Guide to Fresh and Salt Water Species. Golden Press, NY

INSECTS AND OTHER CLOSELY RELATED ARTHROPODS

Arthropods are invertebrate animals which have an external skeleton, bodies usually divided into two or three areas, and paired, jointed appendages.

Insects are arthropods further characterized by having three distinct body areas (head, thorax, and abdomen), a pair of compound eyes, a pair of antennae, three pairs of segmented legs, and usually one or two pairs of wings. See more detailed descriptions of insects on page 296.

Because of the almost overwhelming number of different species of arthropods, including insects, and subject to space limitations, many of the descriptions that follow are of groups (orders, families, genera) rather than of individual species.

Names, both common and scientific, used in the following arthropod and insect sections are those that have been approved and are listed in "Common Names of Insects and Related Organisms" published by the Entomological Society of America, 1989 edition. Names not included in that listing follow the latest available taxonomy published in a variety of scientific sources.

ARTHROPODS OTHER THAN INSECTS

Ticks (Ixodidae)

American Dog Tick *Dermacentor variabilis*   1/8" - 3/8"
Deer Tick   *Ixodes dammini*   1/16" - 1/8"
Both species small, oval, hard-bodied, with unsegmented body broadly joined to smaller protruding head and mouthparts. Both have eight legs. Both reddish-brown. Dog tick is larger, usually with one or two whitish spots on back of body. Deer tick tiny, hardly larger than pin-head, usually does not have any whitish body markings.

American Dog Tick

Deer Tick

Chigger or Harvest Mite

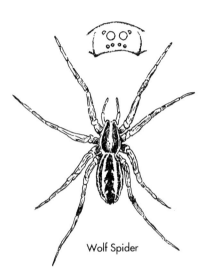
Wolf Spider

Jumping Spider

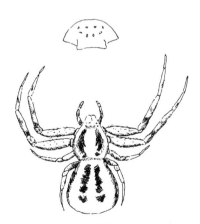
Crab Spider

Arthropods

Ticks are pests of reptiles, birds, and mammals, including humans. People walking through pine barrens vegetation may find one or more ticks on their clothes or body. These should be removed as soon as possible because a small percentage of ticks may transmit a bacterial disease. Dog ticks may transmit Rocky Mt. spotted fever and tularemia. Deer ticks may transmit Lyme disease.

Harvest Mites or Chiggers (Trombidiidae)   1/64" - 3/32"
Reddish, and so tiny these are seldom seen. Immatures crawl on vegetation from which they attach themselves to any passing host. On humans, they prefer areas where clothing is tight. They burrow under the skin and cause considerable irritation. (Illus. on pg. 293)

Spiders (Araneae)
Spiders have bodies separated into two distinct regions: a combined head and thorax (cephalothorax) and abdomen. Spiders have eight legs. All spiders are voracious predators, mainly upon insects.

Wolf or Ground Spiders (Lycosidae)   1/8" - 1 3/8"
Wolf spiders are hunters that chase prey by running over the ground. Most are in shades of brown to dark brown. Eight eyes with distinctive arrangement: four small eyes in front row, two very large eyes behind these, and two medium sized eyes in third row. Female carries egg sac beneath abdomen. When spiderlings emerge, they climb on back of female until ready to be on own. Wolf spiders are largest and most common spiders in pine barrens. (Illus. on pg. 293)

Jumping Spiders (Salticidae)   1/8" - 5/8"
Stout bodied and short legged. Eight eyes arranged in three rows, the two middle eyes in front row much larger than others. Have sharpest vision of all spiders. Are excellent hunters and pounce on prey with spectacular leaps. (Illus. on pg. 293)

Crab Spiders (Thomisidae)   1/16" - 3/8"
Somewhat crab-like in shape with short, rather broad body. Hold legs stretched out to sides. Can walk forward, sideways, or backward. Two forward pairs of legs usually more stout than two hind pairs. Eight small eyes. Forage for prey on ground or climb on vegetation and lie in ambush on flowers waiting for prey. (Illus. on pg. 293)

Black Widow Spider *Latrodectus mactans*   Male 1/8", Female 3/8"
Black. Male with reddish stripes on sides of abdomen. Female with red hourglass mark under abdomen. Rare in open woods. Occasional under boards, in wood piles, or around old sheds and boxes. Poisonous but not aggressive. Usually bites only in self defense. After mating, female may eat male and thus become a "widow."

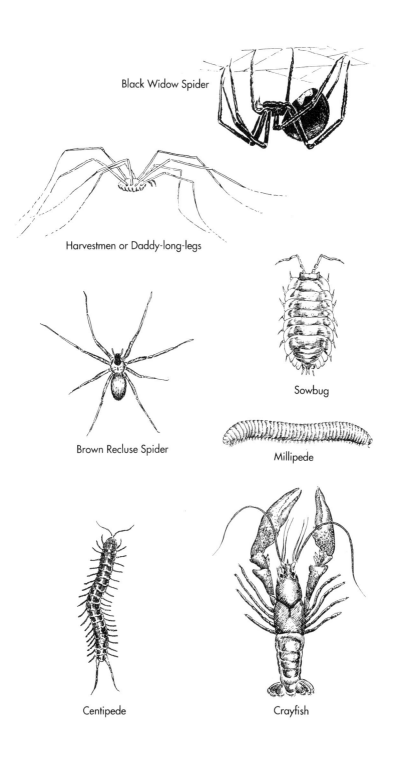

Arthropods

Brown Recluse Spider Loxosceles reclusa   Male 1/4", Female 3/8"
Grayish to reddish brown with dark, fiddle shaped mark on forward part of body. Six eyes in three pairs. In sheltered places. Infrequent in pine barrens but poisonous. (Illus. on pg. 295)

Harvestmen or Daddy-long-legs (Phalangiidae)   1/8" - 1/2"
Single, oval, compact body (no separate regions) with eight extremely long, slender, fragile legs. Two eyes raised on low, turret-like protuberances. Not spiders, though often mistakenly called such. Feed on juices of plants and small invertebrates, both living and dead. Name derives from their abundance during fall harvest season. Abundant in pine barrens.
(Illus. on pg. 295)

Sowbugs or Pillbugs (Isopoda)   1/4" - 1/2"
Light to slate gray, soft bodied, distinctly segmented, with seven pairs of legs. Found in protected places as under boards, bark, and rotting logs. Feed on both living and decaying vegetation. (Illus. on pg. 295)

Millipedes (Diplopoda)   up to 4"
Elongate, cylindrical, worm-like animals with many body segments, most with thirty or more. Most segments with two pairs of short legs. Move with a slow, flowing motion, often curl up when disturbed. Found in and under old logs and forest litter. Feed on decaying vegetation. (Illus. on pg. 295)

Centipedes (Chilopoda)   up to 6"
Elongate, flattened, worm-like animals with many body segments, most with 15 or more. Each segment with only one pair of legs. Usually move rapidly and scurry away when uncovered. Found in soil, under bark, and in and under rotting logs. Predacious and feed on small insects, spiders, and other small invertebrates. (Illus. on pg. 295)

Crayfish Cambarus bartoni (Crustacea: Decapoda)   3"
Two body parts: a combined head and mid-body (cephalothorax) and an abdomen. Two pairs of antennae. Five pairs of walking legs, the fore pair with large pincers used for seizing and tearing food. Aquatic. Breathe by means of gills. Common in and around small streams, ponds, and cranberry bogs. (Illus. on pg. 295)

INSECTS

Insects are invertebrate animals which have an external (exo-) skeleton, segmented body, three distinct body parts (head which contains sensory organs; mid-body (thorax) to which are attached all three pairs of legs; abdomen which contains the digestive and reproductive organs), a pair of compound eyes, a pair of antennae, three pairs of segmented legs, and, usually, two, sometimes only one or none, pairs of membranous wings.

Insects first appeared on this earth about 350 million years ago and since then have spread and adapted to nearly every known habitat on earth, even to such places as the frozen Antarctic (Collembola) and the shores of the oceans (*Anurida maritima*).

Insects form the largest and best known class of animals in the phylum Arthropoda. Nearly 100,000 species are known in North America, and over 10,000 were known to exist in New Jersey back in 1909.[35] Of these, only 280 of the most common and conspicuous species can be described here.

Based upon their different characteristics, insects are grouped into orders. Depending upon which authority is followed, the number of orders varies from 24 to 38. Most authorities, however, agree there are eight or nine principal orders:

ODONATA	Dragonflies, Damselflies
ORTHOPTERA	Grasshoppers, Crickets
DICTYOPTERA	Roaches, Mantids
HEMIPTERA	True Bugs
HOMOPTERA	Cicadas, Hoppers, Aphids, Scale Insects
COLEOPTERA	Beetles
LEPIDOPTERA	Butterflies, Moths
DIPTERA	Mosquitoes, Flies
HYMENOPTERA	Ants, Wasps, Bees

Smaller orders included in the descriptions that follow are:

COLLEMBOLA	Springtails
EPHEMEROPTERA	Mayflies
PHASMATODEA	Walking sticks
ISOPTERA	Termites
DERMAPTERA	Earwigs
MEGALOPTERA	Fishflies, Dobsonflies
NEUROPTERA	Lacewings, Antlions
TRICHOPTERA	Caddisflies

A word is in order about insect growth and development. This occurs in a process called metamorphosis (change in form) of which there are two general types: simple and complete. For the purposes of this book, there are two types of simple metamorphosis: incomplete and gradual.

In incomplete metamorphosis, immatures (naiads) are aquatic and gill breathing and differ considerably in appearance from their winged, air-breathing adults. In this book, this type occurs only in the Ephemeroptera and Odonata.

Insects

In gradual metamorphosis, immatures (nymphs) look very similar to adults except for size, lack of wings and sex organs. These develop gradually through a series of growth periods (instars) and skin sheddings (molts). In this book, this type occurs in the Orthoptera, Dictyoptera, Isoptera, Dermaptera, Hemiptera, and Homoptera.

In complete metamorphosis, immatures and adults usually are quite different in their forms, often living in different habitats, and with different habits. Usually, development occurs in four stages: egg, larva, pupa, and adult. After hatching from their eggs, the young (larvae) feed voraciously and go through a series of growth periods (instars) and skin sheddings (molts) until they transform into a resting state (pupa). Pupae do not feed, often are enclosed in a protective covering such as a cocoon, and many pass the winter in this stage. Following a final molt, pupae emerge transformed into adults. These soon mate and females then deposit their eggs for the start of the next generation. In this book, this type of metamorphosis occurs in the Megaloptera, Neuroptera, Coleoptera, Trichoptera, Lepidoptera, Diptera, and Hymenoptera.

Definitions of a few common terms used in the following descriptions seem in order at this point.

antenna (ae) - a pair of sensory "feelers" arising from fore part of head.

elytron (a) - outer wings, really wing covers, most commonly applied to beetles.

larva (ae) - immature state that differs in form from adult, as grub, maggot, or caterpillar.

naiad - immature, aquatic nymph, as in dragon- and damselflies.

nymph - immature that resembles adult except for size and for lack of wings and reproductive organs.

ovipositor - tubular projection from tip of female abdomen. Used for egg laying. Sometimes modified into an internal organ and used as a "stinger" for defense.

pronotum - shield-like covering over mid-body (thorax), between head and abdomen.

scutellum - small, generally triangular plate between the bases of the outer wings or wing covers.

suture - juncture line of the wing covers (elytra), most noticeable in beetles.

thorax - mid-body, between head and abdomen, to which are attached all three pairs of legs.

Insects

Simple Key to Orders of Insects Described in this Field Guide

1. A. Insects with wings. Forewings may be hard wing covers 2
   B. Wings absent 17
2. A. Only one (fore) pair of wings, or hind wings (if present) small and rounded. Wings membranous 3
   B. Two pairs of wings. Fore pair either membranous or hard wing covers, with membranous hind wings folded underneath 4
3. A. Hind wings small (if present) or vestigial or absent. Tip of abdomen with two or three conspicuous "tails" .... MAYFLIES
   (Ephemeroptera)(page 301)
   B. Hind wings never present. Replaced by pair of halteres. Tip of abdomen without any projecting "tails" .... FLIES, MOSQUITOES
   (Diptera)(page 376)
4. A. Forewings as hard wing covers meeting in straight line down center of back. No veins apparent in wing covers, but striations may be present. Membranous hind wings folded under covers. 5
   B. Forewings not hard. Either parchment-like, membranous, or coated with scales or hairs. Do not meet in a straight line. Wing veins usually evident 6
5. A. Fore wings (wing covers) very short, leathery. Tip of abdomen with a pair of prominent, forceps-like appendages .... EARWIGS
   (Dermaptera)(page 312)
   B. Wing covers covering all or most of abdomen. Tip of abdomen without forceps. .... BEETLES
   (Coleoptera)(page 328)
6. A. Forewings mostly or all parchment-like. Base of forewing may be parchment-like with only tip membranous. 7
   B. Forewings all membranous, hairy, or covered with scales 10
7. A. Mouthparts in the form of a piercing or sucking beak extending below head 8
   B. Mouthparts not as a beak but with mandibles adapted for chewing 9
8. A. Basal portion of forewings parchment-like, with tips membranous. Wings usually flat on back. Beak arises from under front of head .... TRUE BUGS
   (Hemiptera)(page 312)
   B. Forewings all one texture. Usually roof-shaped over back. Beak arises from under rear portion of head, near fore pair of legs. ....
   .... CICADAS, HOPPERS, APHIDS
   (Homoptera)(page 322)
9. A. Hind legs longer and heavier than either fore or middle pair of legs. Adapted for jumping .... GRASSHOPPERS, CRICKETS
   (Orthoptera)(page 304)

Insects

    B. Hind legs similar to fore and middle legs in size and shape. Adapted for running and walking. ....ROACHES, MANTIDS
                                  (Dictyoptera)(page 310)
10. A. Wings covered with scales. Often brightly colored............................
    ....BUTTERFLIES, MOTHS
                                  (Lepidoptera)(page 358)
    B. Wings not covered with scales 11
11. A. Wings membranous with net-like veins 12
    B. Wings membranous, but with few veins, not net-like 14
12. A. Antennae very short, inconspicuous. Larvae aquatic 13
    B. Antennae conspicuous, composed of many segments 15
13. A. Hind wings small. Tip of abdomen with two or three long, hair-like "tails" extending behind ....MAYFLIES
                                  (Ephemeroptera)(page 301)
    B. Fore and hind wings about equal in size and shape. No hair-like "tails" at tip of abdomen. ....DRAGON- and DAMSELFLIES
                                  (Odonata)(page 301)
14. A. Wings covered with fine, long hairs, held roof-like over abdomen when resting. Antennae very long. Aquatic larvae develop inside cases .... CADDISFLIES
                                  (Trichoptera)(page 358)
    B. Wings membranous, usually flat, not roof-like over abdomen when resting. Antennae short ....ANTS, BEES, WASPS
                                  (Hymenoptera)(page 382)
15. A. Wings membranous, with very fine, rather faint veins. Fore and hind wings same size and shape. Abdomen broadly, not thread-like, joined to thorax. .... TERMITES
                                  (Isoptera)(page 312)
    B. Wing veins well developed and apparent. Wings held roof-like, not folded, when at rest. 16
16. A. Hind wings broader at base than fore wings ....FISHFLIES
                                  (Megaloptera)(page 326)
    B. Fore and hind wings similar in size and shape ....
    ....LACEWINGS, ANTLIONS
                                  (Neuroptera)(page 326)
17. A. Small, less than 1/4" long. Appendage under tip of abdomen bends forward under body ....SPRINGTAILS
                                  (Collembola)(page 301)
    B. Larger, more than 1/4" long 18
18. A. Body elongated, stick-like. Antennae long, thread-like ....
    .... WALKING STICKS
                                  (Phasmatodea)(page 312)
    B. Body not so elongated 19
19. A. Abdomen broadly, not thread-like, joined to thorax. Antennae not

elbowed. ....TERMITES
(Isoptera)(page 312)
B. Abdomen joined to thorax by an almost thread-like stem (petiole). Antennae elbowed ....ANTS
(Hymenoptera)(page 382)

COLLEMBOLA (coll=glue; embola=wedge)
Springtails   1/16" - 1/4"
Primitive, minute, wingless. Abdomens have only four to six segments. Common name is from forked, spring-like organ (furcula) which folds forward under insect and is held by a catch. When this "spring" is released, insect is tossed or "jumps" as much as three to four inches into the air. Feed on algae, molds, and decaying vegetation. Included here only because of their great abundance - up to 100,000 individuals per cubic meter of surface soil, even in the pine barrens. Many species. (Illus. on pg. 303)

EPHEMEROPTERA (ephemero=a day; ptera=wing)
Mayflies   1/4" - 5/8"
Brownish, yellowish, or dusky colored, with soft, fragile bodies, delicate transparent wings, and two or usually three filament-like tails which may be twice as long as insect's abdomen. Fore wings triangular, with many veins. Hind wings smaller, more rounded, or absent. At rest, wings held vertically above body. Development by incomplete metamorphosis. Immatures entirely aquatic, are elongate, and have two or usually three tail filaments. Some may live as long as four years. These feed on tiny plants and decaying vegetation. Upon emergence in April and May, adults are first called sub-imagoes, or "duns" by fishermen. After shedding final skin, they become clear winged adults, or "spinners." Adults do not feed, and live for only a day, or part of a day, or, at most, for a very few days. Number of species in pine barrens very limited due to highly acidic conditions of stream waters.[36] Our most common species is likely to be *Leptophlebia cupida*. (Illus. on pg. 303)

ODONATA (odous=tooth)
Dragonflies and Damselflies   3/4" - 5"
Medium to large insects with long, slender abdomens and moveable heads with large, compound eyes and short antennae. Large, well developed, chewing mouthparts. Four membranous, many veined wings, nearly equal in size. Long, hairy legs, not suited for walking, are used in flight like a basket to capture insects. Adults fly on sunny days and most often seen in vicinity of streams, ponds, lakes, and moist meadows.

Dragonflies are larger and more robust than damselflies and are much stronger fliers. When at rest, dragonflies hold their wings out sideways, or horizontally. Damselflies are smaller, more delicate, weaker fliers and, when at rest, most hold their wings together vertically over their body, parallel with their abdomen.

Dragonflies

Immatures of both dragonflies and damselflies are entirely aquatic, are gill breathing, and differ considerably from adults in appearance (incomplete metamorphosis). All are very predacious, feeding on insects, small tadpoles and small fish, and seize their prey by means of a greatly enlarged lower lip which can be projected forward with lightning speed. Both immatures and adults are highly beneficial predators upon mosquitoes and other small insects.

These are ancient insects, with fossil remains dating back 250 to 300 million years. Fossils from the Upper Carboniferous Period show that some dragonfly ancestors had a wingspan of two and one half feet!

Patrick *et al*, 1979 (Table II), records 51 species of dragon- and damselflies as occurring in pine barrens river basins, and further states that the acid waters of pine barrens streams are rich in the odonate fauna.

The following are good examples of pine barrens species. Unless otherwise stated, adults of most species occur from April or May throughout the summer and into early fall.

Dragonflies (Anisoptera)
Generally strong fliers. Wings held sideways, or horizontally when at rest.

Common Green Darner *Anax junius* (Aeshnidae)   2 3/4" - 3 1/8"
Head and thorax green with target-like spot of black, yellow and blue in front of eyes. Abdomen blue to grayish. Wings transparent with iridescent, sometimes amber, sheen. This and the next are two of the largest and fastest dragonflies.

Heroic Green Darner *Epiaeschna heros*   3 1/8" - 3 5/8"
Similar to above but larger and more brownish, with sort of a "T" shaped spot instead of a round spot in front of eyes. (No illus.)

Brown-spotted *Celithemis eponina* (Libellulidae)   1 3/8" - 1 5/8"
Yellow-wing
Head and face yellow or amber. Body amber or brown with darker stripes. Abdomen nearly black with yellow stripes. Wings yellowish with brownish spots and bands.

The Widow *Libellula luctuosa*   1 5/8" - 2"
Body and abdomen dark brown to blackish, striped with yellow. Basal portions of wings brownish or black. Outer portions usually clear but middle portions on old males may become chalky whitish, and tips may be brownish on females.(Illus. on pg. 305)

Ten-spot Skimmer *Libellula pulchella*   2" - 2 1/4"
Body dull brownish, and in old specimens may be covered with a chalky "bloom." Three dark brown blotches on each wing. Spaces between blotches whiter in males than in females. (Illus. on pg. 305)

Springtails

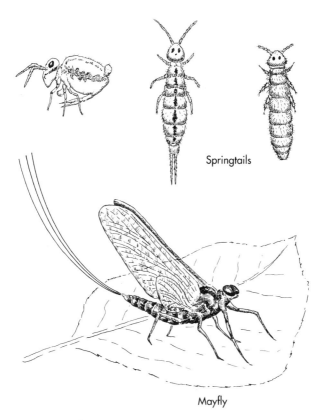

Mayfly

Common Green Darner

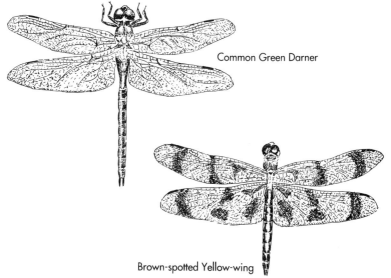

Brown-spotted Yellow-wing

Damselflies

Red Toper or Skimmer *Sympetrum vicinum* 1 1/4" - 1 3/8"
Delicate and thin legged. Head and body red brown, covered with red brown hairs. Abdomen slender, bright red, darker near tip. Underneath lighter. Wings clear, somewhat reddish yellow near base. *Sympetrum* skimmers occur from July through October and are most often seen flying on warm, sunny, autumn days.

Damselflies (Zygoptera)
Generally weak fliers. Wings usually held over body, parallel with abdomen, when at rest.

Black-winged Damselfly *Calopteryx maculata* (Calopterygidae)
Body of male metallic green, of female non-metallic dark brown. Wings of male black, of female brownish and more transparent. Forewings of both sexes have white spot.

Tipped-winged Damselfly *Calopteryx dimidiata* 1 5/8" - 1 3/4"
Very similar to above except that wings are clear with only the outer tips of all four wings black. Both species fly along shady, sandy bottom cedar streams from May through August. (No illus.)

Ruby Spot *Hetaerina americana* 1 1/2" - 1 3/4"
Head and body bronze. Abdomen of male greenish bronze; of female greenish. Wings clear, but bases of all four wings crimson red in males and amber yellow in females. Occurs from late summer into early fall and may be frequently observed around cranberry bogs.

Common Bluet *Enallagma civile* (Coenagrionidae) 1" - 1 1/2"
Entire body in various shades of vivid, light blue with black markings, males more so than females. Wings clear with light, bluish sheen; very narrow (stalked) at base. Several species look nearly alike. Fly about, and rest on vegetation along quiet edges of lakes, ponds, and streams throughout the summer and early fall.

ORTHOPTERA (ortho=straight; ptera=wing)

Grasshoppers and Crickets
Elongate, usually with two pairs of wings. Outer, or front pair long, straight, narrow, more or less leathery, and serve as wing covers (tegmina) for the under, or hind wings. Latter are membranous, larger than fore wings, are used for short flights and are folded, fan-like, under the front wings at other times. Antennae may be short (grasshoppers, locusts) or long, thread-like and many segmented (katydids, crickets). All have well developed mouthparts with strong jaws formed for biting and chewing. A pair of usually short abdominal tails (cerci) is nearly always present. Females possess ovipositors which, in some species, may be very long. Legs

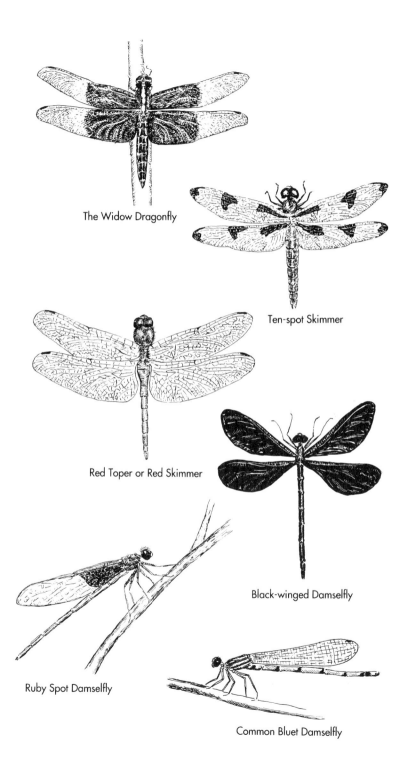

Grasshoppers & Crickets

unequal in size, the upper hind legs (femora) usually larger and thicker, well developed for jumping.

Immatures (nymphs) resemble adults except for size, lack of wings, and sex organs. These develop in gradual stages of growth and skin shedding (gradual metamorphosis).

Males are known for the musical sounds they make, usually at night, and usually by rubbing together (stridulating) portions of their wings and legs. Many of the "songs" these insects "sing" are so characteristic as to enable species identification. Most feed on vegetation. Several are destructive crop pests.

The following are a few examples of pine barrens species.

Crested Pygmy Grasshopper *Nomotettix cristatus* (Tetrigidae)   3/8" - 1/2"
Usually grayish brownish. Pronotum extends back and covers abdomen, narrowing behind. Front wings reduced to small, scale-like structures. Adults most often seen in spring and early summer because individuals hibernate overwinter as adults.

Short-horned Grasshoppers (Acrididae)
Antennae usually much shorter than body.

Bird Grasshopper *Schistocera alutacea*   1 1/8" - 2"
Greenish yellow to greenish brown, with yellow mid-line stripe running along top of entire length of dorsal surface. Outer wings almost blackish yellow. Under (hind) wings transparent yellowish with darker veins. Lower hind legs reddish yellow. Largest of common grasshoppers.

Carolina Grasshopper or Locust *Dissoteira carolina*   1 1/4" - 2"
Mottled grayish brown with small dusky spots. Most conspicuous character is color of hind wings seen in flight: bases black, fading out to pale greenish yellow borders. Makes light buzzing sound with its wings while in flight.

Long-horned Grasshoppers (Tettigoniidae)
Antennae very long, thread-like, usually longer than body.

Fork-tailed Bush Katydid *Scudderia furcata*   1 1/2" - 1 5/8"
Dark leaf green, sometimes tinged with brown. Underneath greenish yellow. Antennae long, thread-like. Fore wings long, narrow, nearly uniform in width, extend beyond abdomen. Tips of hind wings extend beyond fore wings.

Oblong-winged Bush Katydid *Amblycorypha oblongifolia*   1 5/8" - 1 3/4"
Bright leaf green, sometimes tinged with brown or pink. Body lighter green. Underneath greenish yellow. Antennae long, thread-like. Forewings long, narrow but distinctly wider at middle than at tip. Tips of hind wings extend beyond fore wings.

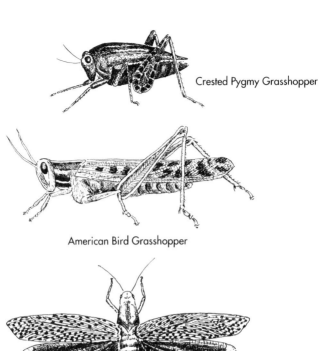

Crested Pygmy Grasshopper

American Bird Grasshopper

Carolina Grasshopper or Locust

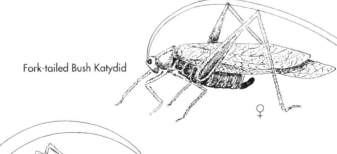

Fork-tailed Bush Katydid

♀

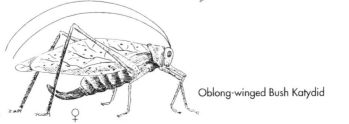

Oblong-winged Bush Katydid

♀

Crickets

Angular-winged Katydid *Microcentrum retinerve*   1 5/8" - 1 3/4"
Pale leaf green. Head and body lighter yellowish green. Underneath yellowish green. Antennae long, thread-like. Forewings broad, obtusely angled.

Northern True Katydid *Pterophylla camellifolia*   1 5/8" - 2"
Dark leaf green. Body lighter green. Underneath yellowish green. Antennae hair-like, longer than wings. Forewings longer than hind wings, very convex, sort of wrap around insect to form an inflated or bulb-like appearance. "Sings" well known "katy-did, katy-didn't" songs from tops of trees summer evenings, usually starting around July 20-28 in oak woods.

Shield-backed Katydid *Atlanticus testaceus*   Head & body 7/8" - 1 1/8"
Female ovipositor 7/8" - 1"
Dull, mottled grayish or brownish. Pronotum extends over first one or two segments of abdomen. Antennae long, thread-like. Nearly wingless, only tiny stubs appearing out from under hind edge of pronotum. Hind legs very long, upper hind legs nearly twice length of pronotum. Female ovipositor spear-like, nearly same length as rest of body.

Crickets (Gryllidae)
Antennae very long, thread-like, usually longer than body.

Field Cricket *Gryllus pennsylvanicus*   1/2" - 3/4"
Gryllus assimilis luctuosus
Two species separated by technical differences. Both stout bodied, black or brownish black. Wing covers sometimes reddish brown. Antennae black, longer than body. Wings shorter than body. Abdomen has pair of "tails." Female has long, straight, spear-like ovipositor. Males produce well known, intermittent, shrill "chirp" or "song." Omnivorous on live and decaying vegetative and dead animal matter. Adults may invade garages and basements in late summer and early autumn.

Narrow-winged Tree Cricket *Oecanthus angustipennis*   1/2" - 5/8"
Very slender. Pale greenish white. Antennae thread-like, twice as long as body. Small black, J-shaped, hooked mark on basal segment of antenna. Black mark on second segment oblong, slightly curved. Forewings of male narrow, held flat over body. Female forewings wrapped closely around body. Occurs in tall weeds, shrubs, and trees. Most common tree cricket in pine barrens.

Snowy Tree Cricket *Oecanthus fultoni*   1/2" - 5/8"
Slender. Whitish to very pale green. Body somewhat flattened. Thread-like antennae over twice as long as body. Small, round black spot on each of first two basal joints of antennae. Male forewings broad, flat, spoon-shaped, held flat over body. Female forewings wrapped closely around body. Fe-

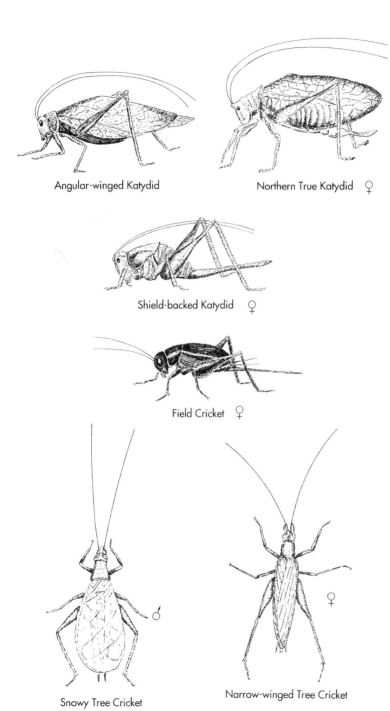

Roaches & Mantids

male with long, straight ovipositor. Males stridulate ("chirp" or "sing") at night, usually in unison with others of its species, at a rate dependent upon the temperature. The number of "chirps" in 13 seconds, plus 40, gives a good estimate of the temperature in degrees Fahrenheit. Thus often called "temperature cricket."

Northern Mole Cricket *Neocurtilla* (= *Gryllotalpa*)   3/4" - 1 1/4"
  hexadactyla (Gryllotalpidae)
Robust. Elongate. Cinnamon or darker brown. Covered with short, fine hairs. Antennae short. Fore wings short. Hind wings longer. Front legs enlarged, "fingers" broad, spadelike, adapted for rapid burrowing in loose sand. Lives in burrows in moist sand, usually 6"-8" below surface.

DICTYOPTERA (dictyo=net; ptera=wing)
Roaches and Mantids
Closely related and quite similar to grasshoppers and crickets but differ mainly in that all three pairs of legs are similar in size and shape, adapted for walking and running, rather than the hind legs being enlarged and adapted for jumping. Fore wings elongate, somewhat thickened, usually with many veins. Hind wings wider than fore wings.

Wood Cockroach *Parcoblatta uhleriana* (Blattellidae)   5/8" - 3/4"
  Parcoblatta pennsylvanica   1/2" - 1"
Two species. Oval, flattened, soft bodied. Head concealed under shield-like pronotum. Antennae long, hair-like. Pronotum and wings brown to dark brown with lighter margins. Wings lie flat over body. Wings of males extend beyond body. Wings of females much abbreviated, reaching only to 2nd abdominal segment. Legs long, spiny. Differences between species are in size and color. *P. pennsylvanicus* is larger and darker brown than *P. uhleriana*. Both found under bark of dead and dying trees in pine woods and both attracted to light. Several other species.

Roaches are among the most ancient of all insects and were very abundant during the Carboniferous Period.

European Mantid *Mantis religiosa* (Mantidae)   1 7/8" - 2 1/2"
Chinese Mantid *Tenodera aridifolia sinensis*   3" - 4"
Two species. Elongate. Body and legs green, greenish yellow, or brownish. Head triangular, loosely joined to thorax and highly moveable. Antennae thread-like. Thorax narrow, much elongated. Wings rounded at tips, reach or exceed tip of abdomen. Forelegs adapted for grasping and holding prey, often held in characteristic predatory pose leading to impression of "praying" mantis.

Two species easily separated by size differences and by differences in size and shape of egg masses. Those of European mantid are oval-elongate and narrow, up to 1 1/2" long by 5/8"-3/4" wide and thick, laid lengthwise on exposed twigs. Those of Chinese mantid also are laid on exposed twigs but are more compact, cylindrical, and conspicuous, ranging about 3/4" by 1".

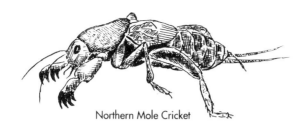
Northern Mole Cricket

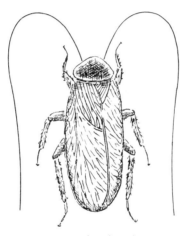
Wood Cockroach

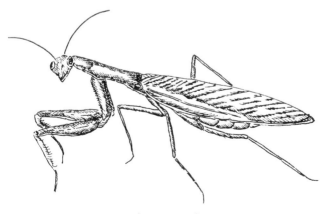
Chinese Mantid

Mantids often are thought of as being beneficial because they feed on other insects, but the fact probably is that they feed on as many beneficial insects, as lady beetles and honey bees, as on injurious ones.

PHASMATODEA (phasmato=spector)
Walkingstick *Diapheromera femorata* 2 5/8" - 4"
Elongate, slender, stick-like. Brownish, sometimes greenish or gray. Antennae very long. Wingless. Three very long, slender, widely separated pairs of legs adapted for walking.

Resemblance to twigs provides camouflage protection from birds and other predators. Sometimes so common in oak-pine woods that one can hear their droppings falling like rain as they chew on oak leaves at night.

ISOPTERA (iso=equal; ptera=wing)
Eastern Subterranean Termite *Reticulitermes flavipes* 1/4" - 3/8"
Small, soft bodied, light colored, social insects with a highly developed caste system. Live in colonies in standing tree trunks, old stumps, buried wood, fallen trees, and logs throughout pine barrens woods. Superficially resemble ants but easily separated by wide waist (abdomen broadly joined to midbody or thorax). Also, termites are soft bodied and usually light colored, thus often erroneously called white ants, while ants are hard bodied and darker colored. Antennae of termites are bead- or thread-like; those of ants usually are elbowed. Termite wings, when present in reproductive caste, are membranous, four in number, similar in size and shape, held flat over body when at rest, and extend beyond tip of abdomen. Hind wings of ants are smaller than forewings and both pair of ant wings are held together, vertically, over body.[37]

Although termites are very destructive when they feed on man's material things, as wood buildings, furniture, and fence posts, in the natural world they are very beneficial for they aid in the conversion of dead trees and other vegetation into humus and other substances that then can be reused by growing plants.

DERMAPTERA (derma=skin; ptera=wing)
European Earwig *Forficula auricularia* 3/8" - 5/8"
Elongate, flattened, with short elytra, exposed abdomen, long thread-like antennae, and a pair of pincers (cerci) (strongly curved in males, nearly straight and parallel in females) at tip of abdomen. Hind wings folded under elytra. Brownish or reddish-brown, shining. Nocturnal, spend daylight hours in dark, dampish crevices, in axils of leaves, and in ground litter. Contrary to old superstitions, earwigs do not get into people's ears and can be no more harmful to humans than a mild pinch with their (male) cerci.

HEMIPTERA (hemi=half; ptera=wing)
True Bugs
Though many people refer to all insects as "bugs," only members of this

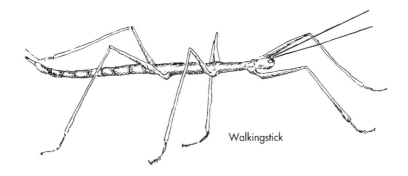

Walkingstick

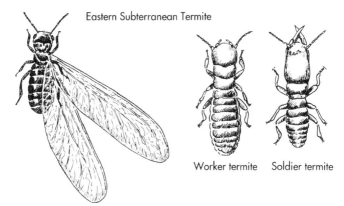

Eastern Subterranean Termite

Winged reproductive termite

Worker termite Soldier termite

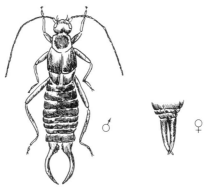

European Earwig

True Bugs

order are correctly called bugs. These are identified by two main characters:

1) Winged species have two pairs of wings: the outer pair, or fore wings, usually thickened and somewhat leathery or horny on their basal half, while the outer, tip ends are membranous and overlap each other. Between the bases of these outer wings (hemi-elytra) there usually is a small, triangular plate, the scutellum. The hind, or inner, wings are entirely membranous and used for flying. When not in use, these are folded beneath the outer wings. Both pairs of wings, when not in use, lie flat over the abdomen.

2) All true bugs have piercing-sucking mouthparts with a jointed beak or tube that arises from under the front of the head and extends back under the body between the legs. These are sucking insects that obtain their food by piercing the surfaces of plants or animals and drawing the sap or blood of their host into their own bodies.

Their bodies often are widest where the abdomen joins the thorax, giving these insects a broad shouldered appearance. Antennae either four or five segmented. Immatures (nymphs) resemble adults except for size, lack of wings, and sex organs. These develop in gradual stages of growth and skin shedding (gradual metamorphosis). Many true bugs are serious pests of cultivated plants, while several others are beneficial because these are predacious and feed on other insects.

Stink Bugs (Pentatomidae)

All "stink bugs," including the following five, have glands on their under surface from which they emit a characteristically pungent odor when touched or disturbed.

Rough Stink Bug *Brochymena quadripustulata*   5/8"
Brochymena arborea

Two species separated by technical differences. Large, broad, oval, flattened, and roughly sculptured above. Mottled grayish brown to dark brown, harmonizing with bark of trees on which they live. Underneath dull yellow. Head extends beyond bases of antennae. Sides of pronotum extend into spinelike projections. Sides of abdomen extend beyond wing covers. Legs with alternating dull red and black rings. Also known as "tree bugs."

One-spotted Stink Bug *Euschistus variolarius*   1/2"

Broadly oval, somewhat flattened above. Shoulder angles acute. Convex beneath. Grayish or greenish yellow with small dark punctures. Underneath greenish-yellow-orange. Antennae light reddish or brownish yellow, fifth and end of fourth segments darker. Legs yellowish with small brownish dots.

Green Stink Bug *Acrosternum hilare*   5/8" - 3/4"

Elongate-oval. Shield shaped. Shining green, with narrow, pale orange or yellowish margins. Underneath greenish yellow. Antennae green with darker areas at ends of last three segments. Feeds on vegetation.

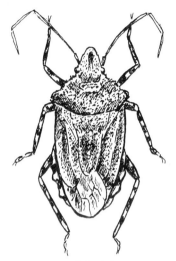

Rough Stink Bug

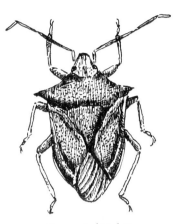

One-spotted Stink Bug

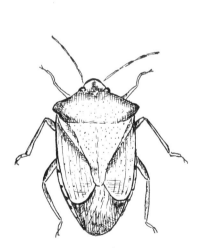

Green Stink Bug

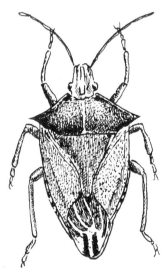

Spined Soldier Bug

True Bugs

Spined Soldier Bug *Podisus maculiventris* 1/2"
Elongate-oval. Shield-shaped, tapering behind. Flattened above. Convex beneath. Shoulders produced outward in slender, projecting spines. Two dark spots near membranous tips of wing covers. Underneath pale yellow. Abdomen with one or two rows of small, dark spots on each side. Sixth ventral segment with median dark spot. A predacious "stink bug."
(Illus on pg. 315)

Leaf-footed Bug *Leptoglossus oppositus* (Coreidae) 3/4"
Elongate oval, flattened above, somewhat convex beneath. Dark brown with small, yellowish markings. Head narrowed and prolonged in front of bases of antennae. Antennae light brown, terminal joint paler. Lower hind legs with leaf-like dilations on each side.

Boxelder Bug *Boisea (= Leptocoris) trivittata* (Rhopalidae) 3/8" - 9/16"
Elongate oval. Flattened above. Dull brownish black with narrow, brick red markings on sides, a mid stripe on top of thorax and on margins of forewings. Feeds on box elder which often has escaped from cultivation. In fall of year, often found on buildings, especially around windows, looking for places to hibernate over winter.

Red Bug *Largus (= Euryophthalmus) succinctus* (Largiidae) 1/2" - 5/8"
Elongate oval. Robust. Brownish black above. Reddish-orange margins of thorax and forewings. Fine, grayish-bluish pubescence beneath.

Large Milkweed Bug *Oncopeltus fasciatus* (Lygaeidae) 1/2" - 3/4"
Elongate oval. Red or orange, and black markings, the black being a large spot covering most of top of pronotum and scutellum, a broad band across middle of closed wing covers, and the overlapping membranes at tip end. Legs black. Red or orange markings on edges of pronotum, scutellum, and wings. Underneath mostly red with black spots. Wing covers extend beyond tip of abdomen. Feeds on foliage and seed pods of milkweed.

Small Milkweed Bug *Lygaeus kalmii* (Lygaeidae) 3/8" - 1/2"
Elongate oval. Black and red markings. Head with oblong red spot. Pronotum black in front with broad red band across back. Scutellum black. Wing covers have two diagonal red lines that cross to form a red "X" mark. Overlapping membrane at tip black with narrow white margin and, usually, two small whitish or grayish spots in middle. Feeds on foliage and seed pods of milkweed.

Ambush Bug *Phymata erosa* (Phymatidae) 3/8"
Broadly oval. Stout bodied. Roughly sculptured. Yellow or greenish yellow with pale or reddish brown markings, including a reddish brown band across widest part of body. Side margins of abdomen expanded and

Leaf-footed Bug

Boxelder Bug

Red Bug

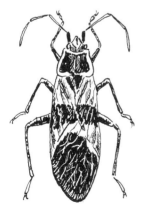
Large Milkweed Bug

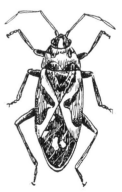
Small Milkweed Bug

True Bugs

upcurved. Wing covers narrow, leaving sides of abdomen exposed. Front legs adapted for catching insects and spiders by being thickened and armed with spines.

These small, predacious bugs conceal themselves in heads of flowers, mainly Compositae, and seize nectar seeking insects that come to visit the flowers.

Assassin Bugs (Reduviidae)
The following three species of assassin bugs have fore legs adapted for seizing and holding prey while drawing out the prey's body fluids with their piercing-sucking beaks. Most attack other insects but a few attack higher animals, even including humans, inflicting painful bites.

Assassin Bug *Stenopoda spinulosa*   3/4" - 1"
Elongate, robust, nearly uniformly straw-yellow with small, squarish, dark brown to black dot just in front of overlapping, membranous wing tips. Upper front legs armed with row of short tubercles or spines.

Kissing Bug *Melanolestes picipes*   5/8" - 3/4"
Elongate, oval. Black, or nearly so. Head elongated in front of eyes. Hind portions of head short, abruptly constricted into a short neck. Thorax somewhat bell shaped and divided into two lobes. Legs short. Uppers of fore and middle legs thickened.

Wheel Bug *Arilus cristatus*   1" - 1 3/8"
Elongate, oval. Stout bodied. Dark brown to grayish black, slightly bronzed. Top of thorax has a high, arched, median, semi-circular crest, like half a cog wheel, with a series of eight or more cog or tooth-like projections, or tubercules. Hind margin of pronotum extends back over scutellum.

Aquatic Bugs
All the following are aquatic or semi-aquatic bugs. All are recognized as true bugs by the overlapping membranous tips of their forewings and by their piercing-sucking beaks. Most are predacious and have fore legs adapted for seizing and holding prey while feeding upon it. Some obtain their air at the water surface by means of posterior breathing tubes. Others carry pockets of air down with them under the water surface.

Water Strider *Gerris remigis* (and others) (Gerridae)   1/2" - 5/8"
Slender, elongate, flattened. Dark to reddish brown to black. Lower sides covered with fine, silvery gray, waterproof pubescence. First joint of antennae distinctly longer than second and third joints combined. Legs very long, stilt-like, fore legs shorter and used for capturing food. Usually wingless. "Skate" or "stride" on top of water surfaces of ponds and streams, aided by tiny pads of short, waterproof hairs on feet.

Ambush Bug

Assassin Bug

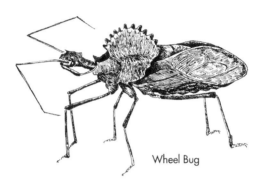
Wheel Bug

Kissing Bug

Water Strider

True Bugs

Creeping Water Bug *Pelocoris femoratus* (Naucoridae) 3/8" - 1/2"
Broadly oval, flattened. Dull yellowish-brown, speckled with browner, whitish, and dull yellowish areas. Fore legs enlarged and adapted for seizing prey. Crawls, rather than swims, on, over, and among aquatic vegetation. Can inflict sharp sting when handled.

Giant Water Bug *Benacus griseus* (Belostomatidae) 1 3/4" - 2 3/8"
Lethocerus americanus
Two species. Both large, elongate-oval, broad, flattened. Underneath somewhat keeled. Dull yellowish brown to dark brown. Both have short, powerful beaks and much enlarged, strong, upper parts of fore legs, adapted for seizing prey. Inside of upper fore legs of *L. americanus* grooved to receive lower legs. No such groove in *B. griseus*.

These are the largest of all true bugs. Are often attracted to electric lights, for which they sometimes, erroneously, are called "flying cockroaches." Can inflict a painful sting.

Giant Water Bug *Belostoma fluminea* 7/8" - 1"
Similar to above but much smaller. Body oval, flattened. Brown. Fore legs enlarged and adapted for seizing prey. Female glues her eggs to back of male where they are carried until they hatch.

Waterscorpion *Ranatra fusca* (Nepidae) 1 1/8" - 1 3/4"
Very elongate, narrow, stick-like. Pale to dark yellowish or reddish brown. Forelegs thickened, armed with stout spines for seizing prey. Middle and hind legs long and slender. A pair of long, tail-like breathing tubes at tip of abdomen are thrust up to water surface to obtain air.

Backswimmer *Notonecta undulata* (and others) (Notonectidae) 1/2"
Oblong. Back of body (dorsal surface) convex, somewhat keeled. Color variable from cream to greenish yellow or even blackish, often with reddish or black markings. Undersurface flat and black. Forelegs adapted for grasping prey. Hind legs long and fringed with hairs for swimming. Swim upside down, on their backs, and use their legs as oars. Both these and water boatmen (next) are efficient predators on mosquito larvae.

Water Boatmen *Corixa* spp. (Corixidae) 1/4" - 1/2"
Oblong-oval. Somewhat flattened. Grayish to brown, back usually with very fine cross banding. Forewings appear to arise immediately behind head. Forelegs short, tips spatulate, or scoop-like. Middle and hind legs elongate, flattened, paddle-like. Hind legs fringed with hairs for swimming. Swim by using legs as oars.

Giant Water Bug

Giant Water Bug
(*B. fluminea*)

Creeping Water Bug

Backswimmer

Backswimmer
(underneath)

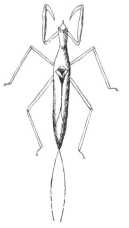

Water Scorpion

Water Boatman

Cicadas

HOMOPTERA (homo=alike; ptera=wing)
Cicadas, Hoppers and Aphids
Closely related to the true bugs (Hemiptera) but with these differences: homopterans also have sucking-piercing, beak-like mouthparts, but these arise at the back of the head (in hemipterans, these arise under the front of the head). Homopterans have four (usually) entirely membranous wings (in hemipterans only the wing tips are membranous). When at rest, wings usually are held roof-like, sloping over sides of body (in Hemiptera these usually are flat over body). All Homoptera are plant feeders and many are serious plant pests.

Cicadas (Cicadidae)
Cicadas are the largest of all Homoptera and often are called "locusts," but this is erroneous as they are not at all like locusts or grasshoppers.

Dog-day Cicada *Tibicen canicularis* 1 1/4" - 2"
Black, with green markings and clear but green veined membranous wings held roof-like over body when at rest. Fore pair of wings twice length of hind pair. Antennae short, bristle-like. Legs slender. Underneath often coated with a chalky white, powdery down. Males have sound producing organs on underside at base of abdomen and, characteristically, emit a loud, high, shrill, ascending and descending buzzing sound, especially during the hot "dog days" of August. Immatures (nymphs) live and develop underground for several (two to five (?)) years sucking on the juices of tree and other plant roots.

Pine-barren Cicada *Tibicen hieroglyphica* 1 1/4"
Similar to dog-day cicadas, above, except for size and an almost transparent abdomen. Our smallest species and a particularly characteristic one to be found in our pine barrens.

Periodical Cicada *Magicicada septendecim* 1" - 1 5/8"
In form, similar to dog-day cicada but smaller. Black with red markings, red eyes and legs, and clear but red veined, membranous wings, often with veins near tips of forewings darkened in form of a faint "W." Undersurface light reddish brown.

Immatures (nymphs) live underground for 17 years (hence, though erroneously, called 17 year "locusts") sucking on the juices of tree and other plant roots. Final nymphs emerge from ground in late May or early June, almost simultaneously, in large population broods, in specific geographic areas, once every 17 years. These final nymphs climb trees and other objects, pause to split down backs, and adults emerge to climb to tops of trees, "sing" and mate.

Best known and largest population brood in New Jersey pine barrens is brood number II (2) which last emerged in 1979 and will next reappear in the year 1996.

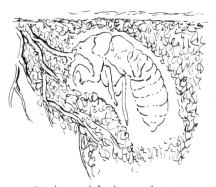
Cicada nymph feeding on plant roots

Cicada adult emerging

Dog-day Cicada

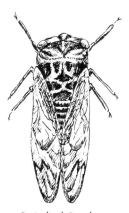
Periodical Cicada

Pine-barren Cicada

Hoppers & Spittlebugs

Leafhoppers (Cicadellidae)
Leafhoppers, the largest family of Homoptera, are common on all types of vegetation and feed on juices which they suck from plants. Hind legs long, adapted for jumping, with one or more rows of small spines on lower leg. A common and conspicuous representative is the:

Red-banded Leafhopper *Graphocephala coccinea* 3/8"
Elongate, slender, tapering toward rear. Head short, blunt but pointed, sort of crescent shaped. Antennae short, bristle-like. Hind legs long, adapted for jumping, with one or more rows of small spines on lower leg. Head yellow, midbody green and yellow, wings bright green with bright red stripes. Underneath yellow. Legs yellow.

Two other species of leafhoppers, the **sharp-nosed leafhopper**, *Scaphytopius magdalensis*, and the **blunt-nosed leafhopper**, *Eusculus striatulus*, are destructive pests on the two most important agricultural crops (cranberries and blueberries) in the pine barrens because they transmit pernicious plant viral diseases.

Treehoppers (Membracidae)
Recognized by large pronotum that covers head and extends back over abdomen, often in varied and odd shapes. Sometimes referred to as the "brownies" of the insect world.

Buffalo Treehopper *Stictocephala bisonia* (= *bubalus*) 3/8"
Green to yellowish green. Lighter beneath. Pronotum does not cover fore wings but, instead, is projected forward on each side to form a short, stout point, or "horn." Feeds on trees and shrubs.

Spittlebugs (Cercopidae)
Spittlebugs, or froghoppers, are small, hopping insects similar to leafhoppers but have only two stout spines on lower legs, not row(s) of small spines as in leafhoppers. Nymphs conceal themselves in protective masses of frothy bubbles which they create by sucking juices from plant tissues and which are found in the axils of leaves and on blades of grass.

Meadow Spittlebug *Philaenus spumarius* 3/8"
Elongate. Vaguely similar to tiny frog in shape. Rounded head. Very short antennae. Wings short, fit closely to body. Two stout spines on lower hind legs. Grayish to yellowish brown with pale spots.

Pine Spittlebug *Aphrophora parallela* 3/8"
Typical spittlebug form, as above, but sloped wings have white bands bordered by parallel black bands on each side. Nymphs found within spittles have scarlet colored eyes and salmon orange bodies. This species is an important pest on pine. (No illus.)

Red-banded Leafhopper

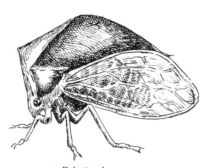

Buffalo Treehopper

Frothy mass on vegetation

Meadow Spittlebug

Spittlebug nymph inside frothy bubbles

Aphids, Fishfly, Lacewing, & Antlion

Planthopper *Acanalonia bivittata* (Fulgoroidea)   1/4"
Green to yellowish green, or brownish yellow. Broadly oval fore wings held almost vertically at rest. Head brown, rounded. Two dark stripes from eyes to tip of fore wings. Feeds on wide variety of herbaceous and woody plants. Most common plant hopper in pine barrens.

Planthopper *Scolops sulcipes*   1/4" - 3/8"
Head prolonged in front into a long, slender, somewhat upturned spine. Head, body, and wings speckled grayish brownish. Prominent longitudinal veins on fore wings. Feeds on grasses.

Aphid *Aphis spiracolae* (Aphididae)   1/16" - 1/8"
One common pine barrens example of a large family of very small, soft bodied, somewhat pear-shaped, plant sucking insects. Most aphids are bright or yellowish green but some occur in a range of other colors. Some adults are winged, others wingless, but nearly all have a pair of slender tubes (cornicles) projecting from near the tip end of abdomen. Aphids often are found in considerable numbers clustered on the tender new growth ends of stems, leaves, and flowers of various plants. They feed on the juices they suck from these plant tissues, thus weakening the plants. Because they reproduce so rapidly, they often occur in such great numbers that they are serious plant pests. Aphids produce a sweet, sticky, anal secretion called "honeydew" on which ants and other insects feed, ants often tending and protecting aphids for this purpose. Aphid populations are kept in check by many predators such as the larvae and adults of lacewing-flies and lady beetles, adult fire-colored beetles, and parasitic wasps.

MEGALOPTERA (mega=great; ptera=wing)

Fishfly *Chauliodes pectinicornis*   1 1/8" - 2"
Soft body. Body and wings gray. Veins on fore wings conspicuously darker. Antennae long with comb-like teeth along one side. Hind wings broader at base than fore wings. At rest, wings fold roof-like over body. Legs weak. Lives near ponds and streams where its aquatic larvae feed on aquatic insects. Metamorphosis complete.

NEUROPTERA (neuro=nerve; ptera=wing)

Green Lacewing *Chrysopa occulata*   3/8" - 5/8"
Soft, delicate, cylindrical. Pale green body. Eyes coppery golden, somewhat iridescent. Antennae thread-like. Wings pale green with green veins. Both larvae and adults are predacious upon aphids and similar small insects as well as their larvae and eggs. Metamorphosis complete.

Antlion *Hesperoleon abdominalis*   1 3/8" - 1 5/8"
Soft, elongate, cylindrical body. Brownish gray to gray. Antennae clubbed. Four wings long, narrow, membranous, and transparent, with many cross

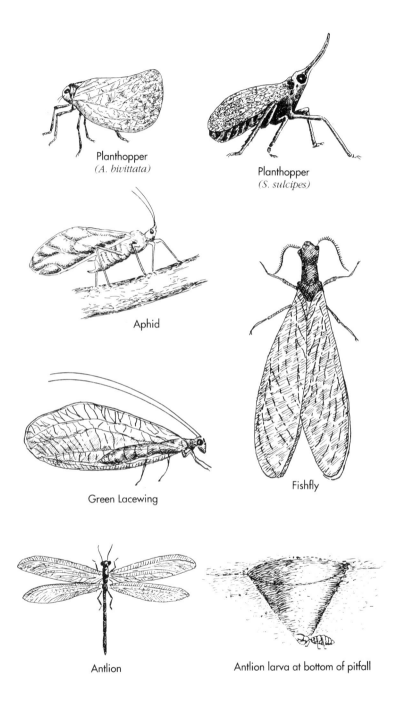

veins. Resemble damselflies in size and shape but hold wings back over abdomen, roof-like, and have clubbed antennae.

Larvae, or "doodlebugs," are oval, flattened, with large heads, long, sickle-shaped jaws, and bristles over bodies. Predacious. Most lie at bottoms of small cone-shaped pit-fall traps they dig in dry sand, then wait for ants and similar insects to fall in, whereupon the antlion larvae seize the prey with their sickle-shaped jaws. Metamorphosis complete.

COLEOPTERA (coleo=sheath; ptera=wing)
Beetles

Adult beetles nearly always have four wings, the outer (fore) pair (wing covers) thickened, hard, and horny, and are called elytra (singular elytron). When closed, these lie over the back and meet down the middle in a straight line (suture). Often, wedged between the bases of the elytra, is a small triangular or rounded segment called a scutellum. The elytra serve as covers for the body and for the membranous hind (under) wings which are the flight wings for the insect and which, when at rest, are folded underneath the wing covers. The elytra are not used for flight but, during flight, are held straight out on the sides so as to allow the hind wings to function. The shield-like segment ahead (in front) of the elytra is the pronotum which covers the mid-body (thorax) of the beetle. Mouthparts are formed for chewing in both adults and larvae and the jaws (mandibles) often are very powerful. The body is usually rather stout and the external skeleton is generally thicker and harder than in insects in most other orders. Metamorphosis is complete: egg, grub or borer (larva), pupa, and adult.

This is the largest order of insects, containing nearly one third of all insect species. Smith, 1909, listed over 3000 species in New Jersey. Because of the great number of common and conspicuous beetle species in the pine barrens, many of the descriptions which follow will, due to space limitations, be of general types (major families) of beetles. In these cases, descriptions of individual species will be restricted to only a few representative examples of each family and will be limited in scope.

Tiger Beetles (Cicindelidae) 3/8" - 3/4"

Slender, oblong or elongate. Run along ground but fly readily when disturbed. Many brightly colored in tones of green, blue, coppery, red, brown, or black, often with metallic or iridescent hues. Others have varying patterns or stripes in black, brown and white. Wing covers (elytra) nearly always banded or spotted with whitish markings. A pair of large, strong, sickle-shaped jaws. Antennae thread-like, eleven segmented, inserted in front of head. Legs long, slender, adapted for running.

Immatures (larvae) live in vertical burrows they construct in soil where they lie in wait to seize passing prey, like ants. Both immatures and adults very predacious.

Diurnal. Open, sunny, sandy areas throughout pine barrens are favored habitats. Sixteen of the 20 New Jersey species occur here. Some examples of common and widespread pinelands species follow. The first six occur from mid-April through May and again in September; the next three occur during summer months. (All illus. on pg. 331)

Cicindela repanda - small, brownish-bronze with light whitish markings; often on or near wet sand. (Illus. No. 1)
C. formosa generosa - robust, reddish-brown to black with broad white bands across and along margins of elytra. (Illus. No. 2)
C. purpurea - coppery-red or purplish green, margins greenish; oblique, yellowish middle band. (Illus. No. 3)
C. patruela consentanea - black with oblique, white middle band. Virtually endemic to New Jersey pine barrens. (Illus. No. 4)
C. scutellaris rugifrons - green or black with white spot, often triangular, on mid-margin of each elytron. (Illus. No. 5)
C. tranquebarica - black with oblique and transverse white bands. May be most common and widespread species along sandy roads throughout the pine barrens. (Illus. No. 6)
C. punctulata - slender, greenish bronze, with inconspicuous white spots. (Illus. No. 7)
C. abdominalis - Shiny black with scattered small white dots. Bright metallic blue underneath. Abdomen brownish red underneath. Our smallest (5/16"-7/16") tiger beetle. This and next are our two red-bellied forms. (Illus. No. 8)
C. rufiventris - Cupreous bronze-brown to dark smoky brown, with white elytral mid band and dots. Underneath metallic blue. Abdomen brownish red underneath. (Illus. No. 9)

Two other species of tiger beetles are unique in that they are either crepuscular or nocturnal and both were thought to be extremely rare but now are known to be quite common in the pine barrens. These are:

Megacephala virginica - larger (3/4"-1") than other tiger beetles. Bright, shining, dark gold-green, lighter on sides, darker to blackish on dorsal surface of elytra. Antennae, legs, and tip of abdomen dull brownish- yellow. Nocturnal. Hides under objects on ground during day. (Illus. No. 10)
Cicindela unipunctata - (5/8"-3/4") Dull, dark bronze-brown. Elytra somewhat flattened and roughly sculptured. Small, triangular white spot midway along margin of each elytron. Underneath dark purplish blue. Crepuscular and nocturnal, only occasionally seen in broad daylight. (Illus. No. 11)

Ground Beetles (Carabidae)

Generally elongate, somewhat flattened. Vary greatly in size, very small to quite large. Most in tones of dull or shiny brown or black, relatively few in brighter colors. Head with pair of large, strong jaws. Antennae thread-like, eleven segmented, inserted under ridges on sides of head. Legs slender, adapted for running.

Live on surface of ground, usually under leaves, bark, boards, or other cover by day, but roam around on ground at night in search of prey. A few climb trees. Most, including their larvae, are predacious. A few feed on seeds. Second largest family of beetles. Smith (1909) listed nearly 400 species in New Jersey.

Snail-eating Ground Beetle *Scaphinotus elevatus* 3/4" - 7/8"
Oval, robust, violet-black, shining, with darker violet tones along upturned edges. Head narrow and elongate to let beetle reach inside shell of a snail and seize and draw out animal as prey. Antennae and legs long, slender.

Numbers limited to abundance of snails which are limited in numbers in pine barrens due to high acid environment and low levels of shell building calcium.[38]

Ground Beetle *Carabus vinctus* 3/4" - 7/8"
Dull black, sometimes with bronze or greenish reflections along margins. Each elytron has three rows of short, dash-like ridges, or broken intervals, like links of a chain, separated by rows of shallow, pitted grooves.

Fiery Caterpillar Hunter *Calosoma scrutator* 1" - 1 1/4"
One of largest and most beautiful beetles. Robust. Pronotum blue or purplish, elytra greenish or bluish, both somewhat metallic, highly iridescent, with reddish margins. Underneath green and red bronze. Legs bluish. Runs over ground and climbs trees in search of caterpillars.

Caterpillar Hunter *Calosoma syncophanta* 1" - 1 1/8"
Similar to above except elytra, including margins, brighter green with an iridescent, coppery sheen. Underneath dark blue. Same feeding habits as above. Introduced as predator on caterpillars of gypsy and browntail moths. (No illus.)

Caterpillar Hunter *Calosoma willcoxi* 3/4" - 7/8"
Similar to *C. scrutator* except for size. Elytra usually darker green with margins sometimes green instead of reddish. Same feeding habits as *C. scrutator*. (No illus.)

Caterpillar Hunter *Calosoma calidum* 7/8" - 1 1/8"
Elongate-oval. Black above and underneath. Elytra with three rows of reddish or copper colored punctures, and with deep, finely punctured grooves. Same feeding habits as *C. scrutator*.

Caterpillar Hunter *Pasimachus depressus* 7/8" - 1 1/8"
Broad, elongate-oval. Black with faintly blue margins. Thorax and abdomen joined by neck-like stem. Pronotum broad with prominent angles. Elytra entirely smooth. Legs stout, fore pair somewhat enlarged for digging. Most common caterpillar hunter type of ground beetle in pine barrens. (Illus. on pg. 333)

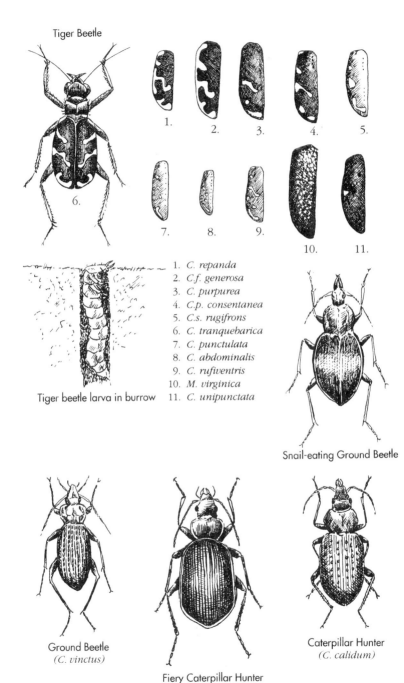

Beetles

Ground Beetle *Scarites subterraneus* 5/8" - 3/4"
Elongate, narrow. Black, shining. Thorax and abdomen joined by neck-like stem. Elytra distinctly grooved, sides parallel. Lower fore legs widened, flattened, toothed on outer side, and fitted for digging.

Canker-worm Hunter or *Galerita janus* 3/4" - 7/8"
False Bombardier Beetle *Galerita bicolor*
Two species, separated by minor differences. Elongate-oval. Head elongate, narrowed in front of eyes, constricted behind eyes into slender neck. Head black. Elytra black or blue-black, covered with short hairs. Legs and bases of antennae reddish brown. Elytra finely grooved.

Ground Beetle *Helluomorphoides bicolor* 9/16" - 3/4"
  Helluomorphoides texana
Two species. Oblong elongate. Head broadly oval with a distinct neck. Pronotum heart shaped, punctured. Elytra shorter than abdomen, squared off at tip, strongly grooved. Elytra of *H. bicolor* considerably darker than head and pronotum. *H. texana* uniformly brown. Previously thought to be rare, now known to be common in pine barrens.

Bombardier Beetle *Brachinus fumans* 1/2"
Head and thorax narrow. Abdomen broad. Head, antennae, thorax, and legs reddish yellow. Abdomen reddish brown. Wing covers bluish or blackish blue, grooved, square cut at tip. (No illus.)
  When disturbed, these beetles emit, from the end of their abdomen, a pungent, volatile liquid which serves as a means of defense. This is ejected, usually, with a sound like a small pop gun. When it contacts the air, it changes to a gas which looks like a small puff of smoke.

Green Ground Beetle *Chlaeneus sericeus* 1/2" - 5/8"
Elongate, oval, convex. Bright green above, black underneath. Elytra bright green, sometimes with bluish tinge, finely grooved, covered with fine hairs. Antennae and legs pale brownish yellow.

Ground Beetle *Harpalus caliginosus* 3/4" - 1"
Robust. Oblong-elongate. Somewhat flattened. Uniformly black, slightly shining, including legs. Antennae and tips of legs reddish brown. Elytra deeply grooved.

Ground Beetle *Harpalus pennsylvanicus* 1/2"
Nearly identical to above, except for size. Moderately shining, both antennae and legs reddish yellow, underneath dark reddish brown. Elytra only moderately grooved. (No illus.)

Caterpillar Hunter
(P. depressus)

Ground Beetle
(S. subterraneus)

Canker-worm Hunter

Ground Beetle
(H. bicolor)

Green Ground Beetle

Ground Beetle
(H. caliginosus)

Beetles

Aquatic Beetles
The next three families are all aquatic or semi-aquatic beetles and have adaptations for swimming and breathing to enable them to live in this environment.

Predacious Diving Beetles (Dytiscidae)
Elongate-oval. Convex above and below. Antennae thread-like. Hind legs long, flattened, oar-like, fringed with hairs, adapted for swimming. Rise to surface for air which they take in by means of breathing pores at end of abdomen while hanging upside down from water surface. Carry air underwater as bubbles and use until need to replenish supply. Usually shining dark olive brown or dark greenish black, often with light yellow markings, especially along margins. Immatures known as "water tigers."

Dytiscus fasciventris   1" - 1 1/8"
Pronotum and elytra with marginal yellow stripe. Female with ten elytral grooves extending beyond middle.

Dytiscus harrisii   1 1/2"
Pronotum and elytra with marginal yellow stripe. Pronotum with yellowish bands across front and hind edges. Elytra with oblique, often indistinct, yellow band across wing tips.

Dytiscus verticalis   1 1/4"
As *D. harrisii* but smaller and without the cross bands on pronotum. (No illus.)

Cybister fimbriolatus   1 1/4"
Pronotum and elytra with marginal yellow stripe. Female with numerous short, fine grooves.

Whirligig Beetles (Gyrinidae)
Two genera: *Dineutes* spp. (over 3/8") and *Gyrinus* spp. (not over 5/16"). Oblong oval, flattened but slightly convex, somewhat tapering at each end. Body smooth, hard, shining. Usually black with metallic reflections. Compound eyes divided into upper and lower parts at water level so they look like they have four eyes. Thus they can see both above and below water surface at same time. Antennae short, clubbed. Front legs long. Middle and hind legs short, flattened, paddle-like. Swim by whirling or gyrating on water surface. Predacious.

Water Scavenger Beetles (Hydrophilidae)
Oval-elliptical, convex above. Dark brownish or olive black, shining. Legs reddish black. Differ from predacious diving beetles by being more convex and having their short antennae end in a distinct club. Certain mouthparts (palpi) are longer than antennae, but not clubbed. Since antennae usually are concealed beneath head, palpi often are mistaken for antennae. Under surface often keeled, produced into a long spine. Middle and hind legs of

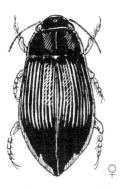

Predacious Diving Beetle
(*D. fasciventris*)

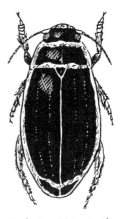

Predacious Diving Beetle
(*D. harrisii*)

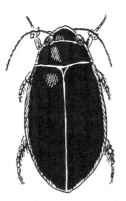

Predacious Diving Beetle
(*C. fimbriolatus*)

Whirligig Beetle

Beetles

aquatic species bear long, hairy fringes for swimming. Aquatic adults obtain air at water surface and carry a film of air down under their wing covers and around their bodies to enable them to breathe under water. Most are aquatic and predacious.

Hydrophilus triangularis   1 3/8" - 1 1/2"
Dark, olive green-black. Our largest species.

Hydrochara obtusata   1/2" - 5/8"
Elytra with four rows of distinct punctures. Often attracted to lights.

Sexton or Burying Beetles (Silphidae)

Elongate, thick bodied, loose jointed. Black, shining, marked with orange-red. Antennae clubbed. Elytra squared off at back, usually shorter than abdomen exposing two or three abdominal segments. Live in, on, and underneath, and feed on dead and decaying animal matter as carcasses of dead birds, mammals, snakes.

Nicrophorus orbicollis   3/4" - 1 1/8"
Disk of pronotum round, smooth, shining, black. Four prominent orange-red spots on elytra. Last three joints of antennal club red.

Nicrophorus tomentosus   5/8" - 1"
Disk of pronotum densely covered with fine silky, yellow hairs. Black, with two orange-red bands across elytra. Antennal club black.

Nicrophorus pustulatus   5/8" - 1"
Pronotal disk smooth, shining. Black with orange bands on elytra reduced to small spots. Last three joints of antennal club orange-red. (No illus.)

Carrion Beetles (Silphidae)

Necrodes (= Silpha) surinamensis   5/8" - 1"
Elongate-oblong, flattened. Black or reddish-black. Disk of pronotum shiny black. Narrow, wavy, yellowish or orange-red band, often reduced to spots or sometimes lacking, across elytra near tips. Elytra gradually broader behind with three distinct ridges on each wing cover. Feed on carrion and small dead animal bodies.

Silpha americana   5/8" - 3/4"
Broadly oval. Elytra "wrinkled," brownish, ridges darker. Pronotum yellow with brown or black spot in center. Found on carrion and decaying fungi.

Silpha novaboracensis   1/2" - 9/16"
Similar to above but smaller. Oblong-oval, flattened. Disk of pronotum brown or black, with wide reddish-yellow margins. Elytra brownish to black with three ridges on each elytron. Found on carrion and decaying fungi. (No illus.)

Water Scavenger Beetle
(H. triangularis)

Water Scavenger Beetle
(H. obtusata)

Sexton or Burying Beetle
(N. orbicollis)

Sexton or Burying Beetle
(N. tomentosus)

Carrion Beetle
(N. surinamensis)

Carrion Beetle
(S. americana)

Beetles

Rove Beetles (Staphylinidae)
Slender, elongate. Elytra short, square-cut, leaving several abdominal segments exposed. Abdomen flexible. When disturbed, insect turns tip of abdomen up over back as if to sting, but is harmless. Found on carrion and decaying vegetative matter. Largest family of beetles but many are very small and difficult to identify so only two of larger, common pine barrens species are described here.

Hairy Rove Beetle *Creophilus maxillosus*   1/2" - 7/8"
Robust, shining, black. Elytra and three segments of abdomen banded with grayish yellow hairs.

Rove Beetle *Staphylinus maculosus*   3/4" - 1"
Dark reddish brown. Elytra and upper side of abdomen fuscous. (No illus.)

Hister Beetles (Histeridae)
Small, oval, short, chunky, convex, hard, compact. Most shining, highly polished, black, some with red markings. Head usually withdrawn under pronotum. Elytra grooved, square-cut, short, leaving tip of abdomen exposed. Found in carrion, excrement, decaying vegetation, garbage. Some in fungi and under bark of trees.

Hister interruptus   1/2"
Hister abbreviatus   1/8" - 1/4"
Both oval, convex, jet-black, shining.

Stag Beetles (Lucanidae)
Elongate. Robust. Mahogany red-brown to black. Antennae usually elbowed with loose, comb-like club of three to four segments. Mandibles of males sometimes very large, antler-like. Both adults and larvae live in and beneath decaying oak and other hardwood stumps.

Pinching Beetle *Pseudolucanus capreolus*   7/8" - 1 1/2"
Elongate-oblong. Smooth, shining. Reddish brown. Mandibles of males as long as head, with a tooth on inner side of each mandible. Female head and jaws smaller.

Antelope Beetle *Dorcus parallelus*   5/8" - 1"
Oblong. Parallel sided. Dark brown to almost black, shining. Thorax and abdomen joined by a neck-like stem. Mandibles of males have large median tooth pointing obliquely inward and upward. Elytra deeply grooved, densely covered with fine punctures.

Horned Passalus *Odontotaenius disjunctus* (Passalidae)   1 1/8" - 1 1/2"
Elongate. Robust. Parallel sided. Thorax and abdomen joined by a neck-like

Hairy Rove Beetle

Hister Beetle

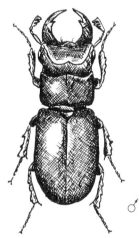
Pinching or Stag Beetle

Antelope Beetle

Horned Passalus or Bess Beetle

stem. Black. Shining. Small forward pointing "horn" or "hook" projects from top of head. Antennae curved, not elbowed, with loose club. Pronotum squarish, smooth, with deep median line. Elytra with deep, finely punctured, lengthwise grooves. These insects are partly social. Live in small colonies in large, rough galleries, they burrow in damp, rotting wood and stumps.

Scarab Beetles (Scarabaeidae)

Body form variable, usually oval to elongate. Often stout and heavy. Antennae end in oval club of 3 to 7 "leaves" which can be closed into a tight club or opened to sense odors. Lower front legs broadened, flattened, toothed on outer edge, fitted for digging in ground. Elytra nearly always hard, often highly polished and attractively marked. Large family. Smith (1909) listed 260 species in New Jersey. Many species quite common in pine barrens.

Tumblebug Beetles *Canthon chalcites*   1/2" - 3/4"
Canthon bispinosus   1/4" - 3/8"

Broadly oval, robust. Dull black. Elytra short, leaving last dorsal segment of abdomen exposed. The two above species separated here only by size. These beetles roll balls of dung along ground before burying ball on which female deposits an egg. Emerged larva feeds on dung. Closest relative here to famous sacred scarab beetle of ancient Egyptians.

Dung Beetle *Dichotomius carolinus*   3/4" - 1 1/8"

Broadly oval, convex, strongly robust. Black, shining. Each elytron with seven shallow, feebly punctured grooves. Male with short, blunt horn on front of head. Our largest and strongest dung beetle. Digs burrow in ground under dung, usually of cattle, which serves as its food supply. Carries excrement down into burrow and forms ball in which female lays its egg. Dung provides food for larva.

Dung Beetle *Phanaeus vindex*   9/16" - 7/8"

Broadly oval, robust, nearly convex. Head bronzed. Pronotum bright coppery red. Elytra metallic green, often slightly bluish. Male has long, curved horn on front of head, female a short tubercle. Elytra grooved, with intervals between grooves ribbed. Our most beautiful and brilliantly colored dung beetle. In same habitats and with same habits as *D. carolinus*.

Earth-boring Dung Beetles *Geotrupes splendidus*   1/2" - 3/4"
Geotrupes egerei

Both broadly ovate, robust, convex, shining. Elytral grooves distinct and strongly punctured. *G. splendidus* varies from bright, metallic green or deep blue-green to metallic purple. *G. egerei* is shining black with traces of iridescent blue or green in punctures. Of six very similar species, these are the two most common in pine barrens. *Geotrupes* species dig burrows in

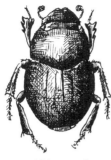

Tumblebug Beetle

Tumblebug beetle rolling ball of dung

Dung Beetle
(*D. carolinus*)

Dung Beetle
(*P. vindex*)

Earth-boring Dung Beetle

Beetles

ground under or near dung or fungi and carry dung into burrows as described for *D. carolinus*.

May Beetles *Phyllophaga* spp. 3/8" - 1"
Most are oblong oval, robust, thick bodied, heavy. Few are small, light. Yellowish to orange brown, to chestnut, to dark reddish brown to brown. Most are smooth, glossy above, without hairs or grooves on elytra. A few are downy or with short, fine hairs. Underneath most have long, dense hairs. All are leaf feeders, usually nocturnally. Larvae are "C"-shaped fat, "white grubs" that live in soil and feed on plant roots. Over 30 species in New Jersey. Two common pine barrens species are:

Phyllophaga fusca 3/4"
Light to dark chestnut brown to piceous. Shining.

Phyllophaga marginalis 3/4"
Reddish or chestnut brown to piceous. Shining. Smooth.

Japanese Beetle *Popillia japonica* 3/8" - 1/2"
Oblong-oval, robust, somewhat flattened. Head, pronotum, and scutellum bright shining green with bronze reflections. Elytra chestnut brown or brownish orange, shining. Suture and margins greenish. Underneath metallic green with white spots along sides. Two white spots on exposed dorsal tip of abdomen. Larvae feed on roots. A pest species of leaf feeding beetle. Introduced from Japan on iris roots in 1916.

Grapevine Beetle or *Pelidnota punctata* 3/4" - 1"
Spotted Pelidnota
Broadly oval, convex, robust. Yellowish to reddish brown, shining. Base of head, scutellum, legs, and underneath greenish black. Ten small, round, black spots on pronotum and elytra. In pine barrens, feeds on leaves of grape vines and virginia creeper.

Goldsmith Beetle *Cotalpa lanigera* 3/4" - 1"
Broadly oval, convex, robust. Head, pronotum, and scutellum yellow with metallic greenish reflections. Elytra yellow. Underneath black, hairy. Feeds on foliage. This is Edgar Allen Poe's famous "gold bug."

Carrot Beetle *Bothynus gibbosus* 1/2" - 5/8"
Oblong, robust, convex. Reddish to dark chestnut brown. Underneath paler. Pronotum with small tubercle on front margin. Elytra with rows of punctures.

Ox Beetle *Strategus antaeus* 1 1/8" - 1 1/4"
Oblong, oval, convex, very robust. Dark cherry red, mahogany or reddish brown. Shining. Male has three erect, curved horns on pronotum, the

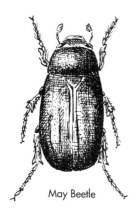
May Beetle

Japanese Beetle

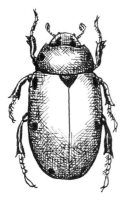
Grapevine Beetle

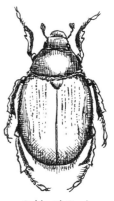
Goldsmith Beetle

Carrot Beetle

Ox Beetle

Beetles

middle one largest and most curved. Female has single tubercle. On ground, under leaves, common around decaying stumps throughout pine barrens.

Rhinoceros Beetle *Xyloryctes jamaicensis*   1" - 1 1/8"
Oblong, convex, very robust. Very dark reddish or blackish brown. Underneath paler, heavily covered with reddish brown hairs. Male with single curved horn arising from middle front of head. Female with small tubercle. Elytra grooved with rows of fine punctures. Larvae are root feeders.

Green June Beetle *Cotinus nitida*   3/4" - 7/8"
Elongate-oblong, somewhat flattened and narrowed toward tips of wing covers. Robust. Pronotum and elytra dull to metallic velvety green with brownish yellow on sides. Head and underneath thorax bright metallic green. Abdomen orange or reddish yellow. Head with small spine or horn. Adults feed on flowers and ripening fruit, especially peaches. Flies with loud buzzing noise. Favors sandy areas as pine barrens. Larvae are root feeders.

Bumble Flower Beetle *Euphoria inda*   1/2" - 5/8"
Oblong oval. Robust. Yellowish brown, feebly bronzed. Hairy. Wing covers mottled with small, black spots and with two feeble ridges which merge behind. Flies with low buzzing similar to bumble bee.

Ant Nest Beetle *Cremastocheilus harrisii*   3/8"
Elongate oval, flattened. Wholly black, shining, roughly sculptured, faintly grooved with dense, shallow, but coarse punctures. Exposed tip of abdomen coarsely punctured. Occurs in nests of large, reddish ants, in sandy locations.

Rough Flower Beetle *Osmoderma scabra*   3/4" - 1"
Hermit Flower Beetle *Osmoderma erimicola*   1" - 1 1/8"
Both species broadly oval, very robust. Flattened above. Pronotum distinctly narrower than elytra. *O. scabra* purplish black, slightly bronzed, roughly sculptured, irregularly pitted. *O. erimicola* dark reddish or mahogany brown to blackish. Shining. Smooth with faint rows of very fine punctures.

Flower Beetle *Trichiotinus piger*   3/8" - 1/2"
            Trichiotinus affinis
Two species separated by technical differences. Both robust. Head and thorax dark greenish, usually covered with yellowish hairs. Elytra smooth, shining, reddish-brown with short, white marginal dash lines. Underneath bronzed with long, white silky hairs. Common on flowers, especially roses. Larvae develop in old oak stumps.

Rhinoceros Beetle

Green June Beetle

Bumble Flower Beetle

Ant Nest Beetle

Rough Flower Beetle

Hermit Flower Beetle

Flower Beetle
(T. piger)

Beetles

Metallic Wood-boring Beetles (Buprestidae)
Hard, cylindrical to flattened, elongate-elliptical, stout, robust. Upper surface usually metallic, shiny, often roughly sculptured. Parallel sided. Elytra usually ends in a pointed tip. Usually metallic or bronzed underneath. Antennae usually saw-toothed. Heads of larvae expanded and flattened.

Flat-headed Pine Borer *Chalcophora virginiensis* 7/8" - 1 1/4"
Elongate-elliptical, robust. Dull black with strong bronze reflections. Upper surface roughened with irregular, lengthwise, often brassy, furrows. Largest of our common buprestids. Larvae burrow under bark of pine trees.

Flat-headed Cherry Tree Borer *Dicerca divaricata* 5/8" - 7/8"
Elongate-elliptical. Robust. Convex. Gray, brown, or black with coppery, brassy, or greenish-bronze reflections. Underneath shiny coppery. Elytra distinctly separated and spreading apart at tips. Breeds in dead and dying hardwoods.

Flat-headed Apple Tree Borer *Chrysobothris femorata* 1/4" - 5/8"
Oblong or elongate-oblong, somewhat flattened. Dark bronze with metallic reflections of green, copper, or brass. Name comes from larvae which bore beneath the bark of virtually all fruit, shade, and forest trees, often completely girdling and killing the trees.

Bronze Birch Borer *Agrilus anxius* 1/4" - 1/2"
Elongate. Bronzed olive green. Elytra covered with granulate scales. Larvae highly destructive to shade and forest trees. (No illus.)

Click Beetles (Elateridae)
Elongate, somewhat convex. Hard bodied. Parallel sided. Antennae usually saw-toothed. Hind angles of pronotum prolonged backwards to form sharp points. Most intriguing character is their ability to snap themselves, from an upside down position, up into the air with a clicking sound and land on their feet. This is due to a loose "hinge" or "clicking" device underneath, between the thorax and abdomen. Larvae are worm-like and have hard coverings, so are known as "wireworms."

Eyed Click Beetle *Alaus oculatus* 1" - 1 3/4"
Elongate. Black. Shining. Pronotum with two large, velvety black "eye spots," each surrounded by a ring of white scales. Elytra grooved lengthwise, speckled with small silvery white scales, a sort of salt and pepper effect. Adults in decaying logs. Is our largest click beetle.

Blind Click Beetle *Alaus myops* 1" - 1 1/2"
Similar to above but smaller. Reddish brown to black. "Eye spots" smaller, indistinctly margined, and more difficult to see. Adults found under bark of dead pine trees. (No illus.)

Flat-headed Pine Borer

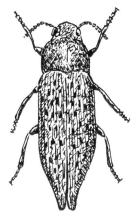

Flat-headed Cherry Tree Borer

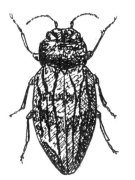

Flat-headed Apple Tree Borer

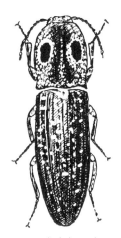

Eyed Click Beetle

Beetles

Fireflies (Lampyridae)
Not flies, but beetles. Elongate or oblong, somewhat flattened. Soft bodied. Soft wing covers. Head concealed under pronotum. Elytra extend beyond abdomen. Most interesting characteristic is ability to produce a "cold" light. Light organs located on one or two segments on underside of abdomen and can be recognized as greenish-yellow spots. Larvae also can produce light and are known as "glowworms." Adult fireflies fly throughout most of early summer.

Ellychnia corrusca 3/8" - 9/16"
Black. Disk and side margins of pronotum black. Space around central disk yellowish tinged with reddish. Does not have light organs. Common on tree trunks in spring.

Photinus pyralis 3/8" - 9/16"
Dark brown to black. Pronotum yellowish with central black spot surrounded with rose-pink. Suture and side margins of elytra pale yellow.

Photuris pennsylvanicus 3/8" - 9/16"
Brownish. Pronotum dull yellowish with central dark stripe surrounded by pale pink. Suture, side margins, and a narrow oblique stripe on elytra pale yellow.

Net-winged Beetles (Lycidae)
Elongate, flattened, roughly wedge-shaped. Soft bodied, soft wing covers. Head concealed under pronotum. Elytra extend beyond abdomen, widen behind. Each elytron with four lengthwise ridges with net-like cross ridges. Closely related to fireflies. Also known as false fireflies.

Banded Net-winged Beetle *Calopteron reticulatum* 5/16" - 3/4"
Golden or orange yellow with black spot in middle of pronotum. Narrow black band across elytra near base. Wide black band at tips of elytra. Antennae saw-toothed.

Banded Net-winged Beetle *Calopteron terminale* 5/16" - 1/2"
Golden or orange yellow with wide black band across tips of elytra.

Soldier Beetles (Cantharidae)
Elongate, slender, rather narrow, parallel sided. soft bodied. Soft wing covers.

Chauliognathus pennsylvanicus 3/8" - 1/2"
Pronotum and elytra dull orange yellow. End of each elytron with large oval black spot bordered with yellow. Pronotum with broad, central black spot. Head and legs black. Abundant in fall on goldenrod flowers.

Firefly
(E. corrusca)

Firefly
(P. pyralis)

Firefly
(P. pennsylvanicus)

Banded Net-winged Beetle
(C. reticulatum)

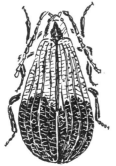

Banded Net-winged Beetle
(C. terminale)

Soldier Beetle
(C. pennsylvanicus)

Beetles

Chauliognathus marginatus 3/8" - 1/2"
Similar to above but smaller and narrower. Black spot on pronotum narrow. Black spots on elytra variable.

Podabrus modestus 3/8" - 1/2"
Pronotum squarish, yellowish, with central dark spot. Elytra dark brownish black with narrow yellow stripes along suture and margins. Superficially looks like a firefly but head is exposed.

Checkered Beetles (Cleridae)

Checkered Beetle *Cymatodera undulata* 3/8" - 1/2"
Slender, elongate, nearly cylindrical. Brownish. Legs dull yellow. Elytra with three (usually) pale crossbars. Feeds on flowers, especially those of prickly pear cactus.

Checkered Beetle *Chariessa pilosa* 3/8" - 1/2"
Elongate. Dull jet black. Pronotum reddish with wide, lengthwise, jet-black stripe on each side of median red line. Elytra with dense, fine punctures and fine hairs. Last three joints of antennae very large, dilated and triangular. Superficially looks like a firefly but is not.

Pleasing Fungus Beetle *Megalodacne heros* (Erotylidae) 9/16" - 7/8"
Elongate oval-elliptical. Convex. Robust. Shining. Black with two orange bands, one at base of elytra, other behind middle, near tips of elytra which taper behind. Black underneath. Occurs on fungi, beneath dead bark, and on rotting wood.

Lady Beetles (Coccinellidae)
Round, convex, sometimes somewhat oval. Head mostly concealed from above. Underneath flat. Usually yellow, orange, or reddish with black spots. A few black with yellow or red spots. Both immatures and adults of most species are valuable predators on aphids, plant lice and scale insects.

Spotted Lady Beetle *Coleomegilla fuscilabris* 3/16" - 1/4"
Oval. Bright red or pink, not orange-red, marked with five black spots on each elytron, one on scutellum and two on pronotum. (Illus. on pg. 353)

Convergent Lady Beetle *Hippodamia convergens* 1/4" - 5/16"
Oval. Light to darker yellowish red with six or less black spots on each elytron, and one on scutellum. Pronotum black with two converging white stripes. (Illus. on pg. 353)

Nine-spotted Lady Beetle *Coccinella novemnotata* 3/16" - 1/4"
Round. Yellowish red marked with four black spots on each elytron, and one on scutellum. (Illus. on pg. 353)

Soldier Beetle
(C. marginatus)

Soldier Beetle
(P. modestus)

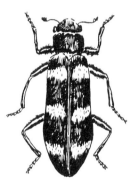

Checkered Beetle
(C. undulata)

Checkered Beetle
(C. pilosa)

Pleasing Fungus Beetle

Beetles

Fifteen-spotted Lady Beetle *Anatis quindecimpunctata* 1/4" - 3/8"
Oval. Very convex. Light to dark yellow, sometimes dark yellowish brown. Eight black spots on each elytron, plus one on scutellum. Two yellowish spots on pronotum.

Two-spotted Lady Beetle *Adalia bipunctata* 1/8" - 3/16"
Round. Orange, with one round black spot centered on each elytron.

Painted Lady Beetle *Mulsantia picta minor* 1/8" - 3/16"
Round. Two black bands across elytra, one across middle, other across and around tips of elytra, reaching up sides to meet with ends of middle band. Common on pine trees.

Plain Lady Beetle *Neomysia pullata* 3/16" - 1/4"
Round. Yellow-orange. Yellow spot with central black dot on each side of pronotum. No spots on elytra.

Black Lady Beetle *Axion plagiatum* 3/16" - 1/4"
Round. One orange-red spot centered on each elytron.

Darkling Beetles (Tenebrionidae)
Larger species elongate-oblong. Somewhat loose jointed. Black. Somewhat like ground beetles but slow and awkward moving. Occur in fungi and under loose bark of dead trees, old stumps and fallen logs. Over 50 New Jersey species.

Xylopinus saperdioides 1/2" - 5/8"
Elongate. Narrow. Somewhat convex. Sides nearly parallel. Dark reddish brown to black. Shining. Elytral grooves deep and distinctly pitted.

Alobates pennsylvanica 3/4" - 7/8"
Elongate oval. Convex. Dull black, feebly shining. Elytra and body gradually widen behind before turning in toward tip. Elytra with rows of fine punctures. Common under loose bark of dead trees.

Tarpela micans 3/8" - 3/4"
Elongate oval. Blackish, bronzed. Elytra metallic with reddish and greenish iridescent reflections. Elytra with finely pitted grooves.

Forked Fungus Beetle *Boletotherus cornutus* 3/8" - 1/2"
Oblong. Robust. Dull dark brown to black, coarsely sculptured above. Male with pair of stout, forward pointing horns on pronotum and one forked horn on head. Female with only tubercle on pronotum. Lives and feeds on woody, bracket type fungi on dead and dying tree trunks.

Spotted Lady Beetle

Convergent L.B.

Nine-spotted L.B.

Fifteen-spotted L.B.

Two-spotted L.B.

Painted L.B.

Plain L.B.

Black L.B.

Darkling Beetle
(*X. saperdioides*)

Darkling Beetle
(*A. pennsylvanica*)

Darkling Beetle
(*T. micans*)

Forked Fungus Beetle

Beetles

Fire-colored Beetle *Neopyrochroa flabellata* (Pyrochroidae) 1/2" - 5/8"
Elongate. Somewhat flattened. Soft bodied with soft wing covers. Head constricted behind into a slender neck. Antennae comb-like. Head and pronotum yellowish. Elytra black.

Blister Beetles (Meloidae)
Elongate, slender, almost cylindrical. Soft bodied with soft wing covers. Loosely jointed. Head constricted behind eyes into a narrow neck. Many species contain cantharidin which can cause blisters on human skin.

Striped Blister Beetle *Epicauta vittata* 1/2" - 3/4"
Dull yellow. Underneath and legs black. Head and pronotum each with two black spots. Each elytron with two longitudinal black stripes.

Margined Blister Beetle *Epicauta pestifera* 1/4" - 5/8"
Dull black. Margins of head and pronotum clothed with grayish pubescence, as also suture and margins of elytra.

Black Blister Beetle *Epicauta pennsylvanica* 1/4" - 5/8"
Uniformly dull black with sparse, fine pubescence. Common on goldenrod.

Long-horned Beetles (Cerambycidae)
Elongate. Cylindrical, sometimes flattened. Antennae very long, often two or three times length of body. Elytra hard. Adults are plant feeders. Larvae, called round-headed borers, tunnel in wood.

Lesser Prionus *Orthosoma brunneus* 7/8" - 1 3/4"
Very elongate. Somewhat cylindrical. Parallel sided. Light to darker reddish brown. Shining. Three short spines on each side of pronotum. Three fine raised lines on each elytron. Larva a destructive borer in wood.

Broad-necked Root borer *Prionus laticollis* 7/8" - 2"
Broad. Very robust and broad shouldered. Reddish black to black. Shining. Two or three short spines on each side of pronotum. Antennae with twelve segments. Elytra wrinkled with three indistinct ridges on each elytron. Elytra taper behind. Immature grubs feed on roots of pine, oak, and blueberry.

Southern Pine Sawyer *Monochamus titillator* 3/4" - 1 1/8"
Elongate. Robust, almost cylindrical. Brownish to blackish. Elytra irregularly mottled with small patches of grayish to blackish pubescence. Antennae extremely long, at least two, usually three times length of body. Occur in pine, immature larvae boring into heartwood.

Fire-colored Beetle

Striped Blister Beetle

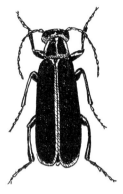
Margined Blister Beetle

Black Blister Beetle

Lesser Prionus Beetle

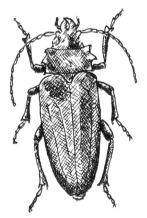
Broad-necked Root Borer

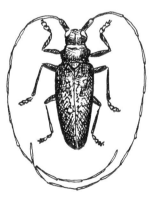
Southern Pine Sawyer Beetle

Beetles

Red Milkweed Beetle *Tetraopes tetraophthalmus* 3/8" - 9/16"
Elongate-oblong. Robust. Orange-red. Pronotum with four black spots. Each elytron with four black spots, the most central one oblong-oval. Feeds and breeds on milkweeds. A similar but slightly smaller (3/8"-1/2") species, *T. melanurus*, is the same orange-red but its elytra has two heart-shaped black spots, one near the middle, the other at the apex, these two sometimes joined along suture. Also on milkweed.

Leaf Beetles (Chrysomelidae)
Oval. Convex. Most small, seldom over 1/2". Antennae less than 1/2 length of body. Many brightly colored. All are leaf and flower feeders. Immatures feed on leaves and roots. Many are destructive pests. Large family. However, because pine barrens has a limited diversity in plant species, the variety of leaf beetles here is not as great as in other habitats.

Dogbane Leaf Beetle *Chrysochus auratus* 3/8" - 1/2"
Oblong. Convex. Brilliant green or blue green. Elytra often with a coppery or brassy reddish sheen. Highly polished and shining. Antennae, legs, underneath bluish-black. Common on dogbane and milkweed.

Milkweed Leaf Beetle *Labidomera clivicollis* 3/8" - 1/2"
Oval. Very convex and robust. Dark blue or blackish blue. Elytra orange yellow with three dark blue or black spots on each, all spots variable in size and shape, the ones nearest suture sometimes merging to form a small "X" or even a band.

Colorado Potato Beetle *Leptinotarsa decimlineata* 1/4" - 3/8"
Oblong-oval, strongly convex. Robust. Dull yellow. Pronotum with numerous black spots, the two central ones often forming a small, black "V." Each elytron with five narrow, black lines, the first joining along the suture, the second and third joining at the apex. In pine barrens, feeds on foliage of ground cherry and black nightshade.

Spotted Cucumber Beetle *Diabrotica undecimpunctata* 1/4"
Oblong-oval, widest behind middle. Greenish-yellow. Each elytron with three pairs of black spots arranged across insect. Feeds on foliage and flowers of members of the gourd or cucumber family and on goldenrods.

Snout Beetles or Weevils (Curculionidae)
Very large family of hard bodied, generally small beetles, most less than 1/2". Most obvious common characteristic is extension of head into a well defined beak, usually long and curved downward. Mouthparts are located at tip ends of these snouts. Antennae usually elbowed, ending in distinct club, basal joint long, often resting in grooves in sides of beak.

Red Milkweed Beetle
(T. tetraophthalmus)

Red Milkweed Beetle
(T. melanurus)

Dogbane Leaf Beetle

Milkweed Leaf Beetle

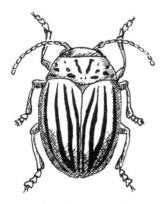

Colorado Potato Beetle

Spotted Cucumber Beetle

Nut and Acorn Weevils *Curculio* spp.
Elongate-oval, sort of egg-shaped, with very long, thin, down-curved snout. Brownish to brown and gray. Often mottled with pale brown, hair-like scales. Adults bore holes and lay eggs, and larvae develop inside nuts. Nut weevils that feed on acorns may destroy 50-90% of an acorn crop and are harmful to reforestation.[39]

Cranberry or Blueberry Blossom Weevil *Anthonomus musculus*   1/32"
Elongate-oval. Dark reddish-brown, thinly covered with whitish pubescence. Feeds on blossoms of blueberry and huckleberry. Abundant and major pest in blueberry fields. (No illus.)

Cockle-bur Billbug *Rhodobaenus tredecimpunctata*   1/4" - 3/8"
Elongate-oval. Red with 13 black spots of varied size on pronotum and elytra. Black underneath. Feeds and breeds on many species of Compositae.

TRICHOPTERA (tricho=hair; ptera=wing)
Caddisflies   1/4" - 1"
Slender, elongate, moth-like, with long, thread-like antennae and four membranous, usually dull colored wings. Although adult caddisflies resemble moths, they do not have the coiled feeding tubes of moths. Also, caddisflies have fine hair rather than scales on their wings and bodies. Most caddisflies are poor fliers and, at rest, hold their wings roof-like over their bodies.

Immatures are most interesting phase of their life cycles. Eggs are laid in water and the caterpillar-like larvae that emerge are entirely aquatic. Many species of larvae construct portable cases of vegetative material or grains of sand and live in and carry these cases around with them underwater. Metamorphosis is complete. Number of species in pine barrens somewhat limited due to highly acid conditions of stream waters.[40]

LEPIDOPTERA (lepido=scale; ptera=wing)
Butterflies and Moths
Usually four membranous wings nearly always, largely or entirely, covered with delicate scales which rub off easily. Adult mouthparts usually consist of a tube used for sucking up liquids but coiled up when not in use. Antennae long, slender, knobbed at tips in butterflies; varied, either thread- or comblike or feathery, never clubbed or knobbed at tip, in moths. Butterflies rest with their wings held vertically over their body. Moths, at rest, usually hold their wings folded flat and roof-like over their bodies, or outstretched in a horizontal position. In general, butterflies are day fliers. Most moths are night fliers but a few moths are day fliers. Larvae of Lepidoptera are called caterpillars. These have chewing mouthparts and most are plant feeders. Metamorphosis is complete. The principal means of separating families of Lepidoptera is based on wing venation.

Acorn Weevil

Cockle-bur Billbug

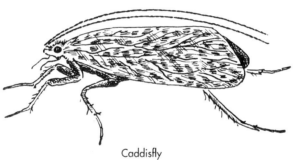
Caddisfly

Caddisfly larva in case

Butterflies
Nymphs, Satyrs and Arctics (Satyridae)
Usually dull brownish, often with eye spots on wings. Front legs reduced, not used for walking.

Wood Nymph *Cercyonis pegala*   Wing span 2" - 2 1/2"
Dark brown with broad yellowish band across fore wings and two bluish eyespots ringed with black. Fly in erratic, bobbing flight, usually near the ground.

Milkweed Butterflies (Danaidae)
Orange or reddish brown, usually marked with blackish veins and small, white speckles.

Monarch Butterfly *Danaus plexippus*   Wing span 3" - 4"
Brownish-orange, distinct dark brown to black on margins and veins. Two rows of small, whitish spots enclosed within black margin. Usually seen in flight or on milkweeds or goldenrods.

Brush-footed Butterflies (Nymphalidae)
Considerable variation in size and color. Front legs greatly reduced, often hairy, brush-like, useless for walking.

Pearl Crescent *Phyciodes tharos*   Wing span 1 1/8" - 1 3/8"
Brownish orange, with numerous scattered dark brown to black markings. Underneath brighter orange. Often on butterfly weed.

Question Mark *Polygonia interrogationis*   Wing span 2" - 2 1/2"
Brownish-orange with numerous, scattered dark brown to black markings, border pale lavender. Wing margins irregular. At rest, holds wings at an angle. Underside of wing resembles a dead leaf with a central silvery mark, something like a question mark in shape.

Comma or Hop-merchant *Polygonia comma*   Wing span 1 1/2" - 2"
Similar to above but smaller. Margins more black than lavender. Silver spot on each hind under wing smaller and more comma shaped. (No illus.)

Mourning Cloak *Nymphalis antiopa*   Wing span 2 1/2" - 3 3/8"
Wings rich deep maroon or dark purplish brown-black, bordered by a black band enclosing numerous bluish spots, and edged with broad, yellow or straw-colored band speckled with brownish spots. Underneath dark brownish with light gray speckling. Hibernates overwinter so one of earliest butterflies to appear in pine woods in spring.

Wood Nymph

Pearl Crescent

Monarch Butterfly

Question Mark

Mourning Cloak

Butterflies

Red Admiral *Vanessa atalanta* Wing span 2" - 2 1/2"
Bases of fore and hind wings brownish, bordered by wide, bright orange band. Tips of fore wings black enclosing several large white spots.

Painted Beauty *Vanessa cardui* Wing span 2" - 2 1/2"
Brownish-orange with distinct pinkish tinge and numerous irregular black spots. Fore wings with black tips enclosing large white spots. Upper hind wings with much pinkish-orange, under hind wings with four eyespots, two small ones between two mid-sized ones.

American Painted Beauty *Vanessa virginiensis* Wing span 1 3/4" - 2 1/4"
Nearly identical to above but smaller, with almost no pink tinge, more just brownish orange. Under hind wings have only two large eyespots.
(No illus.)

Viceroy *Basilarchia (=Limenitis) archippus* Wing span 2 3/4" - 3 1/4
Mimic of monarch butterfly, nearly identical but smaller and hind wings have an additional narrow, curved black band crossing middle of wing.

Red-spotted Purple *Basilarchia* Wing span 2 3/4" - 3 1/4"
arthemis astyanax
Velvety purple-black, tinged with reddish blue or green. Three rows of blue or green spots on outer third of hind wings. Under surface with reddish-orange spots, two near leading edge near base of each forewing, three near base of each hind wing, and seven in a row near edge of hind wing.

Hairstreaks and Elfins (Lycaenidae)
Small, delicate. Antennae often ringed with white. Hairstreaks have one to three hair-like "tails" from the trailing margins of their hind wings, and often have a small red or black eye spot at the inner angle of the hind wing. Elfins are brownish, margins of wings, especially hind ones, somewhat scalloped, and hind wings lack any tails.

Several species prefer habitats of acid, sandy, pine-oak barrens or cedar bogs, and so are considered to be especially characteristic of our New Jersey pine barrens. One should refer to more detailed references for identifications to the following characteristic species. (None illus'd.)

Hessel's Hairstreak *Mitoura hesseli*
Brown Elfin *Incisalia augustinus*
Hoary Elfin *Incisalia polia*
Henry's Elfin *Incisalia henrici*
Frosted Elfin *Incisalia irus*
Pine Elfin *Incisalia niphon*

Red Admiral

Painted Beauty

Viceroy Butterfly

Red-spotted Purple

Butterflies

American Copper *Lycaena phlaeas*   Wing span 1" - 1 1/8"
Fore wings bright orange red, with about eight scattered black spots and wide dark gray to black marginal band. Hind wings grayish black at base with orange margin and a row of black spots along very outer margin.

Bog Copper *Epidemia epixanthe*   Wing span 3/4" - 1"
Small, delicate. Above, male has a blue-purple iridescent gloss, female mouse-gray. Underneath varied from cream yellow to gray-white with round black spots. Limited to acid cranberry and similar bogs.

Eastern Tailed Blue *Everes comyntas*   Wing span 3/4" - 1"
Small, delicate. Above pale blue, gray blue, or grayish, with small tail from trailing edge of hind wing, and two or three small orange spots on edge of hind wing at base of tail. Underneath silvery gray.

Spring Azure *Celastrina argiolus*   Wing span 1" - 1 1/8"
Small delicate. Pale grayish to sky-blue with some iridescence and a darker fringe. Underneath brownish gray with small dark spots. An early indicator of spring. First butterfly to appear in numbers in the pine barrens, sometimes as early as late March.

Swallowtail Butterflies (Papilionidae)
Large, showy, with tail-like projections extending back from outer angles of hind wings.

Tiger Swallowtail *Papilio glaucus*   Wing span 3 1/4" - 4"
Bright yellow with black, tiger-like streaks, and wide black margin enclosing yellow spots.

Black Swallowtail *Papilio polyxenes asterius*   Wing span 3" - 3 1/2"
Black with two rows of yellow spots around wing margins. Outer spots on hind wings crescent-shaped. Spots on males much larger than on females. Area between rows of spots on females suffused with blue.

Spicebush Swallowtail *Papilio troilus*   Wing span 3 1/2" - 4"
Forewings and bases of hind wings blackish. Row of pale yellow spots near outer margin of both wings, crescent-shaped on hind wings. Extensive blue-green areas on rear half of hind wings.

Whites and Sulphurs (Pieridae)
Usually white, yellow, or orange, marked with black. Rounded wings.

European Cabbage Butterfly *Pieris rapae*   Wing span 1 1/2" - 2"
or Imported Cabbage Worm
Yellowish white, with black tips on fore wings. One black spot on fore wing of male, two on female.

Skipper & Moths

Clouded Sulphur *Colias philodice*   Wing span 1 3/4" - 2 1/4"
Yellow, either deep or pale, sometimes whitish, bordered with black. Black margin in fore wings of female encloses a row of small, yellow spots. A black spot centered forward on each fore wing, an orange-yellow spot centered forward on each hind wing.

Skippers (Hesperiidae)
Intermediate between butterflies and moths, with some characteristics of each but generally more like butterflies. Antennae have a small hook at tip rather than just a knob. Flight rapid and erratic. The habitats of four species are primarily sandy oak or pine oak scrub or barrens and thus are especially characteristic of the New Jersey pine barrens. Check with more detailed references for descriptions of the following species.

Sleepy Dusky Wing *Erynnis brizo*
Horace's Dusky Wing *Erynnis horatius* (No illus.)
Cobweb Skipper *Hesperia metea* (No illus.)
Dusted Skipper *Atrytonopsis hianna* (No illus.)

Moths

Sphinx Moths (Sphingidae)
Medium to large with heavy tapering bodies. Wings usually narrow. Fore wings pointed.

Carolina Sphinx *Manduca sexta*   Wing span 3 1/2" - 4 1/2"
Wings grayish, mottled and striped with black and white markings. Abdomen has row of six yellow spots. Larva is tobacco hornworm.

Five-spotted Hawk-moth *Manduca quinquemaculata*   Wing span 4" - 4 3/4"
Similar to above but wings are lighter grayish and abdomen has row of five yellow spots. Larva is tomato hornworm. (No illus.)

Blind-eyed Sphinx *Paonias excaecatus*   Wing span 2 1/2" - 3 1/2"
Fore wing with wide bands of light to deep brown. Outer margin strongly scalloped. Hind wing rosy pink with blue eye-spot surrounded by black.

White-lined Sphinx *Hyles lineata*   Wing span 3 1/2" - 4"
Fore wings dark olive brown to black, with a broad, branched, buff colored band from near bases to tips. Veins outlined in white. Hind wings with much pink, margined with bands of dark brown to black. Body light brownish with paired darker spots on each segment.

Clouded Sulphur Butterfly

Sleepy Dusky Wing Skipper

Carolina Sphinx Moth

Blind-eyed Sphinx

White-lined Sphinx

Moths

Royal and Giant Silkworm Moths (Saturnidae)
Medium to very large, including the largest moths in our area. Body densely hairy. Antennae feathery.

Regal or Royal Walnut Moth *Citheronia regalis* Wing span 3 1/4" - 4 3/4"
Forewings brownish with orange veins and large yellow spots. Hind wings orange with darker veins. Body hairy, orange, with yellow bands. Caterpillar feeds on hickory, walnut, butternut, ash, and sumac, and has black-tipped, orange, backward curved "horns," so is called the "hickory horned devil."

Pine-devil Moth *Citheronia sepulcralis* Wing span 2 3/4" - 4"
Fore wings ash gray to dull brown to brownish-violet with small, basal, rose spot, faint blackish line parallel to border, and central spot. Hind wings deep rose flushed with pink at bases, grayish to brownish on outer halves. Called "pine devil" due to menacing appearance of its "horned" larvae. Feeds on various species of pine.

Imperial Moth *Eacles imperialis* Wing span 3 3/4" - 5 1/4"
Yellow, heavily marked with pinkish-purple-brown spots and bands, including a diagonal band across fore wings to tips, one across middle of hind wings, and a round spot centered on each fore and hind wing.

Rosy Maple Moth *Dryocampa rubicunda* Wing span 1 1/2" - 2"
Body and wings whitish, cream or yellow, with two broad bands of bright pink on fore wings, one at base and one along margin. Larva is green-striped mapleworm.

Spiny Oakworm Moth *Anisota stigma* Wing span 2" - 2 1/2"
Light to dark rusty brown, speckled with black. White spot centered forward on fore wing. Outer margin of hind wing rounded.

Orange-striped Oakworm Moth *Anisota senatoria* Wing span 1 1/4" - 2"
Similar to above but smaller, more reddish brown. Outer margin of hind wing straight. Caterpillar has eight orange-yellow stripes running length of body. Body densely covered with spines. Caterpillars periodically denude large areas of both scrub and white oak. (No illus.)

Buck Moth *Hemileuca maia* Wing span 2" - 2 1/2"
Wings grayish-black at bases and outer edges, with a white median band, and a single eyespot in the white band, on all four wings. Male has brick-red tuft at tip of abdomen. A mid-day flier on sunny days in fall of year, mainly October. A characteristic species of scrub oak barrens, as our pinelands.

Moths

Io Moth *Automeris io* Wing span 2 1/2" - 3 1/4"
Forewings of male yellow, of female reddish-brown. Hind wings yellowish with orange-brown band inside hind margin and dense reddish pubescence near inner margin. Easily recognized by white-centered black and blue "bull's eye" centered on hind wings. Females larger than males.

Polyphemus Moth *Antheraea polyphemus* Wing span 4 1/2" - 5 1/2"
Wings cinnamon to reddish brown, grayish along leading edge of forewings. Faint white line near bases and faint black line parallel to paler outer margin of all four wings. Four large, oval eyespots edged in yellow, one on each wing, those on hind wings surrounded by large blue-black areas.

Luna Moth *Actias luna* Wing span 3 1/2" - 4 3/4"
Wings pale green. Leading margins of fore wings purplish. Hind wings have long, sweeping tails. Transparent eyespot on each wing.

Promethea Moth *Callosamia promethea* Wing span 3 1/4" - 3 3/4"
or Spicebush Silkmoth
Wings dark brown to blackish in males except for faint, whitish, wavy line and pale tan border. Tips of fore wings with eyespot surrounded by pinkish shading. Bases of wings in females light to dark reddish brown with white spot, separated from outer and lighter colored portion by prominent white, wavy line. Margins pale tan. Tip of fore wing eyespots as in males. Females considerably larger than males.

Cecropia Moth *Hyalophora cecropia* Wing span 4 3/4" - 6"
Wings dark grayish brown with red shading in basal area of fore wing. Large, whitish crescent on each wing. White and reddish band parallel to margins on wings. Outer margins light brownish. Bluish-lavender eyespots in tips of fore wings. Body orange red with white rings around abdomen.

Tent Caterpillar Moths (Lasiocampidae)

Eastern Tent *Malacosoma americanum* Wing span 1 1/4" - 1 1/2"
Caterpillar
Stout bodied. Hairy. Wings fawn brown. Fore wings with two oblique, whitish lines nearly parallel to outer margins. Caterpillars spin masses of white webbing ("tents') in shrubs and trees, especially in wild black cherry, in spring.

Tiger Moths (Arctiidae)
Stout, hairy bodied. Moderately broad winged. Usually light colored, often brightly spotted or banded in yellow, orange, red, or black.

Moths

Isabella Tiger or Banded Woolly-bear Moth *Pyrrharctia isabella*   Wing span 2" - 2 1/2"

Fore wings light grayish or orange-brown with numerous black specks and fine wavy lines on fresh specimens. Hind wings pale pinkish orange with a half dozen black spots, at least some near margin along trailing edge. Body orange with a black spot on top of each abdominal segment. Larva is "banded woolly bear."

Salt Marsh Moth *Estigmene acrea*   Wing span 1 3/4" - 2 1/2"

Head white. Body white or yellowish. Abdomen deep yellow with a black spot on sides and top of each segment. Tip of abdomen white in females. Fore wings white with variable scattered black spots. Hind wings same in females but orange-yellow in males. Females larger than males.

Yellow Woolly-bear Moth *Spilosoma virginica*   Wing span 1 3/8" - 2"

Wings almost pure white with few scattered small black specks. Abdomen yellow with a black spot on sides and top of each segment. Fore legs white with tiny orange to yellow brown markings.

Milkweed Tussock Moth *Euchaetes egle*   Wing span 1 3/8" - 1 3/4"

Body and wings unmarked mouse-gray to silvery-grayish. Abdomen yellow with a black spot on sides and top of each segment. Colorful black, white, and orange "harlequin" caterpillar feeds in colonies on milkweed in late summer.

Owlet Moths (Noctuidae)

Noctuids are largest family of Lepidoptera. Most are relatively small, dark colored, and difficult to identify. One group of noctuids, called the owlet or underwing moths, are larger, more conspicuous, and easier to identify. Most of these, in the genus *Catocala* (110 North American species), are strikingly colored, the fore wings a mottled grayish or brownish that camouflages with the bark of trees, the hind wings with concentric bands of red, yellow or orange, or bands of black and white. Eight examples of pine barrens underwings follow:

Graphic Moth *Drasteria graphica*   Wing span 1 1/4" - 1 1/2"

Fore wings banded light and dark gray-brown, with irregular black, wavy markings. Hind wings yellow or light orange with two concentric, wavy black bands. A characteristic pine barrens species.

Widow Underwing Moth *Catocala vidua*   Wing span 2 3/4" - 3 1/4"

Fore wings pale gray with blackish markings. Light, black arc passes just above wing spot over to wing tip. Hind wings dark brown to black, with narrow, white margin.

Isabella Tiger Moth

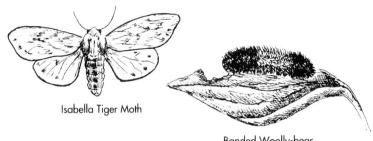

Banded Woolly-bear

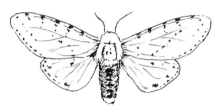

Salt Marsh Moth

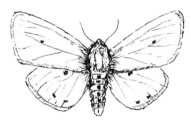

Yellow Woolly-bear Moth

Milkweed Tussock Moth

Graphic Moth

Widow Underwing Moth

Moths

Ilia Underwing Moth *Catocala ilia* Wing span 3" - 3 1/4"
Front wing variable, mottled dark gray to black, usually a lighter band down through middle. Conspicuous white spot centered near forward edge. Hind wings black with two concentric inner, wavy bands of reddish-orange and with narrow cream to orange colored margin. Caterpillar feeds on oaks.

Sordid Underwing Moth *Catocala sordida engelhardi* Wing span 1 1/2" - 1 3/4"
Fore wings dark mottled gray with faint zig-zag whitish line parallel to margin. Hind wing yellow with concentric black bands, the marginal band broken near hind angle. Caterpillar feeds on blueberries.

Graceful Underwing Moth *Catocala gracilis lemmeri* Wing span 1 3/8" - 1 3/4"
Similar to above. Fore wing more mottled, more heavily shaded. Hind wings as above but inner black band does not form a complete loop. Caterpillar feeds on blueberries.

Herodias Underwing Moth *Catocala herodias gerhardi* Wing span 2 1/4" - 2 1/2"
Fore wings grayish with fine wood-grain or dead pine needle pattern of brown and black, leading edges grayish white. Hind wings bright orange red with two concentric black bands and narrow white margin. Prefers scrub oak habitats and is a particularly characteristic New Jersey pine barrens species.

Ultronia Underwing Moth *Catocala ultronia* Wing span 1 3/4" - 2 1/2"

Fore wings brownish with fine wood-grain pattern. Dark brown patch along trailing edge and small dark patch near wing tip. Hind wings orange red with two concentric black bands.

Little Nymph Moth *Catocala micronympha* Wing span 1 5/8" - 1 3/4"
Fore wings grayish mottled with brownish and blackish. Variable. Hind wings yellow with broad marginal black band and narrower inner black band. Caterpillar feeds on oaks.

Tussock Moths (Lymantridae)

Gypsy Moth *Lymantria dispar* Wing span male 3/4"
female 1 1/8" - 2 3/4"
Stout, hairy. Male with feathery antennae. Fore wings of male brownish mottled with both lighter and darker patches and fine wavy lines. Hind wings reddish brown. Both pair of wings of female white, fore wings speckled with black and fine wavy lines. Hind wings clear. Female much larger than male and so heavy can hardly fly and usually does not. Females

Ilia Underwing Moth

Sordid Underwing Moth

Graceful Underwing Moth

Herodias Underwing Moth

Ultronia Underwing Moth

Little Nymph Moth

Moths

lay their eggs in masses on tree trunks, limbs, branches, even leaves as well as on fence posts and other solid surfaces. Masses are recognized by their hairy, light brownish covering. An introduced and serious defoliating pest on oaks, the caterpillars often completely stripping trees of their leaves.

Inchworm Moths (Geometridae)   Wing span 3/8" - 2"
Delicate and slender bodied. Wings broad but delicate. Pattern on fore wings often continued onto the hind wings. Larvae are caterpillars that lack pro-legs in mid-body, so must travel by bringing rear end up to front end in a "humping" or "looping" motion.

Slug Caterpillar Moths (Limacodidae)
Saddleback Caterpillar Moth *Sibine stimulea*   Wing span 1 1/8"
Stout, hairy body. Broad rounded wings dark chocolate brown with darker shading. Hind wings lighter brown. Larva brown at each end, green in middle with an oval purple-brown saddle-shaped mark edged with white, this in the center of the back surrounded by the green. Caterpillar covered with sharp, pointed spines which can inflict a painful sting. Feeds on oaks and other deciduous trees and shrubs.

Bagworm Moths (Psychidae)
Males fly but are rarely seen. Females are flightless and live in spindle-shaped, silken cases, or "bags" covered with bits of leaves, twigs, and other debris. These bags are attached to and hang from twigs with silk strands and hang and swing in the breezes. Males fly to females and mate with them inside these cases.

DIPTERA (di=two; ptera=wing)
Flies and Mosquitoes
A large and important group that have only one, the fore, pair of membranous wings. The other, or hind, pair are reduced to a pair of small, knobbed structures called halteres. In most cases, a hand lens is needed to see these organs which are thought to have something to do with flight stabilization. Antennae are usually short and variable but a few, like mosquitoes, have long and feathery antennae. Compound eyes are relatively large. Mouthparts are adapted for piercing and for sucking liquids. All develop by complete metamorphosis. Immatures often in the form of maggots.

Crane Flies (Tipulidae)   3/8" - 1"
Slender. Soft bodied. Grayish. Like giant, oversized mosquitoes. Very long, slender legs which break off easily. Unlike mosquitoes, crane flies neither buzz nor bite.

Mosquitoes (Culicidae)   1/8" - 1/4"
Slender, soft bodied, with long, sharp beak and hairy to feathery antennae. Wings long, narrow, covered with tiny scales. Females "bite" to obtain a

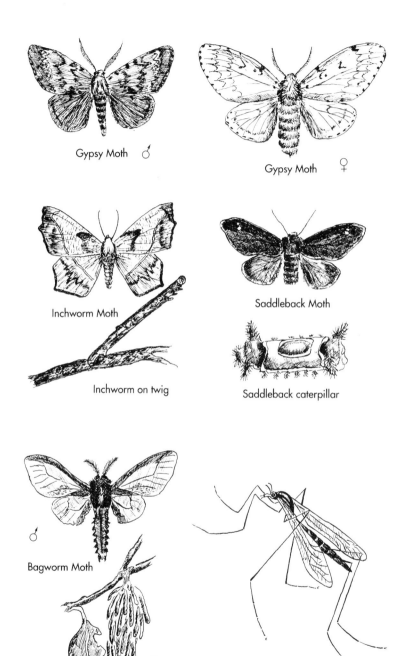

Flies

blood meal so they can produce eggs. Males do not bite. Immatures, called "wrigglers," develop in quiet and often stagnant water. Mosquitoes are carriers of several important diseases.

The following may be among the more common species to be encountered in the pine barrens of New Jersey.

Banded-leg Salt Marsh Mosquito *Aedes sollicitans*
Possibly most common of all pine barrens mosquitoes.

Cedar Bog Mosquito *Culiseta melanura*
May transmit eastern encephalitis disease.

House Mosquito *Culex pipiens*

Tree hole Mosquito *Toxorhynchites rutilus septentrionalis*
Our largest pine barrens mosquito.

Black Flies (Simulidae)   1/8" - 1/4"
Very small, black, somewhat humpbacked flies. Females are pestiferous blood suckers, especially in wooded areas near streams. May occur in large numbers in April and May. Often called "gnats." A common pinelands species is *Prosimulium fuscum*.

Horse and Deer Flies (Tabanidae)

Black Horse Fly *Tabanus atratus*   3/4" - 1"
Stout. Large head with large eyes. Jet black. Wings dark brownish or bluish to black. Females are painful blood sucking "biters." Become most annoying insects in pine barrens after black flies decline in numbers, usually by June.

Deer Fly *Chrysops delicatulus*   1/4" - 3/8"
Yellowish green or brownish. Eyes greenish gold. Abdomen yellow with rows of black markings on segments. Wings shining yellow brownish and often patterned with darker designs and smoky patches. Pestiferous blood sucking "biters."

Robber Flies (Asilidae)   1/4" - 1"
Bodies vary: slender or stout, hairy or nearly bare. Top of head hollowed out between eyes. Dense "beard" of bristles on face. Abdomen usually elongate and tapering. Legs usually large and spiny. Predacious, feeding on other insects with their piercing-sucking mouthparts. A pine barrens example:

Robber Fly *Stichopogon trifasciatus*
Gray with blackish markings. Common in open areas and on bare stretches of sand.

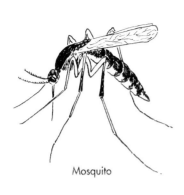
Mosquito

Black Fly

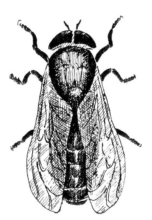
Black Horse Fly

Deer Fly

Robber Fly

Flies

Bee Flies (Bombyliidae)   1/4" - 1/2"
Stout, hairy. Resemble small bumblebees. Many with long slender beak. Many with patterned wings. Hover motionless in air or at flowers. Wings spread sideways when resting.

Hover or Flower Flies (Syrphidae)   1/4" - 5/8"
Many patterned and brightly colored with yellow, brown and black. Others plain brown or black. Many are mimics of bees and yellow jackets. All are good fliers. Most do much hovering. Most visit flowers. Adults do some pollinating. Larvae are predacious on aphids and some scale insects.

Fruit Flies (Tephritidae)
Small, not usually noticed. Included here only to mention the following two common pine barrens examples.

Blueberry Fruit Fly (and Maggot) *Rhagoletis mendax*   1/4"
Shining black to dark tan, with yellowish-white lines on body. Head orange. Eyes red. Wings have black F-shaped bands with curling tails. Most abundant fruit fly in pine barrens. Serious pest of huckleberries and cultivated highbush blueberries. (No illus.)

Goldenrod Round-stem Gall Fly *Eurosta solidagnis*   Adult fly: 3/8"
Gall: 3/4" - 1"
Reddish-brown. Wings brownish with fine network of lines. Female fly lays eggs on stem of goldenrod. Plant develops a thick walled, round, ball-like growth around egg. Egg hatches into a small grub that feeds on gall tissue until adult emerges through tiny hole it bores in side of gall.

House Flies (Muscidae)

House Fly *Musca domestica*   1/8" - 1/4"
Gray with black stripes on sides of body. Abdomen gray with irregular, darker markings. Breeds in filth and may transmit diseases.

Blow Flies (Calliphoridae)
Somewhat larger than house flies. Larvae (maggots) develop in carrion, dung, or in carcasses of dead animals.

Blue Bottle Fly *Calliphora vomitoria*   3/8" - 1/2"
Gray to black with reddish eyes. Abdomen metallic blue. Legs bristly. Wings clear.

Green Bottle Fly *Phaenicia sericata*   3/8" - 1/2"
Brilliant metallic blue-green. Black markings and bristles. Wings clear. (No. illus.)

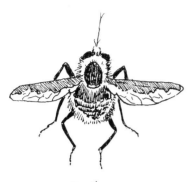
Bee Fly

Hover or Flower Fly

Goldenrod Round-stem Gall Fly and Gall

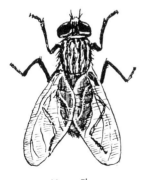
House Fly

Blue Bottle Fly

Sawflies, Wasps, Ants, Bees

HYMENOPTERA (Hymeno=membrane; ptera=wing)
Sawflies, Ants, Wasps, and Bees
Large, important order of most highly developed insects. Hard bodied. Mouthparts chewing, some modified into sucking structures. Antennae fairly long. Female ovipositor sometimes modified into a stinger. Wings, when present, are four in number, membranous, usually with large cells and few veins. Metamorphosis complete.

Conifer Sawflies (Diprionidae)
Small, stout bodied, wasp-like. Antennae fairly long with thirteen or more segments. Abdomen broadly joined to midbody. Larvae look like caterpillars and are much more likely to be noticed than adult wasps. Both of the following feed on pine needles, often clustering near the ends of branches.

European Pine Sawfly *Neodiprion sertifer*   Adult 5/16" - 3/8"
Larva 1" - 1 1/4"
Larva has shiny black head, light green body with two longitudinal gray stripes. (No illus.)

Red-headed Pine Sawfly *Neodiprion lecontei*   Adult 5/16" - 3/8"
Larva 1" - 1 1/4"
Larva similar to above but has a shiny red head. Light, yellow green body with four rows of black spots, a double row down middle of back and a single row on each side.

Ichneumon Wasps (Ichneumonidae)
Large family of mostly stingless wasps that vary greatly in size and color. Most are slender and wasp-like, with antennae usually at least half as long as body. Females of most species have long ovipositors but do not sting. Larvae are parasitic on spiders and small insects.

Long-tailed Ichneumon *Megarhyssa atrata*   1 1/4" - 1 1/2"
Ovipositor up to 3"
Body dark red-brown. Head, face, antennae bright yellow, with black eyes and markings. Wings smoky. Antennae long, thread-like. Abdomen enlarged at tip with two extremely long protruding "tails." Females pierce or bore through several inches of wood with their long ovipositors to lay eggs in the larvae or grubs of a horntail insect.

Short-tailed Ichneumon *Ophion bilineatus*   5/8" - 3/4"
Light brown, darker toward tip of abdomen. Head light brown, eyes darker. Wings clear, iridescent, gauzy. Legs light brown. Antennae long, thread-like. Abdomen flattened. Female ovipositor barely noticeable.

Gall Wasps (Cynipidae)
Adults minute, hump-backed with flattened abdomen. Dark shiny brown

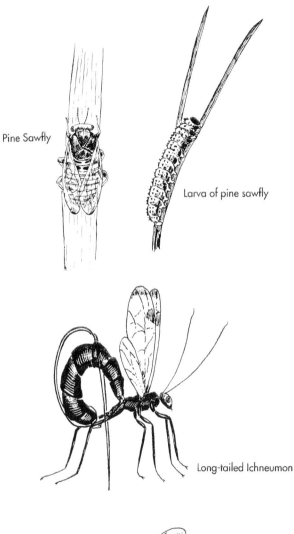
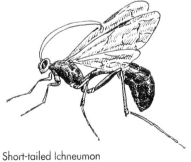

Sawflies, Wasps, Ants, Bees

or black. Adults less noticeable than galls they produce. Female wasps lay eggs on leaves, stems or other plant tissues. Plants produce a growth around each egg to form a gall in which insect hatches into a larva, feeds until adult, then emerges. Each gall shape and host plant is distinctive for each species of gall wasp. Oaks are a favorite host.

Oak Apple Gall *Amphibolips confluenta* Gall up to 2"
Round. Found on leaves of red, black, scarlet, and other oaks.

Pelecinid Wasp *Pelecinus polyturator* (Pelecinidae) 1 3/4" - 2"
Female shining black with very long, shining abdomen. Antennae long, thread-like. Wings clear. Male half size of female and very rare. Female bores hole in ground and lays egg on larva (grub) of a may beetle, *Phyllophaga* spp.

Velvet-Ants (Mutillidae)

Cow Killer *Dasymutilla occidentalis* 1/2" - 1"
Although ant-like, not an ant but a very hairy wasp, with thread-like constriction between midbody and abdomen. Brightly colored in contrasting bands of orange-red and black. Underneath black. Female wingless, most often seen running over ground in hot, sandy areas. Can inflict a painful sting. Male darker than female, with dark, smoky wings, visits flowers.

Ants (Formicidae)
Very common, even abundant insects in the pine barrens, both in terms of numbers and different species. Most distinguishing character is the slender stem (pedicel) between the midbody and the abdomen, usually with one or two lobes on top of the connecting pedicel. Antennae nearly always elbowed, the first segment usually very long. Ants are very successful social insects that live in colonies with differentiation of forms in a caste system.

Black Carpenter Ant *Camponotus pennsylvanicus* 3/8" - 1/2"
Black or dark reddish black. Abdomen with yellow hairs, especially along trailing margins of segments. Pedicel with a prominent hump. Common in all types of dead wood.

Little Black Ant *Monomorium viridum* = *M. minimum* under 1/8"
Tiny, smooth, shiny dark brown to black. Pedicel with two humps

Leaf-cutting Ant *Trachymyrmex septentrionalis* 1/8"
Leaf brown color. Thorax with three pairs of sharp spines. Pedicel with small spines. Colony hill sites identified by crescent rather than circular shape. Most northern of many leaf-cutting ants. Unique to and reaching northern limits of its range in New Jersey pine barrens. These ants cut off pieces of leaves and pine needles, carry them underground, and use the leaves as a culture medium for fungi on which they feed.

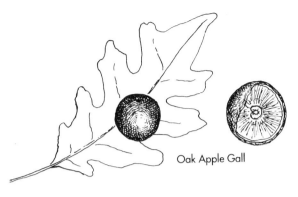

Oak Apple Gall

Pelecinid Wasp

Velvet Ant or Cow Killer

Black Carpenter Ant

Little Black Ant

Leaf-cutting Ant

Sawflies, Wasps, Ants, Bees

Social Wasps (Vespidae)
Obvious wasp-like insects that live in colonies and build nests of papery material they make by chewing wood, bark, or leaves. All have a short petiole connecting abdomen to midbody. All fold their wings lengthwise when at rest. All females have ovipositors at tip of their abdomen. All are pugnacious, resent interference, and can inflict a powerful sting.

Paper Wasp *Polistes fuscatus pallipes*   5/8" - 3/4"
Slender. Reddish brown to blackish with narrow, yellow rings on abdomen. Wings amber to dusky. Female constructs a circular, horizontal layer of cells attached by a short stalk to the underside of an object as the eaves of a building.

Eastern Yellow Jacket *Vespula maculifrons*   1/2" - 5/8"
Stout. Abdomen yellow, banded with black. Wings amber. Builds nests of papery material with the tiers of cells surrounded by an outer, papery covering, often in the ground or in low shrubs. Can be pests around sweet foods at picnics. Can inflict painful sting.

Bald-faced Hornet *Dolichovespula maculata*   3/4"
Stout. Black, patterned with whitish-yellow on face, sides of midbody, and tip of abdomen. Wings amber. Builds large, conspicuous nest, usually out in open, up in a tree, consisting of several tiers of cells surrounded by outer, grayish papery covering.

Thread-waisted Wasps (Sphecidae)
Slender, elongate, with long, thin petiole between midbody and abdomen. Can inflict a painful sting.

Organ Pipe Mud Dauber *Trypoxylon politum*   3/4" - 1"
Shining black. Builds tubular mud nests, side by side, often under open shed roofs. Provisions each of a series of cells with paralyzed spiders on each of which the wasp lays an egg.

Black and Yellow Mud Dauber *Sceliphron caementarium*   3/4" - 1"
Brownish-black with yellowish area on trailing end of midbody. Other smaller yellow markings. Legs yellow black. Wings smoky brownish. Female carries balls of mud with which it builds tubular mud cells. Provisions cells with paralyzed spiders as above. (Illus on pg. 389)

Bees (Apoidea)
Most bees are pollen and honey gatherers. All are covered with short, branched, feathery hairs to which pollen sticks and from which pollen is combed into a pollen basket located on the hind leg or under the abdomen. Most bees are solitary, living under ground or in plant stems, but bumble bees and honey bees are social. Bees are very important in the pollination of plants.

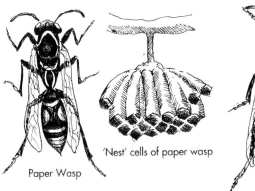

Paper Wasp

'Nest' cells of paper wasp

Yellow Jacket

Bald-faced Hornet

'Paper' nest of Bald-faced Hornet

Organ Pipe Mud Dauber

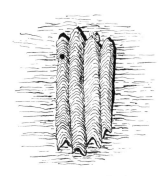

Tubular mud cells of Organ Pipe Mud Dauber

Sawflies, Wasps, Ants, Bees

Carpenter Bee *Xylocopa virginica* 3/4"
Large. Robust. Black or metallic blue-black. Midbody heavily covered with yellow hairs. Resembles a large bumble bee but tip and dorsum of abdomen mostly bare and shining. Female bores tunnel in old, dry wood, then creates a series of cells inside, each cell stuffed with pollen and one of her eggs.

Bumble Bee *Bombus pennsylvanicus* 5/8" - 1"
Bombus vagans
Two of several species, both robust. Densely covered with black and yellow hairs. Most bumble bees nest in ground, often in deserted mouse nests. Bumble bees are important pollinators and can sting.

Honey Bee *Apis mellifera* 5/8"
Light golden brown with darker bands on abdomen. Head, antennae, legs nearly black. Wings clear. Most nest in man made hives. Sometimes swarms escape and nest in hollow trees. These are the most important insects in the pollination of plants. Can sting.

The following are recommended for further identification and reference.

ARTHROPODS, INCLUDING SPIDERS AND INSECTS (GENERAL)

Arnett, R.H., and R.L. Jacques, Jr. 1981. Guide to Insects.

Bland, R.G. and H.E. Jaques. 1978. How to Know the Insects. Wm. C. Brown Co., Dubuque, Iowa

Borrer, D. J. and R. E. White. 1970. A Field Guide to the Insects of America north of Mexico. Houghton Mifflin Co., Boston, MA

Boyd, H. P. and P. Marucci. 1979. Arthropods of the Pine Barrens in Forman, R.T.T., Jr., Pine Barrens Ecosystem and Landscape. Academic Press, N.Y.

Levi, H. W. 1968. A Guide to Spiders and their Kin. Golden Press, N.Y.

Lutz, F. E. 1948. Field Book of Insects. G. P. Putnam's Sons, N.Y.

Milne, L. & M. 1980. The Audubon Society Field Guide to North American Insects and Spiders. A. A. Knopf, N.Y.

Smith, J. B. 1910. Annual Report of the New Jersey State Museum, including a Report of the Insects of New Jersey. Trenton, N.J.

Swain, R.B. 1948. The Insect Guide. Doubleday & Co., NY.

Zim, H.S. and C. Cottam. 1956. Insects. A Guide to Familiar American Insects. Golden Press, N.Y.

GRASSHOPPERS, COCKROACHES, AND TRUE BUGS

Helfer, J.R. 1963. How to know the Grasshoppers, Cockroaches, and their Allies. W.C. Brown Co., Dubuque, Iowa

Slater, J.A. and R.M. Baranowski. 1978. How to know the True Bugs. W. C. Brown Co., Dubuque, Iowa

BEETLES

Arnett, R. H., Jr., N. M. Downie, and H. E. Jaques. 1980. How to know the Beetles. W. C. Brown Co., Dubuque, Iowa

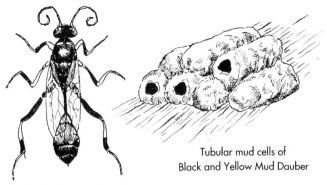

Black and Yellow Mud Dauber

Tubular mud cells of
Black and Yellow Mud Dauber

Carpenter Bee

Bumble Bee

Honey Bee

Dillon, E. S. and L. S. 1961. A Manual of Common Beetles of Eastern North America. Row, Peterson & Co., Evanston, Ill.

White, R. E. 1983. A Field Guide to the Beetles of North America. Houghton Mifflin Co., Boston, MA

BUTTERFLIES AND MOTHS

Covell, C. V., Jr. 1984. A Field Guide to the Moths of Eastern North America. Houghton Mifflin Co., Boston, MA

Klots, A. B. 1951. A Field Guide to the Butterflies of North America, East of the Great Plains. Houghton Mifflin Co., Boston, MA

Mitchell, R. T. and H. S. Zim. 1964. Butterflies and Moths. A Guide to the more Common Species. Golden Press, N.Y.

Pyle, R. M. 1980. The Audubon Society Field Guide to the Butterflies of North America. A. A. Knopf, NY

Shapiro, A. M. 1966. Butterflies of the Delaware Valley. The Amer. Entomol. Soc., Phila., PA

FOOTNOTES

Pine Barrens Areas

1. Harshberger, 1916 pg. 9
2. Collins & Russell, 1988 pg. 5
3. Robichaud & Buell, 1973 pg. 45
4. Office of State Geologist *in* The New Jersey Almanac, 1964-65 pp. 76-78
5. Collins & Russell, 1988 pg. 7
6. Widmer, 1963 pg. 58
7. Rhodehamel, 1979a pg. 42
8. Markley, 1979 pg. 83 Table 1
9. McCormick, 1979 pg. 236
10. McCormick, 1970 pg. 53
11. Havens, 1979 pp. 115-117
12. Rhodehamel 1979b pg. 164
13. N.J. Geological Survey
14. Harshberger, 1916 pg. 314
15. Vivian, pers. comm.
16. McCormick, 1979 pg. 239
17. Good *et al*, 1979 pg. 289
18. McCormick, 1970 pp. 75-81
19. Vivian, pers. comm.
20. McCormick, 1970 pg. 57
21. McCormick, 1970 pg. 56
22. Little, 1979 pg. 297
23. Wacker, 1979 pg. 4

Development & Uses

1. Pierce, 1957 pg. 5
2. Wacker, 1979 pg. 4
3. Wacker, 1979 pg. 6
4. Fairbrothers, 1979 pp. 400-401
5. Mounier, 1984 pp. 93 & 101
6. Pierce, 1957 pg. 5
7. Wacker, 1979 pg. 9
8. Applegate *et al* 1979 pg. 26
9. Wacker, 1979 pp. 9-10
10. Applegate *et al*, 1979 pg. 26
11. N.J. Pinelands Commission 1986 pg. 41
12. Cottrell, 1960 pg. 9
13. Gordon n.d. (1982) pp. 212-213
14. Wacker, 1979 pg. 12
15. Boyer, 1931 pg. 2
16. Wacker, 1979 pg. 12
17. Pierce, 1957 pg. 14
18. Boyer, 1931 pg. 128
19. Pierce, 1957 pp. 16-17
20. Starkey, 1962
21. N.J. Pinelands Commission 1986 pg. 103
22. Starkey, 1962
23. Wacker, 1979 pg. 13
24. Mounier, 1982 pg. 21
25. Beck, 1945 pp. 300-301
26. N.J. Pinelands Commission 1986 pg. 70
27. Ries & Kummel, 1904
28. Cain, 1958 pg. 5
29. Collins, Good & Good *in* Collins & Russell, 1988 pg. 28
30. Patterson, 1969
31. Beck, 1936 pg. 144
32. Pierce, 1957 pg. 47
33. Ewing, 1979 pg. 33
34. Boyer, 1962 pp. 77-83
35. Beck, 1956 pp. 154-155
36. Patterson, 1969
37. Applegate *et al*, 1979 pg. 25
38. Marucci, pers. comm.
39. Michalsky, 1978 pg. 17
40. Applegate *et al* 1979 pg. 31
41. Applegate *et al* 1979 pg. 31
42. Marucci, pers. comm.

Forgotten & Ghost Towns

1. Brown, 1958
2. Cottrell, 1960
3. Ries & Kummel, 1904
4. Beck, 1936 pp. 247-250
5. Bolger *et al*, 1982

6 Michalsky, 1978
7 Beck, 1937 pp. 136-138
8 Boyer, 1931 pg. 86
9 McMahon, 1973 pg. 129
10 Ewing, 1979 pg. 29
11 Ewing, 1979 pg. 33
12 Bisbee & Colesar, 1976 pg. 4
13 Bisbee & Colesar, 1976 pp. 1-9
14 Moore, 1970
15 Bisbee, 1971 pg. 46
16 Fowler & Herbert 1976 pg. 6
17 Dellomo, 1977
18 Fowler & Herbert, 1976
19 Batsto Citizens Gazette, 1966-date
20 Boucher, 1964
21 Ewing, 1986, 1988
22 Kier, 1976
23 Lippincott, 1933
24 Stuckert, 1976
25 Boyer, 1931 pg. 73
26 Beck, 1936 pg. 213
27 Boyer, 1931 pg. 74
28 Mounier, 1982
29 Mounier, 1982
30 Batsto Citizens Gazette, 1981 Vol. XV No. 2 pp. 4-5
31 McMahon, 1973 pg. 210
32 Boyer, 1931 pg. 261
33 McMahon, 1973, pg. 210
34 McMahon, 1973 pg. 219
35 McMahon, 1964 pg. 106
36 Mounier, 1982
37 Boyer, 1931 pg. 65
38 Boyer, 1931 pg. 48
39 Boyer, 1931 pg. 59

Present & Future Uses

1 N.J. Pinelands Commission, 1980 pg. 90
2 Applegate *et al,* 1979 pg. 34

Species Descriptions
Flora & Fauna

1 Johnson, 1983 pp. 930-931
2 McCormick, 1970 pg. 44
3 McCormick, 1970 pp. 44-45
4 Moul & Buell, 1979 pp. 433-439
5 Forman, 1979 pg. 407
6 Hand, 1965, 1984 rev. pg. 1
7 Fairbrothers, 1979 pg. 399, Table 1
8 McCormick, 1979 pg. 239
9 Fairbrothers, pers. comm.
10 McCormick, 1970 pg. 50
11 Stone, 1911 pg. 353
12 McCormick, 1970 pg. 50
13-15 Fairbrothers, 1979 pg. 399, Table 1
16 Stone, 1911 pg. 183
17 Fairbrothers, 1979 pg. 399, Table 1
18 Stone, 1911 pg. 338
19-27 Fairbrothers, 1979 pg. 399, Table 1
28 McCormick, 1970 pg. 18 *and* Boyd & Marucci, 1979 pp. 505-506
29 Smith, 1909 pg. 30
30 Boyle, 1986 pg. 461
31 Leck, 1979 pg. 459
32 Leck, 1979 pg. 459
33 Conant, 1958 pg. 298
34 McCormick, 1970 pg. 24
35 Smith, 1909 pg. 850
36 Patrick *et al,* 1979 pg. 191
37 Borrer *et al,* 1976 pg. 214
38 McCormick, 1970 pg. 18
39 White, 1983 pg. 318
40 Patrick *et al,* 1979 pg. 191

REFERENCES CITED

Anderson, K. 1989. A Checklist of the Plants of New Jersey. Rancocas Nature Center, N.J.A.S.

Applegate, J.E., S. Little, and P.E. Marucci. 1979 Plant and Animal Products of the Pine Barrens *in* Forman, R.T.T., ed., Pine Barrens Ecosystem and Landscape. Academic Press, NY

Batsto Citizens Gazette. 1966-date. Batsto Citizens Committee. Batsto, NJ Vol. 1-date

Beck, H.C. 1936. Forgotten Towns of Southern New Jersey. 1961 reprint. Rutgers University Press, New Brunswick, NJ

Beck, H.C. 1937. More Forgotten Towns of Southern New Jersey. 1963 reprint. Rutgers University Press, New Brunswick NJ

Beck, H.C. 1945. Jersey Genesis. The Story of the Mullica River. Rutgers University Press, New Brunswick, NJ

Beck, H.C. 1956. The Roads of Home. Lanes and Legends of New Jersey. Rutgers University Press, New Brunswick, NJ

Bisbee, H.H. 1971. Sign Posts. Place Names in History of Burlington County. H.H. Bisbee, Burlington, NJ

Bisbee, H.H. and R.B. Colesar. 1976. Martha, 1808-1815. The Complete Martha Furnace Diary and Journal. H.H. Bisbee, Burlington, NJ.

Bolger, W., H.J. Githens, and E.S. Rutsch. 1982. Historic Architectural Survey and Preservation Planning Project for the Village of Whitesbog, NJ. New Jersey Conservation Foundation.

Borrer, D.J., D.M. DeLong, and C.A. Triplehorn 1976 An Introduction to the Study of Insects. 4th ed. Holt, Rinehart, and Winston, NY

Boucher, B.P. (ed.) *et al*. 1963. The New Jersey Almanac. Tercentenary Edition. 1964-1965. Upper Montclair, NJ

Boucher, J.E. 1964. Of Batsto and Bog Iron. Batsto Citizens Committee, Batsto, NJ

Boyd, H.P. and P.E. Marucci. 1979. Arthropods of the Pine Barrens *in* Forman, R.T.T., ed., Pine Barrens Ecosystem and Landscape. Academic Press, NY

Boyer, C.S. 1931. Early Forges and Furnaces in New Jersey. University of Pennsylvania Press.

Boyer, C.S. 1962. Old Inns and Taverns in West Jersey. Camden County Historical Society.

Britton, N.L. 1889. Catalogue of Plants found in New Jersey. N.J. Geological Survey, Final Report of State Geologist. No. 2, pp. 27-649.

Britton, N.L. and A. Brown. 1913, 1952. An Illustrated Flora of the Northern United States, Canada, and the British Possessions. Charles Scribner's Sons, NY

Brown, J. 1958. Allaire's Lost Empire. Transcript Printing House, Freehold, NJ

Cain, O. 1958. The History of Mays Landing and Vicinity (unpublished) *in* the Atlantic County Library, Mays Landing, NJ

Center for Coastal & Environmental Studies. 1978. A Plan for a Pinelands National Preserve. Rutgers University, State of N.J.

Collins, B.R., N.E. & R.E. Good. 1988 The Landscape of the New Jersey Pine Barrens *in* Collins, B.R. & E.W.B. Russell. Protecting the New Jersey Pinelands. Rutgers University Press, New Brunswick, NJ

Collins, B.R. and E.W.B. Russell. 1988. Protecting the New Jersey Pinelands. Rutgers University Press, New Brunswick, NJ

Conant, R. 1979. A Zoogeographical Review of the Amphibians and Reptiles of southern New Jersey with emphasis on the Pine Barrens *in* Forman, R.T.T., ed., Pine Barrens Ecosystem and Landscape. Academic Press, NY

Cottrell, A.T. 1960. The Deserted Village of Allaire. Board of Trustees, Deserted Village of Allaire, NJ

Dellomo, A.N., Jr. 1977. Harrisville. Angelo Publishing Co., Atlantic City, NJ

Ewing, S.W. R. 1979. Atsion. A Town of Four Faces. Batsto Citizens Committee, Batsto, NJ

Ewing, S.W.R. 1986. An Introduction to Batsto. Batsto Citizens Committee, Batsto, NJ

Ewing, S.W.R. 1988. Batsto Lights. The Story of Batsto Glass and the Richards' Dynasty. Batsto Citizens Committee, Batsto, NJ.

Fairbrothers, D.E. 1979. Endangered, Threatened, and Rare Vascular Plants of the Pine Barrens and their Biogeography *in* Forman, R.T.T., ed., Pine Barrens Ecosystem and Landscape. Academic Press, NY

Fernald, M.L. 1950. Gray's Manual of Botany. 8th ed. American Book Co., NY

Ferren, W. B., Jr., J. W. Benton, and L. Hand. 1979. Common Vascular Plants of the Pine Barrens *in* Forman, R.T.T. ed., Pine Barrens Ecosystem and Landscape. Academic Press, NY

Forman, R.T.T. 1979. Common Bryophytes and Lichens of the New Jersey Pine Barrens *in* Forman, R.T.T., ed., Pine Barrens Ecosystem and Landscape. Academic Press, NY

Forman, R.T.T., ed., 1979. Pine Barrens Ecosystem and Landscape. Academic Press, NY

Fowler, M. and W.A. Herbert. 1976. Paper Town of the Pine Barrens. Harrisville, NJ. Environmental Education Publishing Service, Eatontown, NJ

Gasior, J.J. 1986. Jersey Perspective. Selected Topics in Earth Science. Lenape Regional High School District Board of Education, Medford, NJ

Gleason, H.A. 1962. Plants in the Vicinity of New York, NY. Botanical Garden, NY

Gleason, H.A. and A. Cronquist. 1963. Manual of Vascular Plants of northeastern United States and adjacent Canada. D. Van Nostrand Co., NY

Good, R.E., N.F. Good, and J.W. Andresen. 1979. The Pine Barrens Plains *in* Forman, R.T.T., ed., Pine Barrens Ecosystem and Landscape. Academic Press, NY

Gordon, T. (undated, but 1982) Last of the Old-time Charcoal Makers and the Coaling Process in the Pine Barrens of New Jersey *in* Sinton, J.W., ed. History, Culture and Archeology of the Pine Barrens: Essays from the Third Pine Barrens Conference. Center for Environmental Research, Stockton State College, Pomona, NJ

Hand, L.E. 1965. Rev. 1984. Wharton Tract Plant Life. N.J. Division of Parks & Forestry, Department of Environmental Protection, Trenton, NJ

Harshberger, J.W. 1916. Reprint 1970. The Vegetation of the New Jersey Pine Barrens. Dover Publications, NY

Hastings, R.W. 1979. Fish of the Pine Barrens *in* Forman, R.T.T., ed., Pine Barrens Ecosystem and Landscape. Academic Press, NY

Havens, A.V. 1979. Climate and Microclimate of the New Jersey Pine Barrens *in* Forman, R.T.T., ed., Pine Barrens Ecosystem and Landscape. Academic Press, NY

Hitchcock, A.S., 2nd ed., Chase revision. 1971. Manual of the Grasses of the United States. Vols. 1 & 2. Dover Publications, NY

Johnson, L.G. 1983. Biology, William C. Brown Co., Dubuque, IA

Kartesz, J.T. and R. Kartesz. 1980. A synonymized Checklist of the Vascular Flora of the United States, Canada, and Greenland. University of North Carolina Press, Chapel Hill, NC

Kier. C.F., Jr., 1976. Batsto Story a Fascinating One. Batsto Citizens Gazette, Bicentennial Edition. Vol. X, No. 1. Batsto. NJ

Leck, C.F. 1979. Birds of the Pine Barrens *in* Forman, R.T.T., ed., Pine Barrens Ecosystem and Landscape, Academic Press, NY

Lippincott, B. 1933. An Historical Sketch of Batsto, N.J., Batsto Citizens Committee, Batsto, N.J.

Little, S. 1979. Fire and Plant Succession in the New Jersey Pine Barrens *in* Forman, R.T.T., ed., Pine Barrens Ecosystem and Landscape. Academic Press, NY

Markley, M.L. 1979. Soil Series in the Pine Barrens *in* Forman, R.T.T., ed., Pine Barrens Ecosystem and Landscape. Academic Press, NY

McCloy, J.F. and R. Miller. 1976. The Jersey Devil. Middle Atlantic Press, Wilmington, DE

McCormick, J. 1970. The Pine Barrens. A Preliminary Ecological Inventory. New Jersey State Museum, Trenton, NJ

McCormick, J. 1979. The Vegetation of the New Jersey Pine Barrens *in* Forman, R.T.T., ed., Pine Barrens Ecosystem and Landscape. Academic Press, NY

McMahon, W. 1964. Historic South Jersey Towns. Atlantic City Press Publishing Co., Atlantic City, NJ

McMahon, W. 1973. South Jersey Towns. History and Legend. Rutgers University Press, New Brunswick, NJ

McMahon, W. 1980. Pine Barrens Legends, Lore, and Lies. Middle Atlantic Press, Wallingford, PA

McPhee, J. 1967. The Pine Barrens. Farrar, Strauss, and Giroux, NY

Michalsky, B. 1978. Whitesbog, an Historical Sketch. Conservation and Environmental Studies Center, Whitesbog, Browns Mills, NJ

Moore, H. 1943. An Old Jersey Furnace (Martha), a Study. Batsto Citizens Committee, Batsto, NJ

Moul, E.T. and H.F. Buell. 1979. Algae of the Pine Barrens *in* Forman, R.T.T., ed., Pine Barrens Ecosystem and Landscape. Academic Press, NY

Mounier, R.A. 1982. Survey of Historic Glass Factories in southern New Jersey (unpublished) in the library of the Wheaton Historical Association, Millville, NJ

Mounier, R.A. 1984. A Study of Water-powered Sawmills in the Pine Barrens of New Jersey *in* Historic Preservation Planning in New Jersey: Selected Papers on the Identification, Evaluation, and Protection of Cultural Resources. Office of New Jersey Heritage, Department of Environmental Protection, Trenton, NJ

New Jersey Almanac, The. 1963. Tercentenary Edition, 1964-1965. The New Jersey Almanac, Inc., Upper Montclair, NJ,

New Jersey Geological Survey. Trenton, NJ

New Jersey Natural Heritage Program. 1987. Special Plants, Special Invertebrates, and Special Vertebrates of New Jersey. Office of Natural Lands Management, Department of Environmental Protection, Trenton, NJ

New Jersey Pinelands Commission. 1980. New Jersey Pinelands. Comprehensive Management Plan (CMP). N.J. Pinelands Commission, New Lisbon, NJ

New Jersey Pinelands Commission. 1986. Pinelands Cultural Resource Management Plan for Historic Period Sites. N.J. Pinelands Commission, New Lisbon, NJ

Olsson, H. 1979. Vegetation of the New Jersey Pine Barrens: A Phytosociological Classification *in* Forman, R.T.T., ed., Pine Barrens Ecosystem and Landscape. Academic Press, NY

Patrick, R., B. Matson, and L. Anderson. 1979. Streams and Lakes in the Pine Barrens *in* Forman, R.T.T., ed., Pine Barrens Ecosystem and Landscape. Academic Press., NY

Patterson, F.G. 1969. Exploitation of the Pine Barrens. Past, Present, and Future. Conservation and Environmental Studies Center, Whitesbog, Browns Mills, NJ

Pierce, A.D. 1957. Iron in the Pines. Rutgers University Press, New Brunswick, NJ

Pinelands Environmental Council. 1975. A Plan for the Pinelands. Pinelands Environmental Council, Mt. Misery, NJ

Pohl, R.W. 1954. How to Know the Grasses. William C. Brown Co., Dubuque, IA

Rhodehamel, E.C. 1973. Geology and Water Resources of the Wharton Tract and the

Mullica River Basin in southern New Jersey. New Jersey Division of Water Resources, Special Report No. 36.

Rhodehamel, E.C. 1979a. Geology of the Pine Barrens of New Jersey *in* Forman, R.T.T., ed., Pine Barrens Ecosystem and Landscape. Academic Press, NY

Rhodehamel, E.C. 1979b. Hydrology of the New Jersey Pine Barrens *in* Forman, R.T.T., ed., Pine Barrens Ecosystem and Landscape. Academic Press, NY

Ries, H. and H.B. Kummel. 1904. The Clays and Clay Industry of New Jersey. MacCrellish and Quigley, Trenton, NJ

Robichaud, B. and M.F. Buell. 1973. Vegetation of New Jersey. Rutgers University Press, New Brunswick, NJ

Rose, P.S. 1942. Blueberry Queen. (The Whitesbog Story). Abridged from the Saturday Evening Post, Sept. 12, 1942.

Smith, J.B. 1910. Annual Report of the New Jersey State Museum for 1909, including a Report of the Insects of New Jersey. Trenton, NJ

Starkey, J.A. 1962. The Bog Ore and Bog Iron Industry of South Jersey. Bulletin N.J. Academy of Science. Vol. 7 (1)

Stoetzel, M.B., Chr. 1989 Common Names of Insects and Related Organisms. Entomological Society of America

Stone, W. 1911. The Plants of Southern New Jersey. Originally published as Part II of the Annual Report of the New Jersey State Museum for 1910. Reprint 1973. Quarterman Publications, Boston, MA

Stuckert, B.S. 1976. The Batsto Manor House. Batsto Citizens Committee, Batsto, NJ

Tedrow, J.C.F. 1979. Development of Pine Barrens Soils *in* Forman, R.T.T., ed., Pine Barrens Ecosystem and Landscape. Academic Press, NY

Thomas, L.S. 1967 and later reprints. The Pine Barrens of New Jersey. Division of Parks and Forestry, Department of Environmental Protection, Trenton, NJ

United States Department of Interior. 1976. The Pine Barrens Resource. U.S. Dep't of Interior Task Force Report. N.J. Audubon (Magazine) II (6): 1-12.

Vivian, N. 1969. The Jersey Devil. Conservation and Environmental Studies Center, Whitesbog, Browns Mills, NJ.

Wacker, P.O. 1979. Human Exploitation of the New Jersey Pine Barrens before 1900 *in* Forman, R.T.T., ed., Pine Barrens Ecosystem and Landscape. Academic Press, NY

White, R.E. 1983. A Field Guide to the Beetles of North America. Houghton Mifflin Co., Boston, MA

Widmer, K. 1963. The Geology of New Jersey *in* The New Jersey Almanac, 1964-1965. The New Jersey Almanac, Inc. Upper Montclair, NJ

Wolgast, L.J. 1979. Mammals of the New Jersey Pine Barrens *in* Forman, R.T.T., ed., Pine Barrens Ecosystem and Landscape. Academic Press, NY

INDEX

Acanalonia bivittata, 326
Acantharchus pomotis, 290
Accipiter striatus, 244
Acer rubrum, 116
   trilobum (var.), 116
Aceraceae, 116
Acid rain in p.b., 8, 9
Acidity in p.b., 7, 10
Acrididae, 306
Acrosternum hilare, 314
Actias luna, 370
Actitis macularia, 246
Adalia bipunctata, 352
Aedes sollicitans, 378
Aeshnidae, 302
Aetna or Etna Furnace & Forge (Head of River), 58
Aetna or Etna Furnace & Mills (Medford), 48
Agalinis setacea, 204
   virgata, 202
Agelaius phoeniceus, 262
Agriculture, 33
Agrilus anxius, 346
Agrostis altissima, 160
   hiemalis, 160
   hyemalis, 160
Aix sponsa, 244
Alaus myops, 346
   oculatus, 346
Alder, Black, 124
   Common, 118
   Smooth, 118
Aletris farinosa, 176
Algae, 78
Alismataceae, 146
Allaire, 42
Allaire State Park, 62
Alnus serrulata, 118
Alobates pennsylvanica, 352
Amanita muscaria, 79
Amaryllidaceae, 176
Amaryllis family, 176
Amatol, 56
Amblycorypha oblongifolia, 306
Ambush Bug, 316
Ambystoma opacum, 283

Amelanchier canadensis, 122
Amphibians, 282
Amphibolips confluenta, 384
Amphicarpum purshii, 164
Anacardiaceae, 122, 140
Anas platyrhynchos, 242
   rubripes, 244
Anatis quindecimpunctata, 352
Anax junius, 302
Andromeda, Privet, 128
Andropogon abbreviatus (var.), 164
   glomeratus, 164
   scoparius, 164
   virginicus, 164
Angiospermae, 75
Anguilla rostrata, 290
Animal kingdom, 231
Anisoptera, 302
Anisota senatoria, 368
   stigma, 368
Ant Nest Beetle, 344
Ant, Black Carpenter, 384
   Leaf-cutting, 384
   Little Black, 384
Antelope Beetle, 338
Antheraea polyphemus, 370
Anthonomus musculus, 358
Antlion, 326
Aphid, 326
Aphididae, 326
Aphis spiraecolae, 326
Aphredoderus sayanus, 290
Aphrophora parallela, 324
Apiaceae, 196
Apios americana, 140
Apis melifera, 388
Apocynaceae, 198
Apocynum androsaemifolium, 198
   cannabinum, 200
Apoidea, 386
Apple Pie Hill, 5, 47
Apple Tree Borer, Flat-headed, 346
Aquatic habitats/communities, 15
Aquatic plants, 144
Aquifers, 10

Aquifoliaceae, 116, 124
Araceae, 146
Arbutus, Trailing, 136
Archilochus colubris, 248
Arctiidae, 370
Arctostaphylos uva-ursi, 136
Ardea herodias, 242
Area of p.b. in N.J., 2
Arenaria caroliniana, 182
Arethusa, 180
Arethusa bulbosa, 180
Arilus cristatus, 318
Aristida dichotoma, 162
Arney's Mount, 5
Aronia arbutifolia, 120
   melanocarpa, 122
Arrow-Arum, 146
Arrowhead, 146
Arthropods, 292
Arum family, 146
Asclepiadaceae, 200
Asclepias amplexicaulis, 200
   rubra, 200
   tuberosa, 200
Ascyrum hypericoides, 188
   stans, 188
Asilidae, 378
Aspen, Large-toothed, 118
Asphodel, Bog-, 174
   False, 174
Aspleniaceae, 100
Asplenium platyneuron, 100
Assassin Bug, 318
Aster concolor, 216
   dumosus, 216
   gracilis, 216
   linariifolius, 218
   nemoralis, 218
   novi-belgii, 216
   patens, 216
   paternus, 218
   solidagineus, 218
   spectabilis, 216
   undulatus, 216
Aster, Bog, 218
   Bushy, 216
   Golden, 210
   Late Purple, 216
   New York, 216

   Purple, Late, 216
   Showy, 216
   Silvery, 216
   Slender, 216
   Stiff-leaved, 218
   Wavy-leaved, 216
   White-topped, 218
Asteraceae, 206
Atlanticus testaceus, 308
Atrytonopsis hianna, 366
Atsion Forge, Iron Works, Paper &
   Cotton Mill, 49
Aureolaria pedicularia, 204
   virginica, 204
Automeris io, 370
Axion plagiatum, 352
Azalea, Clammy, 126
   Swamp, 126

Backswimmer, 320
Bamber Forge, 42
Baptisia tinctoria, 184
Bard's (Peter) (saw) mill, 44
Bartonia paniculata, 198
   virginica, 198
Bartonia, Twining, 198
   Upright, 198
Basilarchia archippus, 362
   arthemis astyanax, 362
Bass River State Forest, 62
Bat, Little Brown, 234
Batona Trail, 63
Batsto or Batstow Furnace, 53
Batsto-Mullica watershed, 9
Bayberry, 118
   Evergreen, 118
Beach-heather, 134
Beach-plum, 122
Beacon Hill, 5
Beacon Hill gravels, 6
Beak-rush, Small-headed, 168
   White, 170
Bean, Wild, 140
Bearberry, 136
Beardgrass, Broom, 164
   Bushy, 164
Beaver, 236
Bedstraw, Pine-barren, 204

Bee(s), 386
　Bumble, 388
　Carpenter, 388
　Honey, 388
Beech family, 110, 120
Beetle(s), 328
　Ant Nest, 344
　Antelope, 338
　Aquatic, 334
　Blister, 354
　Bombardier, 332
　Burying, 336
　Carrion, 336
　Carrot, 342
　Checkered, 350
　Click, 346
　Cucumber, 356
　Darkling, 352
　Dung, 340
　Earth-boring Dung, 340
　Fire-colored, 354
　Flower, 344
　Forked Fungus, 352
　Goldsmith, 342
　Grapevine, 342
　Ground, 329, 330, 332
　Hister, 338
　Japanese, 342
　June, 344
　Lady, 350
　Leaf, 356
　Long-horned, 354
　May, 342
　Milkweed, 356
　Net-winged, 348
　Ox, 342
　Pinching, 338
　Pleasing Fungus, 350
　Potato, 356
　Predacious Diving, 334
　Rhinoceros, 344
　Rove, 338
　Scarab, 340
　Sexton, 336
　Snout, 356
　Soldier, 348
　Stag, 338
　Tiger, 328, 329
　Tumblebug, 340
　Water Scavenger, 334
　Whirligig, 334
　Wood-boring, 346
Beggar-ticks, 220
Belangee's Sawmill or Skit Mill, 52
Belcoville, 57
Belleplain State Forest, 63
Bellwort, Pine-barren, 174
Belostoma fluminea, 320
Belostomatidae, 320
Benacus griseus, 320
Bentgrass, Tall, 160
Bergen Iron Works, 42
Betula populifolia, 110
Betulaceae, 110, 118
Bidens coronata, 220
Billbug, Cockle-bur, 358
Birch family, 110, 118
　Gray, 110
Birch-Borer, Bronze, 346
Birds, 242
Birmingham Forge, 45
Blackberry, Running, 140
　Sand, 122
　Swamp, 140
Blackbird, Red-winged, 262
Bladderwort family, 150
　Fibrous, 152
　Horned, 152
　Purple, 150
　Swollen, 152
　Zig-zag, 152
Blarina brevicauda, 232
Blattellidae, 310
Blazing-star, Hairy, 210
Blechnaceae, 102
Blister Beetle(s), 354
　Black, 354
　Margined, 354
　Striped, 354
Bloodwort family, 176
Blue Flag, Slender, 176
Blue, Eastern Tailed, 364
Blue-eyed Grass, Eastern, 176
Bluebell family, 206
Blueberry agriculture, 36
Blueberry, Black Highbush, 132
　Highbush, 132
　Low, 130
　New Jersey, 132

Bluebird, Eastern, 256
Bluecurls, 200
  Narrow-leaved, 200
Bluestem, Little, 164
Bluet, Common, 304
Bobwhite, Northern, 246
Bodine's Tavern, 31, 53
Bog iron - see iron
Bog-Asphodel, 174
Boisea trivittata, 316
Boletotherus cornutus, 352
Bombardier Beetle, 332
  False, 332
Bombus pennsylvanicus, 388
  vagans, 388
Bombycilla cedrorum, 256
Bombyliidae, 380
Bonasa umbellus, 246
Boneset(s), 206
  Hairy, 208
  Hyssop-leaved, 208
  Pine-barren, 208
  Rough, 208
  Round-leaved, 208
  White, 208
  White-bracted, 208
Bothynus gibbosus, 342
Bottle Fly, Blue, 380
  Green, 380
Boxelder Bug, 316
Brachinus fumans, 332
Bracken fern, 102
Brake, 102
Branta canadensis, 242
Brasenia schreberi, 150
Breweria caesariense (var.), 142
  pickeringii, 142
Brick making, 28
Broom-Crowberry, 134
Broom-sedge, 164
Brotherton (Indian Mills), 19
Bryophyta, 75
Bryum capillare, 92
Bubo virginianus, 246
Buck Moth, 368
Buckingham and sawmill, 44
Buckwheat family, 182
Budd's Iron Works, 58
Bufo woodhousei fowleri, 284

Bug(s) (true), 312
  Ambush, 316
  Aquatic, 318
  Assassin, 318
  Boxelder, 316
  Kissing, 318
  Leaf-footed, 316
  Milkweed, 316
  Red, 316
  Soldier, 316
  Stink, 314
  Water, Giant, 320
  Wheel, 318
Bugleweed, Sessile-leaved, 202
Bullhead, Brown, 290
  Yellow, 290
Bullhead-lily, 150
Bulltown Glass Works, 54
Buprestidae, 346
Bur-reed family, 144
  Slender, 144
Burying Beetles, 336
Bush-clover, Hairy, 186
  Narrow-leaved, 186
  Wand-like, 184
Butcher's Forge or Works, 42
Buteo jamaicensis, 244
  platypterus, 244
Butorides striatus, 242
Butterflies, 358
  Brush-footed, 360
  Cabbage, 364
  Lycaenid, 362
  Milkweed, 360
  Monarch, 360
  Red-spotted Purple, 362
  Sulphur, 366
  Swallow-tail, 364
  Viceroy, 362
Butterfly-weed, 200
Buttonbush, 132
Buttonweed, Rough, 204
Buttonwood, 118
Buzby's General Store, 46

Cabbage Butterfly, European, 364
Cactaceae, 156
Cactus family, 156
Caddisflies, 358

Calamagrostis cinnoides, 160
Calamovilfa brevipilis, 160
Calico, 52
Calliphora vomitoria, 380
Calliphoridae, 380
Callosamia promethea, 370
Calopogon pulchellus, 180
　tuberosus, 180
Calopteron reticulatum, 348
　terminale, 348
Calopterygidae, 304
Calopteryx dimidiata, 304
　maculata, 304
Calosoma calidum, 330
　scrutator, 330
　syncophanta, 330
　willcoxi, 330
Cambarus bartoni, 296
Campanulaceae, 206
Camping in p.b., 64
Camponotus pennsylvanicus, 384
Candleberry, 118
Canker-worm Hunter, 332
Canoeing in p.b., 64
Cantharidae, 348
Canthon bispinosus, 340
　chalcites, 340
Caprifoliaceae, 132
Caprimulgus vociferus, 246
Carabidae, 329
Carabus vinctus, 330
Cardinal, Northern, 262
Cardinalis cardinalis, 262
Carduelis pinus, 264
　tristis, 264
Carex bullata, 170
　folliculata, 170
　pensylvanica, 170
　walteriana, 170
Carpodacus mexicanus, 264
Carranza Memorial, 47
Carrion Beetles, 336
Carrot Beetle, 342
Caryophyllaceae, 182
Cashew family, 122, 140
Cassandra, 128
Castor canadensis, 236
Catalpa, 118
Catalpa bignonioides, 118

Catbird, Gray, 254
Catbrier, 138
Caterpillar Hunter, 330
　Fiery, 330
Catfoot, 210
Cathartes aura, 244
Catocala gracilis lemmeri, 374
　herodias gerhardi, 374
　ilia, 374
　micronympha, 374
　sordida engelhardi, 374
　ultronia, 374
　vidua, 372
Cecropia Moth, 370
Cedar Creek watershed, 9
Cedar forests/swamps, 14
Cedar, Atlantic White, 14, 108
　Red, 108
'Cedar' water, 10
Celastrina argiolus, 364
Celithemis eponina, 302
Centipedes, 296
Cephalanthus occidentalis, 132
Cerambycidae, 354
Ceratodon purpureus, 90
Cercopidae, 324
Cercyonis pegala, 360
Certhia americana, 254
Ceryle alcyon, 248
Chaetura pelagica, 248
Chaffseed, 204
Chain-fern family, 102
　Narrow-leaved, 102
　Netted, 102
　Virginia, 102
Chalcophora virginiensis, 346
Chamaecyparis thyoides, 14, 108
Chamaedaphne calyculata, 128
Charcoal manufacture, 23
Chariessa pilosa, 350
Chatsworth Country Club, 46
Chauliodes pectinicornis, 326
Chauliognathus marginatus, 350
　pennsylvanicus, 348
Checkerberry, 136
Checkered Beetles, 350
Chelydra serpentina, 272
Cherry Tree Borer, Flat-headed, 346

Cherry, Black, 114
Chickadee, Carolina, 252
Chiggers, 294
Chilipoda, 296
Chimaphila maculata, 134
Chipmunk, Eastern, 234
Chlaeneus sericeus, 332
Chokeberry, Black, 122
　　Red, 120
Chordeiles minor, 248
Chrysemys picta, 274
　　rubriventris, 276
Chrysobothris femorata, 346
Chrysochus auratus, 356
Chrysomelidae, 356
Chrysopa occulata, 326
Chrysops delicatulus, 378
Chrysopsis falcata, 210
　　mariana, 210
Chubsucker, Creek, 290
Cicada(s), 322
　　Dog-day, 322
　　Periodical, 322
　　Pine-barren, 322
Cicadellidae, 324
Cicadidae, 322
Cicindela abdominalis, 329
　　formosa generosa, 329
　　patruela consentanea, 329
　　punctulata, 329
　　purpurea, 329
　　repanda, 329
　　rufiventris, 329
　　scutellaris rugifrons, 329
　　tranquebarica, 329
　　unipunctata, 329
Cicindelidae, 328
Cistaceae, 134
Citheronia regalis, 368
　　sepulcralis, 368
Cladium mariscoides, 170
Cladonia atlantica, 86
　　calycantha, 86
　　chlorophaea, 84
　　coniocraea, 86
　　cristatella, 86
　　incrassata, 86
　　santensis, 86
　　squamosa, 86

　　subtenuis, 84
　　uncialis, 84
Clemmys guttata, 274
Cleridae, 350
Clethra alnifolia, 126
Clethraceae, 126
Clethrinonomys gapperi, 236
Click Beetle(s), 346
　　Blind, 346
　　Eyed, 346
Climate of p.b., 8
Club-moss, Bog, 98
　　Carolina, 96
　　Family, 96
　　Fox-tail, 98
Club-rush, Water, 168
Clusiaceae, 188
Coastal plain, 4
　　Atlantic, 4
　　Inner, 5
　　Outer, 5
Coccinella novemnotata, 350
Coccinellidae, 350
Coccothraustes vespertinus, 264
Coccyzus americanus, 246
Cockroach, Wood, 310
Coenagrionidae, 304
Cohansey aquifer, 10
　　formation, 5, 6
Cohansie Iron Works, 58
Colaptes auratus, 248
Coleomegilla fuscilabris, 350
Coleoptera, 328
Colias philodice, 366
Colicroot, 176
Colinus virginianus, 246
Collembola, 301
Coluber constrictor, 278
Comandra umbellata, 182
Comma (butterfly), 360
Company towns, 39
Compositae, 206
Composite family, 206
Composites, Discoid, 206
　　Ligulate, 206
　　Radiate, 206
Comprehensive Management Plan, 67

Comptonia asplenifolia (var.), 118
   peregrina, 118
Contopus virens, 250
Convolvulaceae, 142
Copper, American, 364
   Bog, 364
Coreidae, 316
Corema conradii, 134
Coreopsis rosea, 218
Corixa spp., 320
Corixidae, 320
Cornus florida, 118
Corvus brachyrhynchos, 252
   ossifragus, 252
Corylaceae, 110, 118
Cotalpa lanigera, 342
Cotinus nitida, 344
Cotton mills, 30
Cotton-grass, Tawny, 168
   Virginia, 168
Cottontail, Eastern, 234
Cow Killer, 384
Cow-wheat, 204
Cowbane, Slender-leaved, 196
Cowbird, Brown-headed, 262
Cranberry agriculture, 33
Cranberry, American, 142
Crataegus uniflora, 122
Crayfish, 296
Creeper, Brown, 254
Cremastocheilus harrisii, 344
Creophilus maxillosus, 338
Crickets, 304
   Field, 308
      Mole, Northern, 310
      Tree, Narrow-winged, 308
      Tree, Snowy, 308
Crotalus horridus, 280
Crow, American, 252
   Fish, 252
Crowberry-, Broom, 134
   Family, 134
Crowleytown Glass Works, 54
Crustacea, 296
Cuckoo, Yellow-billed, 246
Cucumber Beetle, Spotted, 356
Cudweed, Purple, 210
Culex pipiens, 378
Culicidae, 376
Culiseta melanura, 378

Cumberland Forge, 58
Curculio spp., 358
Curculionidae, 356
Curly-grass (Fern), 51, 100
   Family, 100
Cuscuta spp., 142
Cutgrass, Rice, 162
Cyanocitta cristata, 252
Cybister fimbriolatus, 334
Cygnus colunbianus, 242
Cymatodera undulata, 350
Cynipidae, 382
Cyperaceae, 166
Cyperus filiculmis, 166
   grayi, 166
   grayii, 166
   macilentus (var.), 166
Cyperus, Gray's, 166
Cypripedium acaule, 178

Daddy-long-legs, 296
Damselflies, 301, 304
   Black-winged, 304
   Tip-winged, 304
Danaidae, 360
Danaus plexippus, 360
Dandelion, Dwarf, 220
Dangleberry, 130
Danthonia epilis, 160
   sericea, 160
   spicata, 158
Darkling Beetles, 352
Darner, Common Green, 302
   Heroic Green, 302
Darter, Swamp, 292
   Tessellated, 290
Dasymutilla occidentalis, 384
Decapoda, 296
Decodon verticillatus, 192
Deer Fly, 378
Deer, White-tail, 240
Delaware River valley, 5
Dendroica coronata, 258
   discolor, 260
   petechia, 258
   pinus, 260
   striata, 258
   virens, 258

Dennstaedtiaceae, 102
Dermacentor variabilis, 292
Dermaptera, 312
Description of p.b. in N.J., 4, 11
Desmodium obtusum, 184
   rigidum, 184
   strictum, 184
Development & uses of p.b., 19
Dewberry, 140
Diabrotica undecimpunctata, 356
Diapensia family, 136
Diapensiaceae, 136
Diapheromera femorata, 312
Dibotryon morbosum, 79
Dicerca divaricata, 346
Dichanthelium acuminatum, 164
   lindheimeri (var.), 164
Dichotomius carolinus, 340
Dicotyledons, 76
Dicranella heteromalla, 90
Dicranum condensatum, 92
   flagellare, 90
   scoparium, 90
Dictyoptera, 310
Didelphis virginiana, 232
Dineutes spp., 334
Diodia teres, 204
Diplopoda, 296
Diprionidae, 382
Diptera, 376
Dissosteira carolina, 306
Dodder, 142
Dogbane family, 198
   Spreading, 198
Dogwood, Flowering, 118
Dolichovespula maculata, 386
Dorcus parallelus, 338
Double Trouble bogs, 43
Double Trouble State Park, 62
Dove, Mourning, 246
Dover Forge (nr. Bamber), 42
Dover Furnace (Lakehurst), 42
Dragonflies, 301, 302
Drasteria graphica, 372
Dropseed, Late-flowering, 160
   Torrey's, 160
Drosera filiformis, 154
   intermedia, 154
   rotundifolia, 154
Droseraceae, 154

Dryocampa rubicunda, 368
Dryopteris simulata, 100
   thelypteris, 100
Duck, Black, 244
   Wood, 244
Dulichium arundinaceum, 166
Dumetella carolinensis, 254
Dung Beetles, 340
   Earth-boring, 340
Dusky-wing, Horace's, 366
   Sleepy, 366
Dwarf forests, 12
Dytiscidae, 334
Dytiscus fasciventris, 334
   harrisii, 334
   verticalis, 334

Eacles imperialis, 368
Eagle & Eagle Tavern, 30, 47
Eagle, Bald, 244
Earwig, European, 312
Ecology of pine barrens, 11
Ecology, fire, 15
Eel, American, 290
Eft, Red, 283
Elaphe guttata, 278
   obsoleta, 278
Elateridae, 346
Eleocharis olivacea, 168
   robbinsii, 166
Elfin, Brown, 362
   Frosted, 362
   Henry's, 362
   Hoary, 362
   Pine, 362
Ellychnia corrusca, 348
Empetraceae, 134
Enallagma civile, 304
Endangered species, 20, 73
Enneacanthus chaetodon, 290
   gloriosus, 290
   obesus, 290
Ephemeroptera, 301
Epiaeschna heros, 302
Epicauta pennsylvanica, 354
   pestifera, 354
   vittata, 354
Epidemia epixanthe, 364
Epigaea repens, 136

Epilobium angustifolium, 194
Equisetaceae, 96
Equisetum arvense, 96
Ericaceae, 126, 134, 136, 142, 196
Erimyzon oblongus, 290
Eriocaulaceae, 148
Eriocaulon compressum, 148
    decangulare, 148
    septangulare, 148
Eriophorum virginicum, 168
Erotylidae, 350
Erynnis brizo, 366
    horatius, 366
Esox americanus, 288
    niger, 288
Estell Glass Works, 58
Estigmene acrea, 372
Etheostoma fusiforme, 292
    olmstedi, 290
Etna or Aetna Furnace & Forge
    (Head of River), 58
Etna or Aetna Furnace & Mills
    (Medford), 48
Euchaetes egle, 372
Eupatorium album, 208
    hyssopifolium, 208
    leucolepis, 208
    ovatum (var.), 208
    pilosum, 208
    pubescens, 208
    resinosum, 208
    rotundifolium, 208
    saundersii (var.), 208
Euphorbia ipecacuanhae, 188
Euphorbiaceae, 188
Euphoria inda, 344
Eurosta solidagnis, 380
Euryophthalmus succinctus, 316
Euschistus variolarius, 314
Eusculus striatulus, 324
Evening-Primrose, Common, 194
    Cut-leaved, 196
    Family, 194
Everes comyntas, 364
Everlasting, Sweet, 210
    White, 210

Fagaceae, 110, 120
Falco sparverius, 244
False Foxglove, Downy, 204
    Fern-leaved, 204
Fauna, N.J. pine barrens, 231
Federal Forge/Furnace/Works, 42
Fern allies, 95, 96
Fern(s), 95, 98
    Bead, 100
    Bog, 100
    Bracken, 102
    Cinnamon, 98
    Curly-grass, 51, 100
    Family, 100, 102
    Marsh, 100
    Massachusetts, 100
    Royal, 98
    Sensitive, 100
Ferrago Furnace or Forge, 42
Fescue-Grass, Six-weeks, 158
Festuca octoflora, 158
Fetter-bush, 128
Figwort family, 202
Finch, House, 264
Fire ecology, 15
Fire-colored Beetle, 354
Fireflies, 348
Fireweed, 194
Fishes, 288
Fishfly, 326
Flax family, 186
    Ridged Yellow, 186
    Yellow, 186
Flayatem, 48
Flicker, Northern, 248
Flies, 376
    Bee, 380
    Black, 378
    Blow, 380
    Crane, 376
    Deer, 378
    Flower, 380
    Fruit, 380
    Horse, 378
    House, 380
    Hover, 380
    Robber, 378
Floating-heart, 150
Flora, N.J. pine barrens, 77
Flower Beetle(s), 344
    Bumble, 344

Hermit, 344
Rough, 344
Flowering & Fruiting table, 223
Flowering & seed producing plants, 75, 104
Key to, 105
Flowering Fern family, 98
Flyatt or Flyat, 48
Flycatcher, Great Crested, 250
Fomes applanatum, 79
Forests, Cedar, 14
Dwarf (Pygmy), 12
Lowland, 13
Upland, 12
Forficula auricularia, 312
Forgotten towns, 39
Forked River Mts., 5
Formicidae, 384
Fox, Gray, 238
Red, 238
Foxglove, Downy False, 204
Fern-leaved False, 204
Frog, Carpenter, 286
Chorus, New Jersey, 284
Green, 286
Leopard, Southern, 286
Frostweed, 190
Pine-barren, 192
Fruit Fly, Blueberry, 380
Frullania inflata, 88
Fulgoroidea, 326
Fungus Beetle, Forked, 352
Fungus(i), 78
Artist's, 79
Black Knot, 79
Bracket, 79
Earthstar, 79
Mushroom, Fly, 79
Puffballs, 79

Galactia regularis, 140
Galerita bicolor, 332
janus, 332
Galium pilosum, 204
puncticulosum (var.), 204
Gall Wasps, 382
Gall, Goldenrod Round-stem, 380
Oak Apple, 384
Gaultheria procumbens, 136

Gaylussacia baccata, 130
dumosa, 130
frondosa, 130
Geastrum triplex, 79
Gentian family, 198
Pine-barren, 198
Gentiana autumnalis, 198
Gentianaceae, 198
Geology & geologic history of p.b., 4
Geometridae, 376
Geothlypis trichas, 260
Geotrupes egerei, 340
splendidus, 340
Gerardia pedicularia, 204
racemulosa, 202
setacea, 204
virginica, 204
Gerardia, Bristle-leaved, 204
Pine-barren, 202
Thread-leaved, 204
Gerridae, 318
Gerris remigis, 318
Ghost towns, 39
Glass factories, 27
Glaucomys volans, 236
Gloucester Furnace, 56
Glyceria obtusa, 158
Gnaphalium obtusifolium, 210
purpureum, 210
Goat's-rue, 184
Golden Aster, Maryland, 210
Sickle-leaved, 210
Golden Club, 148
Golden-crest, 176
Golden-heather, 134
Goldenrod, Downy, 212
Field, 214
Fragrant, 214
Gray, 214
Pine-barren, 214
Slender, 212
Slender-leaved, 214
Swamp, 212
Wand-like, 212
White, 212
Goldenrods, flat-topped, 212
plume-like, 212
wand-like, 210
Goldfinch, American, 264

Goldsmith Beetle, 342
Goose, Canada, 242
Grackle, Common, 262
Gramineae, 156
Grape family, 142
Grapevine Beetle, 342
Graphic Moth, 372
Graphocephala coccinea, 324
Grass, Blue-eyed, 176
   -Pink, 180
   Cotton-, 168
   Family, 156
   Fescue-, 158
   Indian, 166
   Manna-, 158
   Millet-, 164
   Oat-, 158, 162
   Panic-, 162, 164
   Peanut-, 164
   Poverty-, 162
   Wool-, 168
   Yellow-eyed, 172, 174
Grasses, 156
   Terms used in identification, 158
Grasshoppers, 304
   Bird, 306
   Carolina, 306
   Long-horned, 306
   Pygmy, Crested, 306
   Short-horned, 306
Gratiola aurea, 202
Great Egg Harbor watershed, 9
Green Bank State Forest, 62
Greenbrier, Common, 138
   Glaucous-leaved, 138
   Halberd-leaved, 138
   Laurel-leaved, 140
   Red-berried, 138
'Greens', gathering of, 32
Grist mills, 23
Grosbeak, Evening, 264
   Rose-breasted, 262
Ground Beetle(s), 329
   Green, 332
   Snail-eating, 330
Ground-nut, 140
Grouse, Ruffed, 246
Gryllidae, 308

Gryllotalpa hexadactyla, 310
Gryllotalpidae, 310
Gryllus assimilis luctuosus, 308
   pennsylvanicus, 308
Gum, Black, 116
   Sour, 116
   Sweet, 114
Guttiferae, 188
Gymnospermae, 75
Gypsy Moth, 374
Gyrinidae, 334
Gyrinus spp., 334

Habenaria blephariglottis, 178
   ciliaris, 178
   clavellata, 178
   cristata, 178
   integra, 180
Habitats, major, 11
   Aquatic, 15
   Cedar, 14
   Dwarf (Pygmy), 12
   Lowland, 13
   Upland, 12
Haemodoraceae, 176
Hair-cap Moss, Awned, 92
   Common, 94
   Juniper, 94
   Ohio, 94
Hairgrass, Rough, 160
Hairstreak, Hessel's, 362
Half Moon & Seven Stars Tavern, 31, 48
Haliacetus leucocephalus, 244
Hamamelidaceae, 114
Hampton Gate, Furnace & Forge, 48
Hanover Furnace, 44
Harpalus caliginosus, 332
   pennsylvanicus, 332
Harrisville or Harrisia, 52
Harvestmen, 296
Hawk, Broad-winged, 244
   Red-tailed, 244
   Sharp-shinned, 244
Hawk-moth, Five-spotted, 366
Hawkweed, Hairy, 220
   Vein-leaved, 220
Hawthorn, Dwarf, 122

Hay-scented Fern family, 102
Hazel family, 110, 118
Heath family, 126, 134, 136, 142, 196
Heather, Beach-, 134
  Golden-, 134
  Pine-barren-, 134
Hedge-hyssop, Golden, 202
Helianthemum canadense, 190
  propinquum, 192
Helianthus angustifolius, 218
  divaricatus, 218
Hellumorphoides bicolor, 332
  texana, 332
Helonias bullata, 174
Hemidactylium scutatum, 283
Hemileuca maia, 368
Hemiptera, 312
Hemp, Indian, 200
Hermann City Glass Works, 55
Heron, Great Blue, 242
  Green-backed, 242
Hesperia metea, 366
Hesperiidae, 366
Hesperoleon abdominalis, 326
Hetaerina americana, 304
Heterodon platyrhinos, 276
Hieracium gronovii, 220
  venosum, 220
Hiking in p.b., 63
Hippodamia convergens, 350
Hirundo rustica, 250
Hister abbreviatus, 338
  interruptus, 338
Hister Beetles, 338
Histeridae, 338
Holly, American, 116
  Family, 116, 124
Homoptera, 322
Honeysuckle family, 132
Hop-merchant, 360
Horehound, Water-, 202
Hornet, Bald-faced, 386
Hornworm Moth, Tobacco, 366
  Tomato, 366
Horse Fly, Black, 378
Horsemint, 202
Horsetail, Common, 96
  Family, 96

Field, 96
House Fly, 380
Howell Furnace & Iron Works, 42
Huckleberry, Black, 130
  Blue, 130
  Dwarf, 130
Hudsonia ericoides, 134
  tomentosa, 134
Hudsonia, Heath-like, 134
  Woolly, 134
Humingbird, Ruby-throated, 248
Hunting in p.b., 65
Hyalophora cecropia, 370
Hydrochara obtusata, 336
Hydrophilidae, 334
Hydrophilus triangularis, 336
Hyla andersoni, 284
  crucifer, 284
Hyles lineata, 366
Hylocichla mustelina, 254
Hymenoptera, 382
Hypericum canadense, 190
  densiflorum, 126
  denticulatum, 190
  gentianoides, 190
  stans, 188
  stragulum, 188
  virginicum, 190
Hypoxis hirsuta, 176

Ice age, 6
Ichneumon (Wasp), 382
  Long-tailed, 382
  Short-tailed, 382
Ichneumonidae, 382
Ictalurus natalis, 290
  nebulosus, 290
Icterus galbula, 262
Ilex glabra, 124
  laevigata, 124
  opaca, 116
  verticillata, 124
Imperial Moth, 368
Incisala augustinus, 362
  henrici, 362
  irus, 362
  niphon, 362
  polia, 362
Indian Ann, 49
Indian Grass, 166

Indian-Hemp, 200
Indian-pipe, 196
Indians, American, in p.b., 19
Indigo, Wild, 184
Inkberry, 124
Insectivorous plants, 152
Insects, 296
    definitions of terms used, 298
    Key to orders of, 299
    Metamorphosis of, 297, 298
    Orders of, 297
Io Moth, 370
Ipecac, Wild, 188
Ipecac-Spurge, 188
Iridaceae, 176
Iris family, 176
Iris prismatica, 176
Iron (bog) forges & furnaces, 25
Isopoda, 296
Isoptera, 312
Itea virginica, 120
Ivy, Poison, 140
Ixodes dammini, 292
Ixodidae, 292

Japanese Beetle, 342
Jay, Blue, 252
'Jersey Devil', the, 59
Jointweed, 182
Juglans nigra, 118
Juncaceae, 170
Junco hyemalis, 264
Junco, Dark-eyed, 264
Juncus canadensis, 172
    effusus, 172
    militaris, 172
    pelocarpus, 172
June Beetle, Green, 344
Juniperus virginiana, 108

Kalmia angustifolia, 128
    latifolia, 128
Katydids, 304
    Angular-winged, 308
    Fork-tailed Bush, 306
    Oblong-winged Bush, 306
    Shield-backed, 308
    True, Northern, 308
Kestrel, American, 244

Key to flowering & seed producing plants, 105
    Non-flowering plants, 76
    Orders of insects, 299
Kingbird, Eastern, 250
'Kingdoms' of living organisms, 73
Kingfisher, Belted, 248
Kinglet, Golden-crowned, 256
    Ruby-crowned, 256
Kingsnake, Eastern, 280
Kinosternon subrubrum, 274
Kirkwood aquifer, 10
    formation, 5, 6
Kissing Bug, 318
Krigia virginica, 220

Labiatae, 200
Labidomera clivicollis, 356
Lacewing, Green, 326
Lachnanthes caroliniana, 176
    tinctoria, 176
Ladies'-tresses, Grass-leaved, 180
    Jagged, 180
    Little, 180
    Nodding, 180
    Spring, 180
Lady Beetle(s), 350
    Black, 352
    Convergent, 350
    Fifteen-spotted, 352
    Nine-spotted, 350
    Painted, 352
    Plain, 352
    Spotted, 350
    Two-spotted, 352
Lady's-slipper, Pink, 178
    Stemless, 178
Lambkill, 128
Lamiaceae, 200
Lampropeltis getulus, 280
    triangulum triangulum x *elapsoides,* 280
Lampyridae, 348
Largiidae, 316
Largus succinctus, 316
Lasiocampidae, 370
Latrodectus mactans, 294
Lauraceae, 114
Laurel family, 114

Laurel, Mountain-, 128
　Sheep-, 128
Leaf Beetle(s), 356
　Dogbane, 356
　Milkweed, 356
Leaf-footed Bug, 316
Leafhopper(s), 324
　Blunt-nosed, 324
　Red-banded, 324
　Sharp-nosed, 324
Leather-leaf, 128
Lebanon Glass Works, 44
Lebanon State Forest, 61
Lechea minor, 192
　racemulosa, 192
Lecidea uliginosa, 82
Leersia oryzoides, 162
Leguminosae, 140, 182
Leiophyllum buxifolium, 126
Lemming, Southern Bog, 238
Lentibulariaceae, 150
Lepidoptera, 358
Leptinotarsa decimlineata, 356
Leptocoris trivittatus, 316
Leptoglossus oppositus, 316
Leptophlebia cupida, 301
Leptothrix ochracea, 25
Lespedeza angustifolia, 186
　hirta, 186
　intermedia, 184
Lethocerus americanus, 320
Leucobryum albidum, 92
　glaucum, 92
Leucothoe racemosa, 128
Liatris graminifolia, 210
Libellula luctuosa, 302
　pulchella, 302
Libellulidae, 302
Lichen, Awl, 86
　British Soldier, 86
　Broccoli, 86
　Coastal Plain, 86
　False Reindeer, 84
　Furrowed Shield, 82
　Goat, 84
　Gray Star, 84
　Green Shield, 82
　Lace-tipped, 84
　Ladder, Slender, 86
　Mealy Goblet, 84
　Pyxie-cup, 84
　Reindeer, False, 84
　Rough Shield, 82
　Santee, 86
　Shield, 82
　Slender Ladder, 86
　Squamose, 86
　Swamp, 86
　Tar, 82
　Thorn, 84
　Tunnel, 86
Lichens, 79
　Crustose, 80
　Foliose, 80
　Fruticose, 80
　Terms used in identification, 80
Liliaceae, 138, 174, 176
Lilium superbum, 174
Lily family, 138, 174, 176
　Turk's-cap, 174
Limacodidae, 376
Limenitis archippus, 362
Linaceae, 186
Linaria canadensis, 202
Linum intercursum, 186
　striatum, 186
Lion's-foot, 220
Liquidambar styraciflua, 114
Liriodendron tulipifera, 118
Lisbon mills & forge, 45
Liverwort(s), 88
　Frullania, 88
　Leafy, 88
　Scale 'moss', common, 88
Lizard, Northern Fence, 276
Lobelia canbyi, 206
　nuttallii, 206
Lobelia, Canby's, 206
　Nuttall's, 206
Locust, Black, 118
Locust, carolina, 306
Long-horned Beetles, 354
Loosestrife family, 192
　Swamp, 192
　Swamp-, 196
　Yellow, 196
Lophiola americana, 176
　aurea, 176
Lower Forge, (Burl. Co.) 49
Lower Forge (Ocean Co.), 42

Lowland forests, 13
Loxosceles reclusa, 296
Lucanidae, 338
Ludwigia alternifolia, 194
   sphaerocarpa, 194
Ludwigia, Globe-fruited, 194
Lumbering industry, 21
Luna Moth, 370
Lupine, Wild, 184
Lutra canadensis, 240
Lycaena phlaeas, 364
Lycaenidae, 362
Lycidae, 348
Lycoperdon spp., 79
Lycopodiaceae, 96
Lycopodium alopecuroides, 98
   bigelovii (var.), 98
   carolinianum, 96
   inundatum, 98
Lycopus amplectens, 202
Lycosidae, 294
Lygaeidae, 316
Lygaeus kalmii, 316
Lymantria dispar, 374
Lymantridae, 374
Lyonia ligustrina, 128
   mariana, 128
Lysimachia terrestris, 196
Lythraceae, 192

Madder family, 132, 136, 204
Madtom, Tadpole, 290
Magicicada septendecim, 322
Magnolia family, 114
   Swamp, 114
Magnolia virginiana, 114
Magnoliaceae, 114
Malacosoma americanum, 370
Maleberry, 128
Mallard, 242
Mammals, 232
Manchester Furnace, 42
Manduca quinquemaculata, 366
   sexta, 366
Manna-grass, Blunt, 158
Mantidae, 310
Mantids, 310
   Chinese, 310
   European, 310

Mantis religiosa, 310
Maple family, 116
   Red, 116
   Swamp, 116
   Trident, 116
Marmota monax, 234
Martha Forge (Ocean Co.), 42
Martha Furnace (Burlington Co.), 51
Martha Furnace Diary, 52
Martin, Purple, 252
Mary Ann Forge, 45
May Beetles, 342
Mayflies, 301
McCartyville, 52
Meadow-beauty, 194
   Maryland, 194
Megacephala virginica, 329
Megalodacne heros, 350
Megaloptera, 326
Megarhyssa atrata, 382
Melampyrum lineare, 204
Melanerpes carolinus, 248
   erythrocephalus, 248
Melanolestes picipes, 318
Melastoma family, 194
Melastomataceae, 194
Meloidae, 354
Melospiza georgiana, 266
   melodia, 266
Membracidae, 324
Mephitis mephitis, 240
Metallic Wood-boring Beetles, 346
Metamorphosis, insects, 297, 298
Miarchus crinitus, 250
Microcentrum retinerve, 308
Microtus pennsylvanicus, 236
Milk-Pea, 140
Milkweed Beetle, Red, 356
Milkweed Bug, Large, 316
   Small, 316
Milkweed, Blunt-leaved, 200
   Family, 200
   Red, 200
Milkwort, Cross-leaved, 188
   Family, 186
   Nuttall's, 186
   Orange, 188
   Short-leaved, 188
Millet-grass, Pursh's, 164

Millipedes, 296
Mills, the & forge, 45
Mimus polyglottos, 254
Minerals in p.b., 7
'Mining' of sand & gravel, 28
Mink, 238
Mint family, 200
Minuartia caroliniana, 182
Mitchella repens, 136
Mites, Harvest, 294
Mitoura hesseli, 362
Mniotilta varia, 258
Moccasin-flower, 178
Mockingbird, Northern, 254
Mole, Eastern, 234
Molothrus ater, 262
Monarch Butterfly, 360
Monarda punctata, 202
Monmouth Furnace, 41
Monochamus titillator, 354
Monocotyledons, 76
Monomorium minimum, 384
Monotropa hypopithys, 196
   uniflora, 196
Monroe Forge, 57
Morning-glory, 142
   Family, 142
Mosquito, Banded-leg Salt Marsh, 378
   Cedar Bog, 378
   House, 378
   Tree-hole, 378
Moss(es), 88
   Broom, 90, 92
   Broom, Little, 90
   Bryum, Hairy, 92
   Hair-cap, 92
   Horntooth, Purple, 90
   Peat, 89
   Pin Cushion, 92
   Pohlia, Common, 92
   Sphagnum, 89
   Thelia, Common, 94
Moth, Buck, 368
   Cecropia, 370
   Graphic, 372
   Gypsy, 374
   Imperial, 368
   Io, 370
   Isabella Tiger, 372
   Little Nymph, 374
   Luna, 370
   Milkweed Tussock, 372
   Oakworm, 368
   Pine-devil, 368
   Polyphemus, 370
   Promethia, 370
   Regal, 368
   Rosy Maple, 368
   Royal Walnut, 368
   Saddleback Caterpillar, 376
   Salt Marsh, 372
   Tent Caterpillar, Eastern, 370
   Woolly Bear Caterpillar, 372
   Yellow Woolly Bear, 372
Moths, 366
   Bagworm, 376
   Hornworm, 366
   Inchworm, 376
   Oakworm, 368
   Owlet, 372
   Royal, 368
   Silkworm, Giant, 368
   Slug Caterpillar, 376
   Sphinx, 366
   Tent Caterpillar, 370
   Tiger, 370
   Tussock, 372, 374
   Underwing, 372
Mount, 51
Mountain-laurel, 128
Mountain-Mint, 202
   Short-toothed, 202
Mourning Cloak, 360
Mouse, Meadow, 236
   Meadow Jumping, 238
   Pine, 236
   Red-backed, 236
   White-footed, 236
Mt. Holly Mills & Iron Works, 46
Mt. Misery (saw) mill, 44
Mud-Dauber, Black & Yellow, 386
   Organ Pipe, 386
Mudminnow, Eastern, 288
Muhlenbergia torreyana, 160
   uniflora, 160
Mullica River watershed, 9
Mulsantia picta minor, 352
Musca domestica, 380

Muscidae, 380
Muskrat, 236
Mustela frenata, 238
    vison, 238
Mutillidae, 384
Myotis lucifugus, 234
Myrica heterophylla, 118
    pensylvanica, 118
Myricaceae, 118

Narthecium americanum, 174
Natrix sipedon, 276
Naucoridae, 320
Necrodes surinamensis, 336
Neocurtilla hexadactyla, 310
Neodiprion lecontei, 382
    sertifer, 382
Neomysia pullata, 352
Neopyrochroa flabellata, 354
Nepidae, 320
Net-winged Beetle(s), 348
    Banded, 348
Neuroptera, 326
New Jersey Pinelands Commission, 67
New Jersey State Forests & Parks, 61
New Lisbon mills & forge, 45
New Mills & forge, 45
Newt, Red-spotted, 283
Nicrophorus orbicollis, 336
    pustulatus, 336
    tomentosus, 336
Nighthawk, Common, 248
Noctuidae, 372
Nomotettix cristatus, 306
Non-flowering plants, 75
    Key to, 76
Notonecta undulata, 320
Notonectidae, 320
Notophthalmus viridescens, 283
Notropis chalybaeus, 288
Noturus gyrinus, 290
Nuphar luteum, 150
    variegatum, 150
Nuthatch, Red-breasted, 252
    White-breasted, 252
Nymph Moth, Little, 374
Nymphaea odorata, 150

Nymphaeaceae, 150
Nymphalidae, 360
Nymphalis antiopa, 360
Nymphoides cordata, 150
Nyssa sylvatica, 116
Nyssaceae, 116

Oak(s), 110
    Bear-, 120
    Black, 112
    Black-jack, 112
    Chestnut-, 110
    Dwarf Chestnut-, 120
    Post-, 110
    Scarlet-, 112
    Scrub-, 120
    Southern Red, 112
    Spanish, 112
    White, 110
Oakworm Moth, Orange-striped, 368
    Spiny, 368
Oat-Grass, Black, 162
    Silky Wild, 160
    Smooth Wild, 160
    Wild, 158
Odocoileus virginianus, 240
Odonata, 301
Odontoschisma prostratum, 88
Odontotaenius disjunctus, 338
Oecanthus angustipennis, 308
    fultoni, 308
Oenothera biennis, 194
    laciniata, 196
Old Half Way, 43
Onagraceae, 194
Oncopeltus fasciatus, 316
Ondrata zibethicus, 236
Ong's Hat, 45
Onoclea sensibilis, 100
Opheodrys aestivus, 278
Ophion bilineatus, 382
Opposum, 232
Opuntia humifusa, 156
Orange-grass, 190
Orchidaceae, 178
Orchis, Crested Yellow, 178
    Family, 178
    Green Woodland, 178

Southern Yellow, 180
White Fringed, 178
Yellow Fringed, 178
Oriole, Northern, 262
Orontium aquaticum, 148
Orthoptera, 304
Orthosoma brunneus, 354
Osmoderma erimicola, 344
scabra, 344
Osmunda cinnamomea, 98
regalis, 98
Osmundaceae, 98
Oswego Sawmill, 51
Oswego-Wading watershed, 9
Otter, River, 240
Otus asio, 246
Ovenbird, 260
Ox Beetle, 342
Owl, Great Horned, 246
Screech - see Screech-owl
Oxypolis ambigua (var.), 196
rigidior, 196

Painted Beauty, 362
American, 362
Paisley, 47
Pallavicinia lyelli, 88
Palmer's Sawmill, 42
Panic-Grass, Lindheimer's, 164
Warty, 162
Panicum lanuginosum, 164
lindheimeri (var.), 164
verrucosum, 162
virgatum, 164
Paonias excaecatus, 366
Paper manufacture, 29
Papilio glaucus, 364
polyxenes asterius, 364
troilus, 364
Papilionidae, 364
Parcoblatta pennsylvanica, 310
uhleriana, 310
Parker's mills, 46
Parmelia rudecta, 82
squarrosa, 82
sulcata, 82
Parsley family, 196
Parthenocissus quinquefolia, 142
Partridge-berry, 136

Parula americana, 258
Parula, Northern, 258
Parus bicolor, 252
carolinensis, 252
Parvin State Park, 63
Pasadena, 43
Pasimachus depressus, 330
Paspalum setaceum, 162
Paspalum, Slender, 162
Passalidae, 338
Passalus (Beetle), Horned, 338
Passer domesticus, 260
Pea family, 140, 182
Peanut-grass, 164
Pearl Crescent, 360
Peeper, Northern Spring, 284
Pelecinidae, 384
Pelecinus polyturator, 384
Pelidnota punctata, 342
Pelidnota, Spotted, 342
Pelocoris femoratus, 320
Peltandra virginica, 146
Pencil-flower, 186
Penn State Forest, 62
Pennyroyal, Bastard, 200
Pentatomidae, 314
Pepperbush, Sweet, 126
Perch, Pirate, 290
Peromyscus leucopus, 236
Pewee, Eastern Wood-, 250
Phaenicia sericata, 380
Phalangiidae, 296
Phanaeus vindex, 340
Phasmatodea, 312
Pheucticus ludovivianus, 262
Philaenus spumarius, 324
Phoebe, Eastern, 250
Phoenix Forge, 42
Photinus pyralis, 348
Photuris pennsylvanicus, 348
Phyciodes tharos, 360
Phyllophaga fusca, 342
marginalis, 342
Phymata erosa, 316
Phymatidae, 316
Physcia aipolia, 84
millegrana, 84
Picea abies, 118

Pickerel, Chain, 288
  Redfin, 288
Pickerelweed, 148
  Family, 148
Picoides pubescens, 250
  villosus, 250
Pieridae, 364
Pieris rapae, 364
Pillbugs, 296
Pinaceae, 106
Pinching Beetle, 338
Pine barrens, 1, 2, 4, 11
  Acid rain in, 8
  Acidity of, 7, 10
  Area in N.J., 2
  Areas of, 1
  Camping in, 64
  Canoeing in, 64
  Climate of, 8
  Description of, 4, 11
  Development & uses of, 19
  Ecology of, 11
  Fauna of, 231
  Flora of, 77
  Geology of, 4
  Hiking in, 63
  Hunting in, 65
  Location in N.J., 2
  Minerals in, 7
  Pine 'barrens' vs. 'pinelands', 2
  Precipitation in, 8
  Preservation efforts history, 66
  Preservation future, 67
  Soils of, 6
  Water Reserves, 10
  Watersheds, 9
Pine Crest Sanitarium, 47
Pine family, 106
  Jersey, 108
  Pitch, 12, 108
  Scrub, 108
  Short-leaf, 106
  Virginia, 108
  White, 118
  Yellow, 106
Pinelands Commission, 67
Pine Sawyer (Beetle), Southern, 354
Pine-Borer, Flat-headed, 346
Pine-devil Moth, 368

Pinesap, 196
Pineweed, 190
Pink family, 182
Pink, Grass, 180
Pinus echinata, 106
  rigida, 108
  strobus, 118
  virginiana, 108
Pinweed, Oblong-fruited, 192
  Thyme-leaved, 192
Pipewort family, 148
  Flattened, 148
  Seven-angled, 148
  Ten-angled, 148
Pipilo erythrophthalmus, 264
Piranga olivacea, 262
Pitch distillation, 23
Pitcher-plant, 154
  Family, 154
Pituophis melanoleucus, 278
Pitymys pinetorum, 236
Pityopsis falcata, 210
Plains, East, 13
  Little, 13
  South or Spring Hill, 13
  West, 12
Plant 'Kingdom', 75
Plantanthera blephariglottis, 178
  ciliaris, 178
  clavellata, 178
  cristata, 178
  integra, 180
Planthoppers, 326
Plants, Aquatic, 144
  Composites, 206
  Grass-like, 156
  Herbaceous, 144
  Insectivorous, 152
  Netted-veined, 182
  Parallel-veined, 172
  Unusual structures, 152
Platanus occidentalis, 118
Pleasant Mills, 55
Pleasing Fungus Beetle, 350
Plethodon cinereus, 283
Poaceae, 156
Podabrus modestus, 350
Podisus maculiventris, 316
Pogonia ophioglossoides, 180

Pogonia, Rose, 180
Pohlia nutans, 92
Polistes fuscatus pallipes, 386
Polygala brevifolia, 188
    cruciata, 188
    lutea, 188
    nuttallii, 186
Polygalaceae, 186
Polygonaceae, 182
Polygonella articulata, 182
Polygonia comma, 360
    interrogationis, 360
Polyphemus Moth, 370
Polypodiaceae, 100, 102
Polytrichum commune, 94
    juniperinum, 94
    ohioense, 94
    piliferum, 92
Pond-lily, Yellow, 150
Pondweed, Alga, 146
    Family, 146
Pontederia cordata, 148
Pontederiaceae, 148
Popillia japonica, 342
Populus grandidentata, 118
Potamogeton confervoides, 146
Potamogetonaceae, 146
Potato Beetle, Colorado, 356
Potomac sands, 10
Potomac-Raritan-Magothy formation, 10
Potter's Sawmill, 43
Poverty-grass, 162
Precipitation in p.b., 8
Predacious Diving Beetles, 334
Prenanthes autumnalis, 220
    serpentaria, 220
Preservation of p.b., 66
    Future, 67
    History of efforts, 66
Prickly Pear, 156
Primrose family, 196
Primulaceae, 196
Prionus (Beetle), Lesser, 354
Prionus laticollis, 354
Procyon lotor, 238
Progne subis, 252
Promethea Moth, 370
Pronotaria citrea, 258
Prosimulium fuscum, 378

Prunus maritima, 122
    serotina, 114
Pseudacris triseriata kalmi, 284
Pseudolucanus capreolus, 338
Pseudoparmelia caperata, 82
Pseudotriton ruber, 284
Psychidae, 376
Pteridium aquilinum, 102
Pteridophyta, 75
Pterophylla camellifolia, 308
Pulse family, 140, 182
Pycnanthemum muticum, 202
    verticillatum, 202
Pygmy forests- see dwarf forests
Pyrochroidae, 354
Pyrolaceae, 134, 196
Pyrrharctia isabella, 372
Pyrus arbutifolia, 120
    melanocarpa, 122
Pyxidanthera barbulata, 136
Pyxie, 136

Quaker Bridge, and-Q.B.Tavern, 31,50
Quercus alba, 110
    coccinea, 112
    falcata, 112
    ilicifolia, 120
    marilandica, 112
    montana, 110
    prinoides, 120
    prinus, 110
    stellata, 110
    velutina, 112
Question Mark, 360
Quiscalus quiscula, 262

Raccoon, 238
Racer, Northern Black, 278
Rana clamitans melanota, 286
    utricularia, 286
    virgatipes, 286
Ranatra fusca, 320
Rancocas Creek watershed, 9
Randolph's Mill & Furnace, 47
Rare species, 20, 73
Rattlesnake, Timber, 280
Rattlesnake-root, 220
    Pine-barren, 220
    Slender, 220

Rattlesnake-weed, 220
Realty developments, 31
Red Admiral, 362
Red Bug, 316
Red Lion, 46
Red-spotted Purple, 362
Redroot, 176
Redstart, American, 260
Reduviidae, 318
Reedgrass, Nuttall's, 160
   Pine-barren, 160
Regal Moth, 368
Regulus calendula, 256
   satrapa, 256
Reptiles, 272
Reticulitermes flavipes, 312
Retreat Forge & Factory, 46
Rhagoletis mendax, 380
Rhexia mariana, 194
   virginica, 194
Rhinoceros Beetle, 344
Rhodobaenus tredecimpunctata, 358
Rhododendron glaucum (var.), 126
   viscosum, 126
Rhopalidae, 316
Rhus copallina, 124
   radicans, 140
   vernix, 124
Rhynchospora alba, 170
   capitellata, 168
Rice, Wild, 162
Roaches, 310
Robber Fly, 378
Robin, American, 254
Robinia pseudo-acacia, 118
Rockrose family, 134
Root-borer, Broad-necked, 354
Rosaceae, 114, 120, 140
Rose family, 114, 120, 140
Rosy Maple Moth, 368
Rove Beetle(s), 338
   Hairy, 338
Royal Walnut Moth, 368
Rubiaceae, 132, 136, 204
Rubus cuneifolius, 122
   flagellaris, 140
   hispidus, 140

Ruby Spot, 304
Rush, Bayonet, 170
   Beak-, 168, 170
   Bog-, 170
   Brown-fruited, 170
   Canada, 170
   Chair-maker's, 168
   Club-, 168
   Common, 170
   Family, 170
   Soft, 170
   Spike-, 166, 168
   Swaying, 168
   Three-square, 168
   Twig, 170
Sabatia difformis, 198
Sabatia, Lance-leaved, 198
Saddleback Caterpillar Moth, 376
Sagittaria engelmanniana, 146
Salamander, Four-toed, 283
   Marbled, 283
   Red, Northern, 284
   Red-backed, 283
Salix nigra, 118
Salt Marsh Moth, 372
Salticidae, 294
Sand & gravel 'mining', 28
Sand-myrtle, 126
Sandalwood family, 182
Sandpiper, Spotted, 246
Sandwort, Pine-barren, 182
Santalaceae, 182
Sarracenia purpurea, 154
Sarraceniaceae, 154
Sassafras, 114
Sassafras albidum, 114
Saturnidae, 368
Satyridae, 360
Savannas, 15
Sawfly, Pine, 382
   European, 382
   Red-headed, 382
Saxifragaceae, 120
Saxifrage family, 120
Sayernis phoebe, 250
Scalopus aquaticus, 234
Scaphinotus elevatus, 330
Scaphiopus holbrooki, 284
Scaphytopius magdalensis, 324
Scarab Beetles, 340

Scarabaeidae, 340
Scarites subterraneus, 332
Sceliphron caementarium, 386
Sceloporous undulatus hyacinthinus, 276
Schistocera alutacea, 306
Schizachyrium scoparium, 164
Schizaea pusilla, 51, 100
Schizaeaceae, 100
Schwalbea americana, 204
Scirpus americanus, 168
    cyperinus, 168
    pungens, 168
    subterminalis, 168
Sciurus carolinensis, 234
Scolops sulcipes, 326
Screech-owl, Eastern, 246
Scrophulariaceae, 202
Scudderia furcata, 306
Sedge, Broom-, 164
    Bull, 170
    Button, 170
    Family, 166
    Long, 170
    Pennsylvanica, 170
    Slender, 166
    Three-way, 166
    Walter's, 170
Seedbox, 194
Seiurus aurocapillus, 260
Sericocarpus asteroides, 218
    linifolius, 218
'Serrotinus' pitch pines, 13
Serviceberry, 122
Setophaga ruticilla, 260
Sexton Beetles, 336
Shadbush, 122
Sheep-laurel, 128
Shiner, Ironcolor, 288
Shreveville, 46
Shrew, Masked, 232
    Short-tailed, 232
Shrewsbury Furnace, 41
Sialia sialis, 256
Sibine stimulea, 376
Silpha americana, 336
    novaboracensis, 336
    surinamensis, 336
Silphidae, 336

Silverrod, 212
Simulidae, 378
Siskin, Pine, 264
Sisyrinchium atlanticum, 176
Sitta canadensis, 252
    carolinensis, 252
Skimmer, Ten-spot, 302
Skipper(s), 366
    Cobweb, 366
    Dusted, 366
Skunk, Striped, 240
Smilax glauca, 138
    laurifolia, 140
    pseudo-china, 138
    rotundifolia, 138
    walteri, 138
Smithville, 46
Snake, Corn, 278
    Garter, Eastern, 276
    Green, Rough, 278
    Hognose, Eastern, 276
    Milk, Eastern, 280
    Pine, Northern, 278
    Rat, Black, 278
    Water, Northern, 276
Snakemouth, 180
Snout Beetles, 356
Soils of p.b., 6
Soldier Beetles, 348
Soldier Bug, Spined, 316
Solidago bicolor, 212
    erecta, 212
    fistulosa, 214
    graminifolia, 214
    mexicana (var.), 212
    nemoralis, 214
    odora, 214
    puberula, 212
    sempervirens, 212
    stricta, 212
    tenuifolia, 214
    uliginosa, 212
Sooy's Tavern, 31, 51
Sorex cinereus, 232
Sorghastrum nutans, 166
Sour Gum family, 116
Sowbugs, 296
Sparganiaceae, 144
Sparganium americanum, 144

Sparrow, Chipping, 264
  Field, 266
  House, 260
  Song, 266
  Swamp, 266
  White-throated, 266
Spatter-dock, 150
Species descriptions, 71
Speedwell Mill & Furnace, 47
Spermatophyta, 75
Sphagnum or Peat moss, 89
Sphagnum, gathering of, 32
Sphagnun cuspidatum, 90
  fallax, 89
  flavicomans, 90
  magellanicum, 89
  papillosum, 89
  pulchrum, 90
Sphecidae, 386
Sphingidae, 366
Sphinx, Blind-eyed, 366
  Carolina, 366
  White-lined, 366
Spider, Black Widow, 294
  Brown Recluse, 296
Spiders, Crab, 294
  Ground, 294
  Jumping, 294
  Wolf, 294
Spike-rush, Green, 168
  Triangular-stem, 166
Spilosoma virginica, 372
Spiranthes beckii, 180
  cernua, 180
  laciniata, 180
  praecox, 180
  tuberosa, 180
  vernalis, 180
Spittlebug(s), 324
  Meadow, 324
  Pine, 324
Spizella passerina, 264
  pusilla, 266
Spleenwort, Ebony, 100
Spongs, 15
Spotswood saw- & grist-mills and iron forges, 41
Spring Azure, 364
Springtails, 301

Spruce, Norway, 118
Spurge family, 188
  Ipecac-, 188
Squirrel, Flying, 236
  Gray, 234
  Red, 234
  Southern Flying, 236
St. Andrew's Cross, 188
St. John's-wort, Canada, 190
  Coppery, 190
  Family, 188
  Marsh, 190
  Shrubby, 126
St. Peter's-wort, 188
Stafford Forge, 43
Stag Beetles, 338
Stagger-bush, 128
Staphylinidae, 338
Staphylinus maculosus, 338
Star-flower, 198
Stargrass, Yellow, 176
Starling, European, 256
Stenopoda spinulosa, 318
Sternotherus odoratus, 272
Stichopogon trifasciatus, 378
Stick-tight, 184
Stictocephala bisonia, 324
  bubalus, 324
Stink Bug(s), 314
  Green, 314
  One-spotted, 314
  Rough, 314
Stinkpot, 272
Stipa avenacea, 162
Strategus antaeus, 342
Sturnus vulgaris, 256
Stylisma caesariense (var.), 142
  pickeringii, 142
Stylosanthes biflora, 186
Sulphur, Clouded, 366
Sumac, Dwarf, 124
  Poison, 124
  Winged, 124
Summer-grape, 142
Sundew family, 154
  Round-leaved, 154
  Spatulate-leaved, 154
  Thread-leaved, 154

Sunfish, Banded, 290
   Blackbanded, 290
   Bluespotted, 290
   Mud, 290
   Sphagnum, 290
Sunflower, Narrow
   -leaved, 218
   Woodland, 218
Swallow, Barn, 250
   Tree, 250
Swallowtail, Black, 364
   Spicebush, 364
   Tiger, 364
Swamp Loosestrife, 192
Swamp-candles, 196
Swamp-Loosestrife, 196
Swamp Magnolia, 114
Swamp-pink, 174
Swamps, 14
Swan, Tundra, 242
Sweet Bay, 114
Sweet-fern, 118
Sweetwater, 55
Swift, Chimney, 248
Switchgrass, 164
Sycamore, 118
Sylvilagus floridanus, 234
Sympetrum vicinum, 304
Synaptomys cooperi, 238
Syrphidae, 380

Tabanidae, 378
Tabanus atratus, 378
Tachycineta bicolor, 250
Tamias striatus, 234
Tamiasciurus hudsonicus, 234
Tanager, Scarlet, 262
Tanning & currying, 29
Tanton Furnace & Forge, 48
Tar distillation, 22
Tarpela micans, 352
Taunton Furnace & Forge, 48
Taverns, 30
Teaberry, 136
Tenebrionidae, 352
Tenodera aridifolia sinensis, 310
Tent Caterpillar (Moth), Eastern, 370
Tephritidae, 380

Tephrosia virginiana, 184
Termite, Eastern Subterranean, 312
Terrapene carolina, 274
Tetraopes melanurus, 356
   tetraophthalmus, 356
Tetrigidae, 306
Tettigoniidae, 306
Thallophyta, 75
Thamnophis sirtalis, 276
Thelia asprelia, 94
   hirtella, 94
Thelypteris palustris, 100
   simulata, 100
Thomisidae, 294
Thoroughworts, 206
Thrasher, Brown, 254
Threatened species, 20, 73
Three Partners' sawmill, 42
Thrush, Wood, 254
Thryothorus ludovicianus, 254
Tibicen canicularis, 322
   hieroglyphica, 322
Tick(s), American Dog, 292
   Deer, 292
Tick-trefoil, Rigid, 184
   Stiff, 184
Ticklegrass, 160
Tickseed, Rose-colored, 218
Tickseed-Sunflower, Slender-
   leaved, 220
Tiger Beetles, 328, 329
Tiger Moth, Isabella, 372
Tile manufacture, 28
Tintern Falls Iron Works, 41
Tintern Furnace & Forge
   (Medford), 48
Tinton Iron Works, 41
Tipulidae, 376
Titmouse, Tufted, 252
Toad, Fowler's, 284
   Spadefoot, Eastern, 284
Toadflax, Bastard, 182
   Blue, 202
   Old-field, 202
   Star, 182
Tofieldia racemosa, 174
Toms River watershed, 9
Toper, Red, 304
Towhee, Rufous-sided, 264

Toxicodendron radicans, 140
　vernix, 124
Toxorhynchites rutilus
　septentrionalis, 378
Toxostoma rufum, 254
Trachymyrmex septentrionalis, 384
Treefrog, Pine Barrens, 284
Treehopper(s), 324
　Buffalo, 324
Trees, 106
Triadenum virginicum, 190
Trichiotinus affinis, 344
　piger, 344
Trichoptera, 358
Trichostema dichotomum, 200
　setaceum, 200
Trientalis borealis, 198
Troglodytes aedon, 254
Trombidiidae, 294
Trypoxylon politum, 386
Tuckahoe Furnace & Forge, 58
Tuckahoe River watershed, 9
Tuckerton Road, the, 48
Tulip tree, 118
Tumblebug Beetles, 340
Tupelo, 116
Turdus migratorius, 254
Turkeybeard, 174
Turpentine distillation, 22
Turtle, Box, Eastern, 274
　Mud, Eastern, 274
　Musk, 272
　Painted, Eastern, 274
　Red-bellied, 276
　Snapping, Common, 272
　Spotted, 274
Tussock Moth, Milkweed, 372
Twig-rush, 170
Tyrannus tyrannus, 250

Umbelliferae, 196
Umbra pygmaea, 288
Underwing Moth, Graceful, 374
　Herodias, 374
　Ilia, 374
　Sordid, 374
　Ultronia, 374
　Widow, 372
Union Clay Works, 43

Union Works or Forge, 46
Unknown Mill, 48
Upland forests, 12
Upper Mill, 45
Upton/Upton Station, 44
Urocyon cinereoargenteus, 238
Uses of p.b., present & future, 61
Utricularia cornuta, 152
　fibrosa, 152
　inflata, 152
　purpurea, 150
　subulata, 152
Uvularia nitida (var.), 174
　puberula, 174
　pudica, 174

Vaccinium atrococcum, 132
　caesariense, 132
　corymbosum, 132
　macrocarpon, 142
　pallidum, 130
　vacillans, 130
Vanessa atalanta, 362
　cardui, 362
　virginiensis, 362
Vascular plants, 75, 94
Velvet-ants, 384
Vespidae, 386
Vespula maculifrons, 386
Viburnum cassinoides, 132
　nudum, 134
Viceroy, 362
Vine family, 142
Viola lanceolata, 192
　pedata, 192
　primulifolia, 192
Violaceae, 192
Violet, Birdfoot, 192
　Family, 192
　Lance-leaved, 192
　Primrose-leaved, 192
Vireo griseus, 256
　olivaceus, 256
Vireo, Red-eyed, 256
　White-eyed, 256
Virginia Creeper, 142
Virginia-willow, 120
Vitaceae, 142
Vitis aestivalis, 142

Vole, Meadow, 236
   Pine, 236
   Red-backed, 236
Vulpes fulva, 238
Vulpia octoflora, 158
Vulture, Turkey, 244

Wading River Forge & Slitting Mill, 52
Wading River watershed, 9
Walker's Forge, 57
Walkingstick, 312
Walnut, Black, 118
Warbler, Black & White, 258
   Black-throated Green, 258
   Blackpoll, 258
   Pine, 260
   Prairie, 260
   Prothonotary, 258
   Yellow, 258
   Yellow-rumped, 258
Washington & Washington Tavern, 31, 51
Washington Furnace, 42
Wasp, Paper, 386
   Pelicinid, 384
Wasps, 382
   Gall, 382
   Ichneumon, 382
   Social, 386
   Thread-waisted, 386
Water Boatmen, 320
Water Bug, Creeping, 320
   Giant, 320
Water reserves in p.b., 10
Water Scavenger Beetles, 334
Water Strider, 318
Water-Horehound, 202
Water-lily family, 150
   Fragrant, 150
   White, 150
Water-plantain family, 146
Water-shield, 150
Waterford Works Glass Factory, 56
Waterscorpion, 320
Watersheds in p.b., 9
Wax-myrtle, 118
   Family, 118
Waxwing, Cedar, 256

Weasel, Long-tailed, 238
Weevil(s), 356
   Acorn, 358
   Blueberry Blossom, 358
   Cranberry, 358
   Nut, 358
West Creek Forge, 43
Westecunk Forge, 43
Weymouth Furnace, Forge, & Paper Mill, 57
Wharton State Forest, 11, 61
Wharton, Joseph, 11, 61
Wheatlands, 43
Wheel Bug, 318
Whip-poor-will, 246
Whirligig Beetles, 334
White Alder family, 126
White Horse, 47
White-topped Aster, Narrow-leaved, 218
   Toothed, 218
Whitesbog (White's bogs), 44
Widow, The, 302
Williamsburg Forge, 42
Willow, Black, 118
Willow-herb, Great, 194
Winslow Glass Works, 56
Winslow Junction, 56
Winterberry, 124
   Smooth, 124
Wintergreen family, 134, 196
   Spotted, 134
Witch-hazel family, 114
Witherod, 132
   Naked, 134
Wood-boring Beetles, metallic, 346
Wood Nymph, 360
Woodbine, 142
Woodchuck, 234
Woodland Country Club, 46
Woodpecker, Downy, 250
   Hairy, 250
   Red-bellied, 248
   Red-headed, 248
Woodwardia areolata, 102
   virginica, 102
Wool-grass, 168
Woolly-bear Moth, Banded, 372
   Yellow, 372

Wren, Carolina, 254
    House, 254
Wright's Sawmill & Forge, 42

Xerophyllum asphodeloides, 174
Xylocopa virginica, 388
Xylopinus saperdioides, 352
Xyloryctes jamaicensis, 344
Xyridaceae, 172
Xyris caroliniana, 174
    torta, 172

Yellow Jacket, Eastern, 386
Yellow-eyed Grass, Carolina, 174
    Family, 172
    Slender, 172
    Twisted, 172
Yellow-poplar, 118
Yellow-winged, Brown-spotted, 302
Yellowthroat, Common, 260

Zanaida macroura, 246
Zapus hudsonius, 238
Zizania aquatica, 162
Zonotrichia albicollis, 266
Zosteraceae, 146
Zygoptera, 304

NOTES

NOTES

NOTES

NOTES

NOTES

NOTES

NOTES

NOTES

NOTES

NOTES

NOTES

/ = Richard Stockton College
RSAE = 2 Crestridge Dr, Bridgeton NJ 08302 (note All CAPS)
LAC = Lower Alloways Creek

NOTES

9-12-2007
    Saw Great Egret at RSC

9-19-2007
    Saw Laughing gull @ LAC
    Saw Double Crested Cormorant @ LAC